Lecture Notes in Physics

Springer-Verlag Berlin Heidelberg GmbH

The Editorial Policy for Proceedings

The series Lecture Notes in Physics reports new developments in physical research and teaching – quickly, informally, and at a high level. The proceedings to be considered for publication in this series should be limited to only a few areas of research, and these should be closely related to each other. The contributions should be of a high standard and should avoid lengthy redraftings of papers already published or about to be published elsewhere. As a whole, the proceedings should aim for a balanced presentation of the theme of the conference including a description of the techniques used and enough motivation for a broad readership. It should not be assumed that the published proceedings must reflect the conference in its entirety. (A listing or abstracts of papers presented at the meeting but not included in the proceedings could be added as an appendix.)
When applying for publication in the series Lecture Notes in Physics the volume's editor(s) should submit sufficient material to enable the series editors and their referees to make a fairly accurate evaluation (e.g. a complete list of speakers and titles of papers to be presented and abstracts). If, based on this information, the proceedings are (tentatively) accepted, the volume's editor(s), whose name(s) will appear on the title pages, should select the papers suitable for publication and have them refereed (as for a journal) when appropriate. As a rule discussions will not be accepted. The series editors and Springer-Verlag will normally not interfere with the detailed editing except in fairly obvious cases or on technical matters.
Final acceptance is expressed by the series editor in charge, in consultation with Springer-Verlag only after receiving the complete manuscript. It might help to send a copy of the authors' manuscripts in advance to the editor in charge to discuss possible revisions with him. As a general rule, the series editor will confirm his tentative acceptance if the final manuscript corresponds to the original concept discussed, if the quality of the contribution meets the requirements of the series, and if the final size of the manuscript does not greatly exceed the number of pages originally agreed upon. The manuscript should be forwarded to Springer-Verlag shortly after the meeting. In cases of extreme delay (more than six months after the conference) the series editors will check once more the timeliness of the papers. Therefore, the volume's editor(s) should establish strict deadlines, or collect the articles during the conference and have them revised on the spot. If a delay is unavoidable, one should encourage the authors to update their contributions if appropriate. The editors of proceedings are strongly advised to inform contributors about these points at an early stage.
The final manuscript should contain a table of contents and an informative introduction accessible also to readers not particularly familiar with the topic of the conference. The contributions should be in English. The volume's editor(s) should check the contributions for the correct use of language. At Springer-Verlag only the prefaces will be checked by a copy-editor for language and style. Grave linguistic or technical shortcomings may lead to the rejection of contributions by the series editors. A conference report should not exceed a total of 500 pages. Keeping the size within this bound should be achieved by a stricter selection of articles and not by imposing an upper limit to the length of the individual papers. Editors receive jointly 30 complimentary copies of their book. They are entitled to purchase further copies of their book at a reduced rate. As a rule no reprints of individual contributions can be supplied. No royalty is paid on Lecture Notes in Physics volumes. Commitment to publish is made by letter of interest rather than by signing a formal contract. Springer-Verlag secures the copyright for each volume.

The Production Process

The books are hardbound, and the publisher will select quality paper appropriate to the needs of the author(s). Publication time is about ten weeks. More than twenty years of experience guarantee authors the best possible service. To reach the goal of rapid publication at a low price the technique of photographic reproduction from a camera-ready manuscript was chosen. This process shifts the main responsibility for the technical quality considerably from the publisher to the authors. We therefore urge all authors and editors of proceedings to observe very carefully the essentials for the preparation of camera-ready manuscripts, which we will supply on request. This applies especially to the quality of figures and halftones submitted for publication. In addition, it might be useful to look at some of the volumes already published. As a special service, we offer free of charge LaTeX and TeX macro packages to format the text according to Springer-Verlag's quality requirements. We strongly recommend that you make use of this offer, since the result will be a book of considerably improved technical quality. To avoid mistakes and time-consuming correspondence during the production period the conference editors should request special instructions from the publisher well before the beginning of the conference. Manuscripts not meeting the technical standard of the series will have to be returned for improvement.

For further information please contact Springer-Verlag, Physics Editorial Department II, Tiergartenstrasse 17, D-69121 Heidelberg, Germany

Harald Friedrich   Bruno Eckhardt (Eds.)

Classical, Semiclassical and Quantum Dynamics in Atoms

Springer

Editors

Harald Friedrich
Physik-Department T30,
Technische Universität München
James-Franck-Strasse
D-85747 Garching, Germany

Bruno Eckhardt
ICBM, Carl von Ossietzky Universität
Postfach 2503
D-26111 Oldenburg, Germany

Cataloging-in-Publication Data applied for.

Die Deutsche Bibliothek - CIP-Einheitsaufnahme

Classical, semiclassical and quantum dynamics in atoms / Harald Friedrich ; Bruno Eckhardt (ed.).

(Lecture notes in physics ; 485)
ISBN 978-3-662-14152-6 ISBN 978-3-540-69055-9 (eBook)
DOI 10.1007/978-3-540-69055-9

ISSN 0075-8450
ISBN 978-3-662-14152-6

Originally published by Springer-Verlag Berlin Heidelberg New York in 1997
Softcover reprint of the hardcover 1st edition 1997

Typesetting: Camera-ready by the authors/editors
Cover design: *design & production* GmbH, Heidelberg
SPIN: 10550706 55/3144-543210 - Printed on acid-free paper

Preface

Dieter Wintgen was born on 11 June 1957 in the Westphalian town of Arnsberg. He graduated with a Diploma in Physics from Münster University in 1982 and received his doctoral degree from the Technical University Munich in 1985. He joined the atomic physics group at Freiburg University in 1986, and returned there in 1992 after three years at the Max Planck Institute for Nuclear Physics in Heidelberg. He died tragically in a mountaineering accident on 16 August 1994.

Dieter became acquainted with atomic physics while working for his doctorate in Munich and he remained faithful to this field in the following years. He was quick to realize that small atomic systems provided ideal testing grounds for modern classical and semiclassical theories including all the exciting new developments we might summarize under the heading "quantum chaos". In his all too short career he was tremendously active, bringing many conventional atomic physicists into contact with the exciting field of nonlinear dynamics, and also showing many a more mathematically oriented colleague the beauty of physics in small atoms. Dieter's "Weltanschauung" was radically democratic with little or no regard for hierarchical structures. His clear and straight thinking and speaking earned him respect and affection among his colleagues and students. His untimely death has left a deep wound in the atomic physics community.

A number of the world's most prominent physicists in the fields of nonlinear dynamics and atomic physics have contributed to produce this volume, dedicated to the memory of Dieter Wintgen. We are grateful to them all for their part in making this project a success. We are also grateful to Professor W. Beiglböck, Mrs. U. Beiglböck and Mrs. B. Reichel-Mayer from Springer-Verlag and especially to Sabine Henne, Bruno Hüpper and Dipl.Phys. Petra Meerwald for their help in editing and handling of the electronic files.

Garching, Oldenburg, March 1997 Harald Friedrich, Bruno Eckhardt

Contents

Introduction

1 Background

The relation between classical mechanics and quantum mechanics has been a central theme of physics ever since quantum mechanics was discovered at the beginning of this century. For several decades most efforts focussed on trying to understand counterintuitive concepts associated with the quantum mechanical wave function. The formal translation of the classical equations of motion into the quantum mechanical Schrödinger equation was, and is, seen as straightforward for an ordinary mechanical system. The emergence of classical behaviour from the quantum dynamics was studied within the WKB formalism which gives a reliable prescription for approximately deriving quantum mechanical wave functions and level spectra from classical trajectories in the semiclassical limit, where wave lengths are small compared to typical lengths and classical actions are large in comparison to $\hbar$ [Berry and Mount (1972), Fröman and Fröman (1996)]. Little notice was taken of the fact that this formalism is applicable to integrable systems only [Einstein (1917)].

Integrability is, however, the exception rather than the rule in classical mechanics, even for seemingly simple systems with few degrees of freedom Although this has been known in principle since the work of Poincaré and others around the turn of the century, the far-reaching implications have only recently become generally realized and accepted. A tangible consequence of this realization is the explosively developing field of nonlinear dynamics, "chaos", which has now permeated into virtually all fields of physics, and beyond.

The continuing progress in the understanding of the rich and diverse behaviour in classical dynamics has presented new challenges for the old problem of how classical dynamics is related to corresponding quantum mechanics. One crucial question is how the nonlinearity of classical evolution survives the transition to the strictly linear quantum mechanical description [Haake (1991)]. A milestone was reached in the 1970's, when Gutzwiller (1971) and Balian and Bloch (1972) discovered that the quantum mechanical energy level density of a physical system could be expressed in terms of its periodic classical orbits in the semiclassical limit. The central formula expressing this relation is obtained by taking the trace over the semiclassical Green's function and is known as Gutzwiller's trace formula. In contrast to earlier quantization rules, viz. WKB quantization for one-dimensional systems and EBK quantization for higher-dimensional but integrable systems,

Gutzwiller's trace formula applies also to non-integrable systems with non-linear, chaotic classical dynamics. It took almost two decades to understand the deep and beautiful properties of the trace formula. Considerable improvements of our understanding of the organization of classical chaos and of the analytical properties of the quantum trace were required in order to turn the trace formula into the useful and practical tool it has become within the last few years [Gutzwiller (1990), Chaos (1992), Eckhardt (1993)].

Advances in the theoretical understanding of Gutzwiller's trace formula and its various representations have been accompanied by extensive numerical investigations of the classical and quantum behaviour of simple model systems, such as billiards of various shape, the three- or more-disc scattering system, the kicked rotor and the kicked top. Beyond such simple model systems, one- and two-electron atoms have provided important examples of naturally occurring and experimentally accessible systems in which the quantum manifestations of classical chaos can be studied. Of all such systems, simple atoms play an outstanding role in this context. They have few degrees of freedom, their properties are accurately described in detail by non-relativistic quantum mechanics and modern experimental techniques such as high resolution laser spectroscopy can accurately measure finest details of spectra and cross sections. Furthermore the corresponding classical dynamics is (except for the hydrogen atom and genuine one-electron ions in the absence of external fields) non-integrable.

The whole field now frequently referred to as "quantum chaos" has received and still is receiving significant stimulation from the study of simple atoms [Gay (1991)]. The discovery in 1974 of the ionization of hydrogen atoms by microwaves with small frequencies [Bayfield and Koch (1974)] and its classical explanation [Jensen (1984)] made it clear, that chaotic classical motion was something real with observable consequences, even in systems on an atomic scale where quantum mechanics is undisputedly the theory governing the dynamics. The study of one-electron atoms in time-oscillating fields has since become an important and active subfield of atomic and optical physics [Cederbaum *et al.* (1997)].

Atomic examples for conservative Hamiltonian systems with non-integrable classical dynamics are provided e.g. by one-electron atoms in an external magnetic field or by atoms or ions with more than one, in the simplest case two, electrons in the absence of external fields. The hydrogen atom in a uniform magnetic field was the first system in nature for which the relation between periodic classical orbits and quantum mechanical spectra was analysed and understood on the basis of Gutzwiller's trace formula [Pullen (1981), Wintgen and Friedrich (1987), Wintgen (1987)]. It provides a beautiful example of how long ranged fluctuations in the experimentally measured quantum spectra are related to the classical actions of (unstable) short periodic orbits, while short ranged statistical fluctuation measures tend to follow the universal behaviour expected for a system whose classical dynamics undergo a transition from regularity to chaos [Friedrich and Wintgen (1989), Hasegawa *et al.* (1989)].

For two-electron atoms, modern semiclassical theory has lead to breakthroughs in two different regimes. On the basis of the new understanding of the role of (unstable) periodic orbits, Richter and Wintgen in 1991 presented a satisfactory semiclassical description of low-lying states of the helium atom [Richter (1991), Wintgen *et al.* (1992)]. At that time, this had been one of the central unsolved problems of semiclassical theory for more than sixty-five years [Born (1925)]. More recently, significant progress has been made in understanding the dynamics of two-electron atoms close to the break-up or fragmentation threshold, where both electrons are in highly excited states — or in the continuum. The spectrum of doubly excited resonant states just below the threshold is characterized by increasing numbers of overlapping and interacting Rydberg series, and the investigation of semiclassical theories in this regime has only recently become meaningful, as extensive numerical calculations, in particular by Bürgers and Wintgen, have produced reliable quantum mechanical spectra [Bürgers (1995), Bürgers *et al.* (1995)]. At the same time, more and more doubly excited states of helium are becoming accessible to experiment, and the energy region where a detailed comparison between calculation and measurement is possible, is gradually approaching the really interesting regime very close to the threshold [Domke *et al.* (1996) and references therein].

The usefulness of semiclassical methods in atomic physics reaches, of course, beyond the examples of one- and two-electron atoms sketched above, see e.g. Englert (1988). However, it is in these simple atoms that modern semiclassical theories, formulated for small numbers of degrees of freedom, find an ideal testing ground. Numerical calculations both for the classical as well as for the quantum dynamics are challenging, but they are possible *ab initio* without model assumptions. The variety of interesting physical effects is tremendous and they are all, at least in principle, experimentally accesible. Thus the study of simple atoms has in recent years reclaimed at least part of its role as a pioneering field at the forefront of theoretical physics. Dieter Wintgen played a major part in this development, and this volume is a tribute to his achievement and to his enthusiasm and dedication.

2 Outline

The body of this book consists of two parts. Part I reports advances in general semiclassical theory while Part II describes recent progress in applications to simple atomic systems.

Martin **Gutzwiller** begins with a detailed recollection of the developments around 1970 which led to the trace formula, derived via the semiclassical approximation to the quantum propagator. In the following chapter, Predrag **Cvitanović**, Gábor **Vattay** and Andreas **Wirzba** formulate an alternative "quasiclassical" quantization scheme for Hamiltonian dynamics; this is based on a "quasiclassical approximation" to the quantum evolution operator which, in contrast to the conventional semiclassical approximation, is multiplicative along the flow in an extended phase space. This leads to a quasiclassical alternative to Gutzwiller's trace formula and the corresponding zeta function with improved analyticity properties.

In the chapter by Bruno **Eckhardt** and Bruno **Hüpper**, the semiclassical formalism is applied to the concrete example of the photodissociation of small molecules. They are able to identify certain limitations to the applicability of leading order semiclassics. Jonathan **Keating**'s contribution focusses on the ability of appropriately resummed expressions for the zeta function to accurately reproduce exact quantum mechanical eigenvalues. Contributions of tunnelling trajectories to the semiclassical propagator and the relation between the energy-domain representation and the time-domain representation are the subject of the chapter by Neepa **Maitra** and Eric **Heller**. Tunnelling corrections also play a mayor role in the contribution by Fritz **Haake** and Peter **Gerwinski**. They quantize kicked tops by semiclassically estimating matrix elements between coherent states. Since initial and final points are specified, the dominant contributions come from complex or 'ghost' orbits. This method is efficient and makes it possible to systematically estimate and improve the errors involved in the semiclassical approximation.

In the final chapter of part I, a field theoretical approach to semiclassics is applied by Oded **Agam** and Shmuel **Fishman** to the two-dimensional disordered Lorentz gas. The combination of techniques from solid state theory with semiclassics has lead to important insights into the fluctuation properties of spectra and cross sections, some of which have already been translated into atomic and molecular physics (see e.g. the contributions by Blümel, by Delos and Schwieters and by Burgdörfer et al).

The theoretical understanding of the dynamics of the kicked rotor and the kicked top has developed over the years in close cooperation with experimental and theoretical studies of one-electron atoms in time-oscillating external fields, as provided by microwaves or lasers [Gay (1991)]. These atomic systems played the pioneering role in demonstrating that manifestations of classical chaos were actually observable in naturally occurring quantum systems, and they are the subject of the first three chapters of Part II.

Reinhold **Blümel** uses a comparison of the kicked rotor and the kicked hydrogen atom to discuss chaotic ionization and dynamical localization for one-electron Rydberg atoms in a microwave field. Derek **Richards** presents a comprehensive overwiew of the enormous richness and diversity of behaviour of a periodically driven hydrogen atom in the various regimes of (scaled) frequency and field strengths. The chapter by Dominique **Delande** and Jakub **Zakrzewski** on hydrogen atoms in *circularly polarized* microwave fields gives a comprehensive account of the current state of knowledge in the field, and of some misconceptions which have now been eliminated. The occurrence of non-dispersive wave-packet eigenstates stabilized by the oscillating field is worth a special mention as an example of the exotic effects observable in such seemingly simple systems. The discussion of fluctuations in the threshold field strengths for ionization establishes a connection to the subject of conductance fluctuations, which is a topic of some interest in the condensed matter community, to where a substantial amount of the effort devoted to the study of "quantum chaos" has moved [Richter *et al.* (1996)]. A comparison of interference effects observed in the spectra of atoms in external fields and those important in the conductance properties of microscopic junctions is made in the contribution by John **Delos** and Charles **Schwieters**.

The hydrogen atom in an external magnetic field was the first physically real example to which the semiclassical theory based on Gutzwiller's trace formula was applied, and Dieter Wintgen was the pioneer of this development. Jörg **Main**, working as a theorist in the experimental group of K.H. Welge, was active in this effort from the beginning, and his contribution gives a summary of the state of the art in the field which has developed out of these studies. An interesting and appealing result of recent work is the extension of periodic orbit theory to describe contributions due to real and ghost orbits near bifurcations and the successful formulation of uniform semiclassical approximations for the neighbourhoods of various types of bifurcations observed. A further recent advance has been the extension of periodic orbit theory developed for the example of a pure hydrogen atom in a magnetic field to more general atoms with a non-Coulombic core in addition to the one valence electron.

After the quick success of the semiclassical theory based on Gutzwiller's trace formula applied to atoms in a magnetic field, Dieter Wintgen was the first to realize the power of these concepts for the hitherto unresolved problem of describing two-electron atoms semiclassically. He and his students started a true renaissance of activity in this field. The state of the art today is described in the chapters by Jan-Michael Rost and Gregor Tanner and by Joachim Burgdörfer, Yanghui Qiu and Jörg Müller. Among other results **Rost** and **Tanner** emphasize the power of modern semiclassical theory by demonstrating that the information extracted from the classical orbits alone can reproduce fine details of the quantum spectra including, e.g., the effects of interference of interacting Rydberg series. They also discuss the interesting prospect of relating the lifetimes of doubly excited states just below the break-up threshold to ionization cross sections just above the threshold. The intriguing regime of doubly excited states just below the break-up threshold, where all approximate systematic assignments of quantum numbers fail, is also a main theme in the chapter by **Burgdörfer**, **Qiu** and **Müller**. These authors impressively demonstrate how we can utilize our knowledge of periodic classical orbits to classify and understand, to a certain extent, observable features of the quantum spectra without actually giving a complete description in detail, as is possible for integrable systems with complete sets of quantum numbers. There is one reason in particular, why the region just below the fragmentation threshold breaks out of the frame in which conventional atomic physics is usually pursued: most doubly excited states in this regime may not be observable as individual lines, but, if the line widths of these resonant states become larger than their mean separation, the line spectrum should merge into a fluctuating continuum as the break-up threshold is approached from below. As mentioned in the last two chapters, evidence is accumulating that, except for a few isolated states or series of states associated with exotic stable configurations of the three-body system, this dissolution of the line spectrum into a fluctuating continuum actually happens. Atomic physics here makes a connection to developments in solid state physics and nuclear physics.

References

Balian R and Bloch C (1972), Ann. Phys. (N.Y.) **69**, 76.
Bayfield JE and Koch PM (1974), Phys. Rev. Lett. **33**, 258.
Berry MV and Mount KE (1972), Rep. Prog. Phys. **35**, 315.
Born M (1925), *Vorlesungen über Atommechanik* (Springer-Verlag, Berlin) p. 341.
Bürgers A (1995), *Doppelt-Angeregte S-Zustände im Heliumatom*, doctoral thesis, University of Freiburg (unpublished).
Bürgers A, Wintgen D and Rost J-M (1995), J. Phys. B **28**, 3163.
Cederbaum LS, Kulander KC and March N (eds.) (1997), *Atoms and Molecules in Intense Fields*, Structure and Bonding **86** (Springer-Verlag, New York).
Chaos **2** (1) (1992), special focus issue on *Periodic Orbit Theory*.
Domke M, Schulz K, Remmers G, Kaindl G and Wintgen D (1996), Phys. Rev. A **53**, 1424.
Eckhardt B (1993), in: *Quantum Chaos*, G. Casati, I. Guarneri and U. Smilansky (eds), (North Holland, Amsterdam), 77.
Einstein A (1917), Verh. Dt. Phys. Ges. **19**(9/10), 82.
Englert B-G (1988), *Semiclassical Theory of Atoms*, Lecture Notes in Physics **300** (Springer-Verlag, New York).
Friedrich H and Wintgen D (1989), Phys. Reports **183**, 37.
Fröman N and Fröman P-O (1996), *Phase Integral Method*, Springer Tracts in Modern Philosophy **40**, ed. C. Truesdell (Springer-Verlag, New York).
Gay JC (ed.) (1991) *Irregular Atomic Systems and Quantum Chaos*, special volume of: Comments on At. Mol. Phys. **25**.
Gutzwiller MC (1971), J. Math. Phys. **12**, 343.
Gutzwiller MC (1990), *Chaos in Classical and Quantum Mechanics* (Springer-Verlag, New York).
Haake F (1991), *Quantum Signatures of Chaos*, Springer Series in Synergetics **54** (Springer-Verlag, New York).
Hasegawa H, Robnik M and Wunner G (1989), Prog. Theor. Phys. suppl. **98**, 198.
Jensen RV (1984), Phys. Rev. A **30**, 386.
Pullen RA (1981), *Semiclassical Treatment of the Quadratic Zeeman Effect*, PhD thesis, University of London (unpublished).

Richter K (1991), *Semiklassik von Zwei-Elektronen-Atomen*, doctoral thesis, University of Freiburg (unpublished).
Richter K, Ullmo D and Jalabert RA (1996), Phys. Reports **276** 1.
Wintgen D (1987), Phys. Rev. Lett. **58** 1589.
Wintgen D and Friedrich H (1987), Phys. Rev. A **36** 131.
Wintgen D, Richter K and Tanner G (1992), *CHAOS* **2**, 19.

The Origins of the Trace Formula

Martin C. Gutzwiller

Thomas J Watson Research Lab, IBM, Yorktown Heights

Abstract. The trace formula establishes a mathematical relation between the spectrum of energy levels of a dynamical system in quantum mechanics on one hand, and the set of periodic orbits in the corresponding system in classical mechanics on the other hand. Such a formula was first derived by Atle Selberg in the context of geodesics on a surfaces of constant negative curvature for the purposes of number theory, while other mathematicians and some physicists were mostly inspired by Weyl to get the asymptotic distribution of eigenvalues of differential operators. Eventually, the author obtained a more general trace formula on the basis of Feynman's path integral, and a similar result was obtained by Colin de Verdiere shortly thereafter on more geometric grounds. The use of various zeta-functions in differential geometry and statistical physics is mentioned at the end.

0.1 Introductory Remarks

Ever since Max Planck introduced his famous constant h at the end of the year 1900, the connection between classical and quantum mechanics has become one of the everlasting conundrums in physics. Is there any useful information to be found in the classical trajectories of an arbitrary dynamical system, if we want to learn something about the wave functions in the corresponding quantum system?

In this general form, the question has sofar eluded any global answer. The trace formula, however, seems to cover quite nicely a more restricted issue: Is the spectrum of the quantum system in any way related to some special feature in the corresponding classical system? The answer is very simple and clean, and therefore particularly convincing: The special feature is the collection of the periodic orbits.

Moreover the relation is expressed quite succinctly in a formula for a function $g(E)$ of the energy E that can be calculated in two ways. The first way (or left-hand side of the formula) is a sum over all the eigenvalues E_j in the required spectrum, while the second way (or right-hand side of the formula) is a sum over the lengths (more exactly the action integral $S(E)$) of all the smoothly closing, periodic orbits as provided by the classical mechanics. Both sides of the formula depend on the choice of an arbitrary function that allows us to separate out certain parts of the spectrum for special attention.

One can hardly expect that such a general scheme can be made to function without some difficulties. A good deal of hard work has, therefore, gone into finding the mathematical procedures as well as good examples, if not some useful applications of the trace formula in physics. An abridged history of all these efforts will be given below, but it is only fair to mention first, and discuss at some length, the ideas that prompted the mathematicians to wrestle with the trace formula in their own way and for their own purposes.

1 Mathematical Precursors

1.1 Mathematical Physics Before Quantum Mechanics

The problems of classical mechanics have a very simple, purely geometric analog in the geodesic lines on a two-dimensional surface that can be embedded in the three-dimensional Euclidean space like the surface of an ellipsoid. This connection between mechanics and geometry can be generalized to embrace classical optics where the light rays act like the motion on geodesics whose speed depends on the local coordinate.

It is then only a small jump to include wave optics as well as wave mechanics in the discussion. The simplest wave equation on a curved surface involves no more than the metric, i.e., the quadratic differential for the distance of two neighboring points in terms of the internal coordinates. Waves $\psi(x, y, z, t) = \phi(x, y, z)\exp(i\omega t)$ with a purely harmonic time-dependence become eigen-functions of the Laplace-Beltrami operator Δ in the simple form,

$$\Delta\phi + \lambda\phi = 0, \tag{1}$$

where $\lambda = k^2$, and $k = \omega/c$ with $c =$ speed of light. The formal relation between geodesics and the Laplace operator entered mathematics well before quantum mechanics because of its importance in optics.

The mathematical investigation of the wave-operator led quite naturally to the study of the propagator, also called Green's function $G(x, y, z; x', y', z'; \lambda)$, which gives the solution to the inhomogeneous wave-equation,

$$\Delta G + \lambda G = \delta(x - x'); \tag{2}$$

the right-hand side provides a local excitation in (x', y', z').

Green's function can be written as a sum over the eigenfunctions $\phi_j(x, y, z)$ from (1) with the eigenvalues λ_j ,

$$G(x, y, z; x', y', z'; \lambda) = \sum_j \frac{\phi_j(x, y, z)\phi_j^\star(x, y, z)}{\lambda - \lambda_j}. \tag{3}$$

Various special cases of this formula can be found in the classic treatise of Courant and Hilbert, *Methoden der Mathematischen Physik*, whose first volume appeared at the beginning of 1924, almost two years before Schrödinger's first paper on the new wave mechanics. The physical applications are mostly related to the problems in acoustics like the vibration of strings, rods, membranes, and plates.

Eventually the mathematicians, if not the physicists, should have arrived at the idea of relating the set of smoothly periodic geodesics with the eigenvalues of the Laplacian on the same Riemannian manifold. But the mathematicians were not interested nearly as much in mathematical physics as they were in the more abstract number-theory.

1.2 The Temptations of Number Theory

The Riemann zeta-function is defined by the formula,

$$\zeta(s) = \sum_n \frac{1}{n^s}, \tag{4}$$

where $s = \sigma + it$ is a complex number. This function has a long and honorable history: it was first used by Euler to prove that there are infinitely many prime numbers. Then a simple generalization was used by Dirichlet in 1837 to show that any arithmetic series contains its fair share of prime numbers. Finally, in his short article of 1859, Riemann represented the number $F(x)$ of primes below some positive real number x as a sum over the roots of the equation $\zeta(s) = 0$ in the strip $0 < \sigma < 1$.

What intrigued the mathematicians is a short statement by Riemann about half way into his article: "It is very probable that all the roots have $\sigma = 1/2$. A rigorous proof of this claim, however, is desirable; but meanwhile, I have abandoned the search for it after some preliminary unsuccessful efforts, because it is not required for the next purpose of my investigation." This sofar unproven conjecture is known as Riemann's hypothesis, and remains the most sought-after result in mathematics. Needless to say that it has been completely confirmed by extensive numerical calcualtions, even before C.L.Siegel found in 1931 a very efficient method for computing the zeta-function in Riemann's notes.

A superficial similarity of Riemann's formula for $F(x)$ with some form of trace in (3), i.e., $\int d^3x G(x, y, z; x, y, z; \lambda)$, was recognized by Hilbert, among others. Therefore, he suggested that a proof of Riemann's hypothesis might involve finding a mechanical system whose eigenvalues coincide with the roots of the zeta-function. This idea had a short-lived moment of glory in the 1980's when such a system was believed to be found in the geodesics on the modular surface. But the mistaken identity was eventually traced back by Dennis Hejhal to a computing error.

1.3 Selberg's Trace Formula

In 1948, Atle Selberg stunned the world with an elementary proof of the socalled prime-number theorem: the quotient $F(x) \log x / x$ approaches 1 when x goes to infinity. This result can be restated by looking at the logarithms $q = \log p$ of the prime number p, and asking for their number $f(y)$ below y.

Clearly the quotient $f(y)y/\exp(y)$ goes to 1 as y goes to infinity. Sometimes around 1950 Selberg made the crucial association of the prime numbers with the primitive periodic orbits on a surface of constant negative curvature, because their lengths ℓ have the asymptotic distribution $f(\ell)$.

Selberg succeeded in making this close analogy very evident in his trace formula for a compact and complete surface of constant negative curvature. The derivation is not difficult, but it requires some caution and a good knowledge of group theory. We will write down this important result because it has become the model for all the other trace formulas. The eigenvalues of the Laplacian are written in the form,

$$(\Delta + \kappa^2 + \frac{1}{4})\phi = 0, \tag{5}$$

where the constant curvature has been normalized to -1. The primitive periodic orbits (ppo), i.e., geodesics that close smoothly without repetitions, are designated by p. Their length is called $\ell(p)$; their number proliferates in the manner described in the two paragraphs above if we put $y = \ell(p)$.

Selberg's trace formula involves two arbitrary functions $P(u)$ and $\Lambda(\kappa)$ that are Fourier transforms of each other,

$$\Lambda(\kappa) = \int_{-\infty}^{+\infty} du\, e^{i\kappa u} P(u). \tag{6}$$

$P(u)$ is assumed symmetric in the variable u, and quite strongly decaying for large $\mid u \mid$. Selberg proved the formula,

$$\sum_{j=0}^{\infty} \Lambda(\kappa_j) = \int_{-\infty}^{+\infty} \nu \Lambda(\nu) \tanh(\nu) d\nu + \sum_p \sum_{n=1}^{\infty} \frac{\ell(p)}{2\sinh[n\ell(p)/2]} P(n\ell(p)). \tag{7}$$

The derivation of this formula depends critically on the high degree of symmetry of these compact surfaces of constant negative curvature. Each surface provides a tiling of the Poincare upper half-plane, i.e., a complete covering without overlap or gaps. Each tile can be transformed into any other by an element of an infinite discrete subgroup of $PSL(2, R)$, i.e., the continuous group of 2 by 2 matrices with real entries and determinant = 1. A periodic orbit is associated with each conjugacy class of this discrete subgroup, and its length is given by the trace of the corresponding 2 by 2 matrix in $PSL(2, R)$.

This formula can be greatly generalized to include non-compact surfaces of constant negative (Gaussian) curvature. Many physicists may not realize the tremendous variety of such surfaces in contrast to the surfaces whose curvature is either constant positive (only the spheres) or zero (only the flat tori). Even the simplest special geometry for (7), a sphere with two handles or double torus, requires the choice of six real parameters in addition to the value of the curvature.

Moreover, in contrast to the motion on a surface with positive or vanishing curvature, the motion on a surface with strictly negative Gaussian curvature was shown by Hadamard in 1896 to be totally unstable. This kind of motion was studied extensively by the mathematicians throughout this century, including G.D.Birkhoff, Artin, Hedlund, Anosov, and Arnold. It fits all the descriptions of a chaotic motion, and has a well defined Kolmogoroff entropy.

1.4 Weyl's Law

The physicists at the turn of the century inspired the mathematicians to solve a problem that also bears on the connection between classical and quantum mechanics. Planck made the great discovery in 1900 that energy is parceled out in discrete steps, at least in the case of the electromagnetic radiation inside an isolated cavity. Already the earlier, unsuccessful discussions of this problem were based on estimating the number of electromagnetic modes of vibration in the limit where the wavelength of the vibrations is small with respect to the dimensions of the cavity. Only the volume, but not the shape, of the cavity turned out to enter into the final result.

In 1912, Peter Debye proposed a theory for the specific heat of solid bodies that worked very much along the lines of Planck's crucial result, and with the same success. What counted at low temperatures was the asymptotic number of mechnical modes of vibration, and again only the volume and the elastic constants, but not the shape was important. He issued a challenge to the mathematicians to prove this empirical fact on the basis of the mathematical theory of elasticity. Hilbert thought that a general proof would be very difficult to find.

Nevertheless, already in 1912, Hermann Weyl provided such a proof on the basis of Hilbert's theory of integral equations. In the special case of the Laplace operator $-\Delta$ the number $N(E)$ of eigenvalues λ in (1) with $\lambda < 2mE/\hbar^2$ is given by the formula,

$$N(E) \cong \frac{A}{2\pi}\frac{2mE}{\hbar^2}, \quad \frac{V}{6\pi^2}\frac{(2mE)^{3/2}}{\hbar^3}. \tag{8}$$

in the limit of E going to infinity. (In quantum mechanics the operator $-2\hbar^2\Delta/2m$ becomes the kinetic energy E of a particle with mass m, and $\hbar = h/2\pi$.) The right-hand side gives the estimate for the eigenvalues of (1) in two (three) dimensions where A (V) is the area (volume) of the domain where (1) is solved.

Weyl's Law (8) was widely discussed by the mathematicians, and an extensive discussion can be found in the 1924 monograph of Courant and Hilbert. An important unresolved issue was the next term in what was clearly an asymptotic expansion in powers of $\hbar/\sqrt{2mE}$. The history of this subject and some partial answers can be found in the well-known 1967 article of Mark Kac, "Can you hear the shape of a drum?".

No effort is made in any of this work to refine Weyl's law in order to find an approximate expression for some special eigenvalue such as the lowest. Obviously, that would have been contradictory to the purpose of of the asymptotic formula (8), and therefore, nothing worthwhile could be expected. Nevertheless, Weyl's law became very important in physics, and was sometimes used to estimate particular eigenvalues for arbitrary systems in quantum mechanics where no other method was available.

With the help of some rather crude but effective arguments, L.H.Thomas and Enrico Fermi in 1927 reduced Weyl's law to a very general statement about the eigenvalues of Schrödinger's equation. Their number $N(E)$ with energy less than E is proportional to the volume of phase space with energy less than E, i.e., the $2n$-dimensional space whose coordinates are the n position coordinates q_i and n momentum coordinates p_i. The constant of proportionality is Planck's constant h raised to the power n. One can hardly ask for a more concise statement: phase space is divided into cells of volume h^n each of which accomodates one eigenvalue of Schrödinger's equation.

2 Semiclassical Mechanics

2.1 The Old Quantum Theory

After Planck's and Einstein's work on the electromagnetic radiation, it was Niels Bohr who first proposed a viable theory for the energy spectrum of the hydrogen atom. This story is so fundamental for the development of modern physics that it is told in every physics textbook. Tremendous ingenuity was put into explaining atomic and molecular physics on the basis of some shaky principles connecting classical mechanics and quantum theory; a particularly useful idea was Bohr's correspondence principle. In 1917 Einstein pointed out that these principles could not possibly work in the majority of cases where several degrees of freedom were badly interfering with one another.

When the big breakthrough came in 1925/26, particularly with the Schrödinger equation, the previous efforts were collectively called "old quantum theory" in contrast to the new "quantum mechanics". The textbooks usually allocate one short chapter to the pre-1925 ideas because they had a good deal of intuitive appeal. But very little is said or done to justify mathematically the famous correspondence principle of Bohr, except either in systems with only one degree of freedom or in systems that could be separated into independent degrees of freedom by an appropriate transformation of the coordinates.

The method of choice for this limited endeavour is tied to the names of Leon Brillouin, Hendrik Kramers, and Gregor Wentzel. The main idea goes back to Liouville in the last century, as well as to Harold Jeffreys in this century, and even to Werner Heisenberg in his Ph.D. thesis on hydrodynamics in 1923. The WKB method can be used in many problems, but works best for a particle that is bound in a one-dimensional potential $V(x)$. The spectrum of eigenvalues is then given by the condition,

$$\int_{x_1}^{x_2} dx[2m(E-V)]^{1/2} = (\nu + \frac{1}{2})h, \tag{9}$$

where the integral goes from x_1 to x_2 where $V(x) = E$; these are the points where the classical orbit turns around. The integer ν numbers the bound states starting at 0. The extra 1/2 on the right-hand side is the important correction that comes from the wave function spilling over into the classically forbidden region where $E < V(x)$.

2.2 Forty Years in the Desert

The time from 1926 to 1966 was an era of unprecedented success in physics, particularly in everything that required quantum mechanics for its explanation, such as atomic, molecular, condensed-matter, chemical, and optical physics on one side of the ledger, as well as nuclear and elementary particle physics on the other. With computing resources becoming available at an ever accelerating clip, there was little incentive to improve our understanding of the relations between the old-fashioned classical and the only 'correct' quantum mechanics.

Nevertheless, a steady flow of isolated work during this time is concerned with this problem. Presumably, some individuals did not like to be submerged in an approach that reduced everything to the solution of one partial differential equation. Of course, many schemes were invented to carry out this last job under different circumstances, and some intuitive insight was required to pick the most likely method to give results in a particular project. Nevertheless, the main task was always to prove Schrödinger's equation correct in every single case; everybody was expected to act like a good engineer and get on with the job, since the basic recipe for success was known.

The WKB approximation is limited to systems with one degree of freedom, and does not provide a good picture of systems with at least two freedoms, even if they can be separated. Einstein's paper in 1917 provides better insight to the origin of the quantization condition (9) by considering the situation in phase space. The trajectories of integrable systems with a given energy E lie on tori whose dimension equals the number of freedoms. As many topologically independent, but otherwise arbitrary closed loops C_i are chosen on such a torus, and each gives rise to a quantization condition,

$$\int_{C_i} \mathbf{p}\, d\mathbf{q} = (\nu_i + \frac{\alpha_i}{4})h, \tag{10}$$

where ν_i is an non-negative integer. According to the WKB condition (9), the integer $\alpha_i = 2$ for a vibration; but it has to vanish when the relevant degree of freedom describes a rotation.

The issue of the integer versus half-integer quantum numbers was settled by Joseph Keller in 1958 after he had rediscovered Einstein's paper of

1917. In contrast to WKB, he constructed a smooth wave function on the n-dimensional torus of a classical trajectory in phase space. This wave function has to be projected into the position space of coordinates q_i. If this projection leads to the singularities, such as the classical return points x_1 and x_2 in (9), then α_i is generally an even positive number differing from 0.

These corrections to the original quantization conditions of the old quantum theory were first discussed in the context of integrable dynamical systems by Viktor Maslov, in papers and books in Russian that were translated into western languages only in the 1970's. The idea really goes back to the 1920's, and to Marston Morse in his work on the variational principle of classical mechanics. Unfortunately, Morse's results were completely ignored by the physicists, and the writers of textbooks in classical mechanics. Cf. below for more on this topic.

2.3 Feynman's Path Integral (PI)

In 1948 the Reviews of Modern Physics published an article by Richard Feynman, "Space-Time Approach to Non-Relativistic Quantum Mechanics", which some people consider his finest and most important work. Taking off from a hint in the second edition of Dirac's Principles of Quantum Mechanics, Feynman offers an explicit solution for the time-dependent Schrödinger equation in terms of an integral over all paths in position space that connect the source in q' at time t' with the detector in q'' at the time t''.

Again there are antecedents in mathematics to this idea which were ignored by the physicists. Norbert Wiener had shown in the 1920's how to solve the classic equation for heat flow in this manner, which was close in spirit to the idea of random walks in statistical mechanics. After Feynman's first paper, the mathematicians under the leadership of Mark Kac generalized Wiener's approach. It has the great virtue of being amenable to rigorous mathematical proofs because heat dissipates exponentially in time, whereas the solutions of Schrödinger's equation propagate like waves that are subject to the conservation of particles.

Feynman's PI was immediately generalized to cover not only systems with infinitely many particles, but even second-quantized fields that describe the creation and annihilation of particles. Most endeavours in elementary-particle theory nowadays start out with an expression in terms of a path integral over all possible fields in some Hilbert space. The use of this starting point, however, is purely formal, and is made to insure the internal consistency of the resulting field equations. The idea comes to fruition as a method for computing in various Monte Carlo schemes for both chemistry and lattice gauge theory, although only in its heat-flow version where the time variable is purely imaginary.

The PI calculates the probability amplitude $K(q'', t''; q', t')$ for a particle to leave its initial position in q' at time t' and arrive at the final position in q'' at time t'', where the symbol q stands for a vector in any number of dimensions. This propagator K is given by the integral over all continuous paths $q(t)$ connecting the two endpoints in the stipulated time-interval,

$$\lim\,[\frac{m}{2\pi i\hbar\tau}]^{N/2}\int dq_1...\int dq_{N-1}\exp(\int dt L(\dot{q},q)), \tag{11}$$

as N goes to ∞ . For the purpose of defining the limit, the time interval from t' to t'' is divided into N equal pieces of length $\tau = (t'' - t')/N$. The intermediate stations in the path are called (q_j, t_j) with $j = 0, 1, ..., N$, where $(q_0, t_0) = (q', t')$ and $(q_N, t_N) = (q'', t'')$. Notice that the Lagrangian appears in the exponent of the integrand, not the Hamiltonian! The velocities $\dot{q}$ in the Lagrangian are expressed in terms of finite differences. The value of the each position q_j for $j = 1, ..., N-1$ is allowed to vary without bound in the domain of the Schrödinger equation.

Already in his original article, Feynman made it clear that the PI was ideally suited to make the transition from quantum back to classical mechanics. If the integral of the Lagrangian over the classical trajectory between the endpoints is large compared to Planck's constant, only the paths in its neighborhood will contribute to (11) because the other paths cancel one another by destructive interference. But he did not pursue this thought beyond the most obvious application, and limited himself mostly to discussing the harmonic oscillator. Even in his book of 1965 with Hibbs, he never adresses the general problem of using the results of classical mechanics to get an explicit value for the propagator in a more general context

Meanwhile, there were some scattered efforts in this direction, the earliest by Cecile Morette-DeWitt in 1948. Philippe Choquard wrote his Ph.D. thesis in 1954 with Pauli on this subject, and succeeded in deriving a systematic expansion of the propagator in powers of the time difference $(t'' - t')$. Such an expansion converges only for a limited time-interval as will be explained in the next chapter. It does yield Weyl's formula, however, because large energies are compatible with short times. Kac in 1966 as well as McKean and Singer in 1967 found the two first correction terms to (8) in this way on the basis of the heat-flow equation.

3 The General Trace Formula

3.1 The Systematic Use of the Path Integral

The story sofar gives only a bare sketch of all the work until 1967 that went into establishing the relation between classical and quantum mechanics. My first article on this subject was published in that year, but I was blissfully unaware of most of these earlier attempts. I had become familiar, however, with some of the theoretical work in molecular and solid-state physics. I was disappointed with having only two methods of solving quantum-mechanical problem: either assembling ever larger sets of basis functions from which to find the ground-state wave-function by the variational principle, or calculating ever higher terms in a perturbation series.

Somehow I reached the conclusion that the semiclassical approach had not been worked out under sufficiently general circumstances. In particular, the PI had not been tried seriously enough, since not even the hydrogen atom had been done directly on the basis of the PI. Therefore, this looked like a good start on a new enterprise, although many of the pieces in the puzzle were missing.

Feynman's PI gives the primary information which is the propagator K; it is valid even for time-dependent systems. The spectrum can only be gotten for a time-independent system, however, and the transition requires a Fourier transform with respect to the time-variable $t = t'' - t'$. The result is Green's function,

$$G(q'', q', E) = \frac{1}{i\hbar} \int_0^\infty dt K(q'', q', t) \exp(iEt/\hbar), \tag{12}$$

which converges for complex values of E with a positive imaginary part. The expansion in eigenfunctions takes the form analogous to (3),

$$G(q'', q', E) = \sum_j \frac{\phi_j(q'')\phi_j^*(q')}{E - E_j}, \tag{13}$$

where we assume quite generally that $H_{op}\phi_j = E_j\phi_j$ with the quantum-mechanical Hamilton-operator H_{op} for the particular problem.

Most people don't seem to realize it, but there is no path-integral for Green's function! There are some poor substitutes, however, such as integrals over "histories in phase-space" with some global subsidiary conditions. Also, Feynman's PI is meant for Cartesian coordinates, and even the transition to polar coordinates has to be approached with great caution. The use of the PI for the Laplacian on a Riemannian manifold is, therefore, fraught with pitfalls that were discussed by Bryce DeWitt. In any case, the classical limit in the PI requires another feature in classical mechanics that seemed to be taboo in the discussion of the variational principle of classical mechanics by the physicists: the second variation of the integral over the Lagrangian. The subject had been treated extensively by Marston Morse, as mentioned earlier.

Classical mechanics requires the first variation to vanish, and thereby defines the ordinary differential equations for the classical trajectory. The classical limit of the PI requires the integration over all possible deviations in the lowest order, i.e., the second variation. The two features yield the semiclassical aproximation K_c of the propagator as a sum over the classical trajectories between the given endpoints,

$$K_c(q'', q', t) = \sum_{class.traj.} \frac{\sqrt{C}}{(2\pi i\hbar)^{n/2}} \exp[iR(q'', t''; q', t')/\hbar - i\kappa\pi/2], \tag{14}$$

where $R(q'',t'';q',t')$ is the integral over the Lagrangian along the particular trajectory, and C is the determinant of the second derivatives with respect to q'' and q' of R. Van Vleck had written down this expression in 1929, but he had not considered the sum over several trajectories which requires the additional phase $-\kappa\pi/2$. The Morse index κ is the number of negative eigenvalues in the second variation, and also the number of conjugate points along the trajectory. If several trajectories are allowed to start in q' at t' with slightly different momenta, they will cross one another again at well defined later times in the conjugate points.

In order to get the classical approximation to Green's function G_c, one has to carry out the Fourier transform (12) on the expression (14). The stationary-phase method seems to be compatible with the classical limit (and can be used also for the transition from position variables q to the momentum variables p). The resulting expression for $G_c(q'',q',E)$,

$$\frac{2\pi}{(2\pi i\hbar)^{(n+1)/2}} \sum_{class.traj.} \sqrt{(-1)^{n+1}D}\exp[iS(q'',q',E)/\hbar - i\lambda\pi/2], \qquad (15)$$

resembles (14) except that the exponential contains the function $S(q'',q',E)$ which is the integral over $\mathbf{p}\,d\mathbf{q}$ along the trajectory at the given energy E. The phase-loss $\lambda\pi/2$ now comes from the number of conjugate points along the trajectory at the constant energy E, i.e., how many times trajectories starting at the same place, but in slightly different directions, cut into the original trajectory.

This whole program is carried out in the two papers of the Journal of Mathematical Physics (JMP), 1967 and 1969, for the case of a particle in a rotationally symmetric potential. All that was known about such problems in the semiclassical approximation came out quite naturally, including the WKB approximation for the radial motion and the integer quantum numbers for the Kepler problem. Surprisingly, the exact eigenfunctions for the hydrogen atom appear, when momentum coordinates are used, as Ian Percival found simultaneously and independently.

3.2 Periodic Orbits

The formula (14) for the classical propagator and (15) for the classical Green's function can be obtained without the PI. Van Vleck's expression in either case satisfies the appropriate inhomogeneous Schrödinger equation with an error of order h^2. The phase loss due to a conjugate point comes up quite naturally because Van Vleck's expression diverges in its neighborhood, and has to be completed with a solution of the full wave-equation near the conjugate point or focus. Arguments of this kind are well known in the theory of optical diffraction.

In order to calculate the trace $g(E)$ of Green's function, or of the propagator, the two endpoints have to be identified, i.e., $q' = q'' = q$, and the integration over q is carried out,

$$g(E) = \int dq\, G(q,q,E), \tag{16}$$

where we have used (13). In the classical approximation G_c or K_c, the classical trajectories now have to begin and end at the same place q. In general, the classical trajectories become closed loops whose tangent (or momentum) at the begininng does not coincide with the tangent (or momentum) at the end. Each loop contributes an integral to the classical trace $g_c(E)$.

As the common start-end point is dragged around to get the integral, the exponent in (14) or (15) changes according as,

$$\frac{\partial}{\partial q} S(q,q,E) = [\frac{\partial S(q'',q',E)}{\partial q''} + \frac{\partial S(q'',q',E)}{\partial q'}]_{q''=q'=q} = p'' - p', \tag{17}$$

i.e., as the change in momentum at q. The stationary-phase argument now tells us that the contribution of a particular closed loop is large whenever there is no such change of momentum, i.e., the trajectory closes smoothly as if the particle continued on the periodic orbit. The classical trace $g_c(E)$ becomes a sum over periodic orbits in lowest order of an expansion in powers of Planck's constant h.

This argument came up for the first time in my paper in JMP 1970 which was submitted on December 4, 1969. The trace is calculated and the corresponding spectrum is obtained for several integrable systems, including the rectangular billiard and a spherically symmetric potential. After explaining the effect of (17) on the trace, I comment: "This rule should be generally valid ... the importance of the closed geodesics for the spectrum becomes a very attractive feature." Therefore, the next paper was devoted entirely to this idea. It was submitted 27 August 1970 and published in JMP March 1971, with the title "Periodic Orbits and Classical Quantization Conditions."

The main job in writing down the complete expression for $g_c(E)$ is to reduce the amplitude to some reasonable quantities that are associated with a periodic orbit in classical mechanics, and to carry out the integral in the framework of the stationary-phase approximation. The monodromy matrix appears eventually, or more explicitly, the local transverse-area-preserving map whose fixpoint is the periodic orbit. There are different cases depending on the eigenvalues of this linear map, parabolic (integrable), elliptic (locally stable), hyperbolic (unstable), loxodromic (two transverse degrees of freedom are coupled). The most interesting is the (directly) hyperbolic case which had never been discussed before (except in the Selberg trace formula),

$$g_c(E) = g_0(E) + \frac{1}{i\hbar} \sum_{\substack{prim.\\ per.orb.}} \sum_{n} \frac{T}{2\sinh(n\chi/2)} \exp[in(S(E)/\hbar - \mu\pi/2)], \tag{18}$$

where T is the period, $S(E) = \int \mathbf{p}\, d\mathbf{q}$, χ is the instability exponent, and μ is the number of conjugate points, all calculated for the one (primitive) trip. The first term $g_0(E)$ on the right comes from the closed orbit of length

0, and gives the smoothly increasing number of energy levels according to Weyl's law. The second term generates the poles in E.

The example for this general trace formula was the anisotropic Kepler problem (AKP), where I knew only one periodic orbit at the time so that the sum over the primitive periodic orbits consisted of only one term. Although the resulting spectrum, when properly interpreted, turned out to be in fairly good agreement with quantum-calculations, a number of colleagues protested against this gross simplification. The objections were effectively answered in the next paper, submitted on 6 April 1972 and published in JMP January 1973, where the AKP was shown to have a symbolic dynamics in terms of binary sequences.

The development of the general trace formula in my papers some 25 years ago has been described with some historical detail, because it is entirely different, independent, and earlier than the significant work that came out of France shortly thereafter. Also the papers in JMP try to follow through on a more consistent and exhaustive approach. At the end, the AKP is a new example that represents an almost ideal case of a totally chaotic motion. In spite of having only unstable periodic orbits, the AKP has a relatively simple structure of conjugate points.

3.3 The Input from Nuclear Physics

Among the most productive areas in the sciences, research in nuclear physics has enjoyed the general support of many governments. However, in contrast to the more down-to-earth problems concerning atoms, molecules, condensed matter, optics, and so on, the basic ingredients into an understanding of the nucleus are still unknown, fifty years after its first large-scale applications in generating violent as well as non-violent energy. Perhaps for this reason, theoreticians in this area were less tempted by any systematic approach, and more willing to try out novel ideas to gain new insights into physics in general.

Roger Balian and Claude Bloch of the Centre d'Etudes Nucleaires in Saclay submitted three large papers to the Annals of Physics in 1970 (received 20 March, 18 August, and 7 October) on the distribution of eigenfrequencies and the asymptotic evaluation of Green's function for large quantum numbers. The various improvements of Weyl's law are reviewed very conscientiously, and Carleman's paper for the Congress of Scandinavian Mathematicians in 1934 is credited with being the first to use Green's function. Balian and Bloch also adopt this method, but they immediately assume that the energy E has a positive imaginary part γ. Therefore, all sums over eigenvalues are smoothed out, and only the gross asymptotic features are preserved as E becomes large.

This is the first time the number of eigenstates $N(\Lambda)$ for $\lambda = k^2 \leq \Lambda$, as in (8), is calculated to higher order in $1/k$ for an arbitrary three-dimensional cavity with a smooth boundary. An arbitrary linear relation $\partial\phi/\partial n = \kappa\phi$ is postulated to couple the boundary value ϕ with its normal derivative $\partial\phi/\partial n$.

Weyl's formula (8) gets two more terms that depend on the value of the angle $\delta = \arctan(\kappa/k)$,

$$N(\Lambda) \cong \frac{Volume}{6\pi^2}k^3 + \frac{Surface}{4\pi^2}k^2(\frac{\pi}{4} - \delta) + ..., \tag{19}$$

where the term in k is left out because it involves a fairly complicated integral over the mean (not the Gaussian) curvature of the surface, as well as a factor that depends on δ.

The general method that led to these results is applied to the electromagnetic radiation inside a conducting cavity, to a general domain with a Riemannian metric, and finally to Schrödinger's equation. The main idea is always the same as in the paper of Mark Kac (and other papers), although we now deal with exponentially damped waves rather than particles or heat diffusing. The short-time solutions of the wave equation are considered locally, first near a plane boundary, and then near a boundary with a constant curvature. Ridges and corners at the boundary present problems that require special treatment.

A long paper by Balian and Bloch appeared in the Annals of Physics of 1972 (received 3 May 1971) with the title "Eigenfrequency Density Oscillations". It discusses first the spectrum of vibrations in parallelepipeds and spheres, and then treats the expansion of Green's functions for general shapes of cavities into multiple reflections of rays. When taking the trace, the periodic orbits appear quite naturally in the form of closed polygons whose vertices lie on the boundary. The stability of orbits and the phases connected with the conjugate points are not mentioned. The generalization to Schrödinger's equation is discussed toward the end with a number of references to work in semiclassical mechanics including my two papers in JMP in 1970 and 1971. In the last paragraph one reads that "it is easy to guess that they (density oscillations) are associated with closed classical trajectories in the considered potential." There is no trace formula.

The last paper in this series was published 1974 again in Annals of Physics. Schrödinger's equation is solved with the help again of a multiple scattering expansion for Green's function. The separation of variables is no longer necessary, but the energy still has a positive imaginary part to enforce convergence, and complex coordinates are used for the classical paths. Although no examples are treated, some generalizations such as multi-component wavefunctions are considered without any further details. The whole problem was already in the court of the mathematicians!

3.4 The Mathematicians at Work

Selberg's trace formula remained an isolated achievement because it was based on the high degree of symmetry in the spaces of constant negative curvature. It also claimed an equality, rather than an asymptotic relation between the two sides of the formula, valid only for large energies. It was viewed, therefore, more as a statement concerning the symmetry group involved than a general feature for the motion of a particle or a wave.

This group-theoretical interpretation can also be given to the trace formula for an integrable dynamical system. As mentionned above, the trajectories in phase space lie on a torus of n dimensions, where n equals the number of freedoms. Each periodic orbit is described by n integers, and its length at a fixed energy E is given by a homogeneous function of these integers of degree 1. The singularities in the sum over all periodic orbits are most easily obtained with the help of Poisson's formula: Choose any pair of functions $\Lambda(\kappa)$ and $P(u)$ that are Fourier transforms of each other, as in (6), then,

$$2\pi \sum_{n=-\infty}^{+\infty} P(2\pi n) = \sum_{\nu=-\infty}^{+\infty} \Lambda(\nu). \tag{20}$$

This formula was used in the papers of JMP 1969 and 1970, and it was also the main tool in the papers by Michael Berry and Michael Tabor of 1976 and 1977 where they also treated only the integrable systems. It gives immediately the trace formula for the heat-flow equation on a circle of radius 1 where the physical constants have been normalized to 1,

$$\sum_{\nu=-\infty}^{+\infty} \exp(-\nu^2 t) = \sqrt{\pi/t} \sum_{n=-\infty}^{+\infty} \exp[-(2\pi n)^2/4t]. \tag{21}$$

The eigenvalues of the energy are simply ν^2 while the lengths of the periodic orbits are $2\pi n$. On the right-hand side, we have the propagator for this problem $K(x,t;0,0) = \sqrt{\pi/t}\exp(-x^2/4t)$. Notice that both sides of this equation are convergent for positive t.

Yves Colin de Verdiere published two short notes in the Comptes Rendus of the French Academy of Sciences for 1972, and then a full account in the Compositio Mathematica 1973. His declared intention is to generalize Poisson's formula, and he expects to do so in some form resembling Selberg's trace formula, but for more general Riemannian manifolds. He proves a formula that looks exactly like (21) with the eigenvalues of the Laplacian replacing ν^2 on the left, and the lengths of the closed geodesics replacing $2\pi n$ on the right. Moreover, there is a term for Weyl's law on the right and, most impressive, an estimate for the error provided t remains small. The proof uses Wiener's idea of integrating over all continuous paths, but then picks the stationary ones which are obviously the geodesics.

These results are probably not independent of the work in Saclay by Balian and Bloch, but the exact flow of ideas is not quite evident. There is no doubt, however, that a number of mathematicians were inspired to try their own methods to get new versions of the trace formula. Jacques Chazarain dealt with the wave equation on a Riemannian manifold. He constructs the propagator with the help of Hormander's Fourier-integral operators which spread the waves along the geodesics. The trace to be calculated is given by the sum,

$$S(t) = \frac{1}{2} \sum_{\ell \geq 0} \exp[\pm i\omega_\ell t], \tag{22}$$

where the frequencies ω_ℓ are the square-roots of the eigenvalues λ_ℓ, and both signs are taken in the sum in order to achieve symmetry with respect to the time t. The above sum is shown to have singularities for isolated values of the time which coincide with the periods of the periodic orbits.

This last claim seems quite strange at first, but can be explained qualitatively at least on the basis of the earlier trace formula (18). The inverse of the Fourier transform (12) gets us back to the propagator, or to its trace if it is applied to (18). The left-hand side then looks like (22) with the frequencies essentially equal to the eigenvalues of the Hamilton operator. On the right-hand side of (18), however, we use the stationary-phase argument since the energy E is large: the action integral $S(E)$ in the exponent is effectively a linear function of the energy, and its rate of change is the period T of the periodic orbit. Therefore, the Fourier transform of each term becomes a δ-function in time with the peak at $t = T$.

This result of Chazarain has a fascinating consequence: A measurement of the spectrum in an atom or molecule by the absorption of a photon gives essentially the left-hand side of (18). The experimental data consist of a series of spikes in energy that can be transformed numerically. The result should again be a series of spikes in the complementary variable which is time. Thus the experiment will give information on the periodic orbits of the corresponding classical system, or equivalently, the correlations in the experimental data are explained by the periodic orbits.

4 Classical and Quantum Mechanics

4.1 The Phase Portrait

Until the end of the 19-th century, mechanics was dominated by its celestial branch. The discovery of Uranus and Neptune, as well as the exploration of the asteroid belt between Mars and Jupiter, had generated new interest in finding better methods for predicting the motions in the solar system. The emphasis was on improving perturbation theory in order to get higher precision and find solutions for longer time-intervals.

Henri Poincare was the first mathematician-physicist to move mechanics away from its exclusive concern with the construction of better algorithms for the computation of particular trajectories. His early work on the theory of ordinary differential equations had convinced him that special solutions of the equations of motion are generally deceptive. Trajectories with almost identical initial conditions can have entirely different behavior for long times, a situation that is now described by the ubiquitous word chaos. Poincare proposed to study the overall appearance of the flow in phase space, the phase portrait.

His most useful trick in this task was the surface of section, which is still the tool of choice in the study of dynamical systems with few degrees of freedom. The map of the surface of section into itself, area conserving if there is no friction, is most easily understood in terms of its fixed points, so that one is immediately led to the search of periodic orbits. Poincare considered them to be "the only opening through which we might try to penetrate into the fortress which has the reputation of being impregnable".

George William Hill had the crucial idea of explaining some of the most difficult features in the motion of the Moon around the Earth in this manner. He had found a special solution for the full three-body system Moon-Earth-Sun that closes itself smoothly after one synodic month, although the Moon does nothing of the kind. But then he was able to construct the real trajectory of the Moon by investigating the stability of this periodic orbit. Poincare recognized the importance of this new approach, and he made his appreciation quite clear when he wrote the biography for the publication of Hill's Collected Works.

In spite of this auspicious beginning, the investigation of periodic orbits in dynamical systems remained the province of mathematicians, astronomers, and engineers until the second half of the 20-th century. The work of Hill and Poincare came shortly before the great breakthrough in physics of 1900, and the physicists were still caught in the earlier lessons of Lagrange, Hamilton, and Jacobi. Although quantum mechanics presented a very different picture of nature, it was quickly adapted to the methods that had been invented more than a century before. Herbert Goldstein's textbook of Classical Mechanics, even in its new edition of 1980, is a typical expression of this spirit which has dominated the post-Second-World-War generation of physicists in the US.

4.2 Dynamic Zeta-Functions

The use of Riemann's zeta-function in number theory seemed to be the main incentive for studying Selberg's trace formula, as was emphasized in section 1.2. At some point, however, maybe in the early sixties, the idea of a zeta-function as a book-keeping device in geometry and classical mechanics became popular. The integer n in (4) was replaced by the exponential of the length $\ell(p)$ of a periodic orbit p.

If the number of the periodic orbits increases as discussed in section 1.3, the corresponding formula (4) defines a complex-valued function of s. It exists at least in a halfplane which is defined by some lower bound for the real part of s. The sigularities of this zeta-function give more detailed information about the proliferation of the number of periodic orbits as their length increases. In exactly the same way, the singularities of the Riemann zeta-function were supposed to be the keys to account exactly for the proliferation of the prime numbers.

The singularities themselves did not seem to exist on their own ground, but they derived their existence from the periodic orbits. A long survey article by Steven Smale on Differentiable Dynamical Systems in the Bulletin of the American Mathematical Society of 1967 proposes such a function; he does not sound very confident of the use to be expected from this tool, because he calls it rather disparagingly "another zeta-function", as if there were already enough of them on the market.

In the early seventies, the statistical mechanicians under the leadership of David Ruelle explored a similar approach, again as a method for dealing with the probabilities involved in classical phase-space. This idea has a precursor in the use of generating functions in the theory of probability, and goes back to a long paper of Laplace in the Memoirs of the French Academy of Sciences of 1782, where he also proposes the method of steepest descent. This is the real version of the stationary-phase argument which plays a crucial role in both sections 3.1 and 3.2.

Although the example of Selberg's trace formula was available by then, all these zeta-functions missed an essential point: the denominator in (7) and (18) increases exponentially only with half the stability exponent $n\chi$, because each term represents a probability-amplitude whose square gives the probability in quantum mechanics.

In the case of a billiard where $S(E) = k\ell(p)$ and $k = \sqrt{2mE}/\hbar$, the general trace formula (18) converges absolutely only when the imaginary part of k is sufficiently large to beat the stability exponent $\chi/2$. But there are good arguments to believe that the trace converges conditionally all the way to the real k-axis because the phase factor in (7) causes destructive interference between the many terms. Of course, such a phenomenon is to be expected since quantum mechanics involves waves rather than numbers of particles.

4.3 Outlook

The idea of establishing a direct connection between the eigenvalues of Schrödinger's equation and the classical periodic orbits was not well received for about ten years, until 1980. On one hand the general trace formula presents so many technical difficulties that it could not be trusted; in particular the exponential proliferation of the periodic orbits in chaotic systems appeared to make any practical use almost impossible. On the other hand, there were some trivial examples like a few coupled harmonic oscillators that don't have any periodic orbits at all.

In all these objections, the basic idea of the trace formula is badly misunderstood. Its purpose is not to offer a universally valid recipe for cooking up and distilling out a series of approximate eigenvalues. There is no doubt, however, that a good grasp of the periodic orbits in a particular classical system will teach us a lot about the spectrum in the corresponding quantum system. The trace formula serves as a guide that is subject to careful scrutiny before any of its predictions can be accepted at face value.

This is not the place to review all the important work in the last 15 years that has helped to explain and justify the general trace formula. Among the many collections of articles on this subject I will only mention the "Chaos Focus Issue on Periodic Orbit Theory" which appeared at the beginning of volume 2 of the the journal CHAOS. It contains two contributions by Dieter Wintgen and his coworkers, the first on "The Semiclassical Helium Atom", and the second on "The Quantization of Chaotic Systems" where the anisotropic Kepler problem is discussed.

As the authors of the first article correctly emphasize, the spectrum of the helium atom represents the ultimate test for our understanding of the connection between classical and quantum mechanics. It was Dieter Wintgen's great merit to have recognized the historical importance of this problem, and to have devoted his great energy and skill to this endeavour. Of course, I have a special liking for the second article because it settles a problem that I had proposed some twenty years before, and on which I had used up a lot of my time, not always with the desired success.

The general trace formula has finally become a useful tool in making our intuitive knowledge of classical mechanics available to the solution of problems in quantum mechanics. That will become particularly valuble in the development of mesoscopic physics, as can be seen in the special issue on "Quantum Complexity in Mesoscopic Systems" that appeared in 1994 as volume 83 of Physica D. Unfortunately, Dieter Wintgen is no longer with us to participate in this large project where his unusual personality and great talents would have been extremely helpful.

References

Balian R and Bloch C, Ann Phys (New York) 60 (1970) 401, 63 (1971) 592, 64 (1971) 271 with Errata 84 (1974) 559, 69 (1972) 76, 85 (1974) 514

Berry M and Tabor M, Proc Roy Soc London A 349 (1976) 101 and A 356 (1977) 375, J Phys A 10 (1977) 371 and 10 (1977) 2083

Bishop A R, Ecke R E, and Mainieri R (editors), Quantum Complexity in Mesoscopic Systems, North-Holland Elsevier, Pysica D 83 (1995) 1-311

Carleman T, Eighth Cong Math Scand Stockholm 1934, p 34

Chazarain J, Inventiones Math 24 (1974) 65

Choquard P, Helv Phys Acta 28 (1955) 89

Colin de Verdiere Y, Compositio Mathematica 27 (1973) 83 and 159

Debye P, Annalen der Physik 39 (1912) 789-839

Dewitt B S, Rev Mod Phys 29 (1957) 377-397

Dirichlet G L (1837), Abhand Ak Wiss Berlin, Jahrgang 1837; cf Vorlesungen ueber Zahlentheorie, edited by Dedekind, Supplement VI

Einstein A, Verh Dtsch Phys Ges 19 (1917) 82

Fermi E, Zeits f Physik 48 (1928) 73 and 49 (1928) 550

Feynman R P, Rev Mod Phys 20 (1948) 367-387

Feynman R P and Hibbs A R, Quantum Mechanics and Path Integrals, McGraw-Hill 1965

Gutzwiller M C, J Math Phys 8 (1967) 1979, 10 (1969) 1004, 11 (1970) 1791, 12 (1971) 343, 14 (1973) 139

Goldstein H, Classical Mechanics, Addison-Wesley 1950 and 1980

Hadamard J, J Math Pure Appl 4 (1898) 27

Hejhal D A, Some Observations Concerning the Eigenvalues of the Laplacian and Dirichlet L-Series. Recent Progress in Analytic Number Theory, vol. 2, H. Halberstam and C. Hooley, eds. (Academic Press, London, 1981) 95–110

Hill G W, privately published 1877, then reprinted in Acta Math 8 (1886) 1-36; full account in Am J Math 1 (1878) 5, 129, and 245

Jeffreys H, Asymptotic Expansions, Oxford UP 1961 (and references there)

Kac M, Am Math Monthly 73 (1966) 1-23

Kac M, Probability and Related Topics in Physical Sciences, Interscience Publishers 1959; Integration in Function Spaces and Some of its Applications, Lezioni Fermiane, Pisa 1980; cf also, L Schulman, Techniques and Applications of Path Integration, Wiley-Interscience, 1980

Keller J B, Ann Phys (New York) 4 (1958) 180

Kolmogoroff A N, Dokl Akad Nauk SSSR 124 (1959) 754-755

Laplace P S, Histoire et Memoires de l'Academie Royale des Sciences, Annee 1782, Paris 1785, p 1-88

Liouville J, cf his biography "Joseph Liouville 1809-1882 Master of Pure and Applied Mathematics" by Jesper Luetzen, Springer-Verlag 1990

Maslov V P, Theorie des Perturbations et Methodes Asymptotiques, Paris, 1972; translation of original edition in Russian of 1965

McKean H P and Singer I, J Diff Geom 1 (1967) 43-69

Morette C, Phys Rev 81 (1951) 848

Morse M, The Calculus of Variations in the Large, Am Math Soc Pub 18, 1934

Poincare H, The Collected Works of George William Hill, vol 1, Introduction VII-XVIII, Carnegie Foundation, Washington 1905

Poincare H, Journal de Math 3 serie 7 (1881) 375-422 and 8 (1882) 251- 296, 4 serie 1 (1885) 163-244 and 2 (1886) 151-217; this series of articles is reprinted in volume 1 of Oevres, Gauthier-Villars 1928

Ruelle D, Statistical Mechanics and Thermodynamic Formalism, Addison- Wesley, 1978

Riemann B (1859), Gesammelte Mathematische Werke, p 145

Thomas L H, Proc Camb Phil Soc 23 (1927) 542

Selberg A (1948), Canad J Math 2 (1950), 66-78

Selberg A, J Indian Math Soc 20 (1956), 47-87

Siegel C L (1931), Quellen und Studien zur Geschichte der Mathematik, Astronomie und Physik, Abteilung B Studien, Berlin, Band 2, p 45

Smale S, Bull Am Math Soc 73 (1967) 747-817

Van Vleck J H, Proc Natl Acad Sci USA 14 (1929) 178

Weyl H, Nachr Akad Wiss Goettingen (1911) 110, Math Annalen 71 (1912) 441-479, J Reine Angew Math 141 (1912) 1-11, 163-181 and 143 (1913) 177- 212; these articles are reprinted in volume 1 of Gesammelte Abhandlungen, Springer-Verlag 1928

Wiener N, several articles in volume 1 of Collected Works, MIT Press 1976; cf the article "Wiener and Integration in Function Spaces" by Mark Kac, Bull Am

Math Soc 72 (1966) 52-68, in particular the appreciation of Wiener's pivotal role at the end of this article.
Wintgen D, Richter K, and Tanner G, The Semiclassical Helium Atom, CHAOS 2 (1992) 19-33
Wintgen D and Tanner G, Quantization of Chaotic Systems, CHAOS 2 (1992) 53-59

Quantum Fluids and Classical Determinants

Predrag Cvitanović[1], Gábor Vattay[1,2] and Andreas Wirzba[3]

[1] Center for Chaos and Turbulence Studies, Niels Bohr Institute
Blegdamsvej 17, DK-2100 Copenhagen Ø, Denmark
[2] Eötvös University, Department of Solid State Physics
Múzeum krt. 6-8., H-1088 Budapest, Hungary
[3] Institut für Kernphysik, TH Darmstadt
Schloßgartenstraße 9, D-64289 Darmstadt, Germany

Abstract. A "quasiclassical" approximation to the quantum spectrum of the Schrödinger equation is obtained from the trace of a quasiclassical evolution operator for the "hydrodynamical" version of the theory, in which the dynamical evolution takes place in the extended phase space $[q(t), p(t), M(t)] = [q_i, \partial_i S, \partial_i \partial_j S]$. The quasiclassical evolution operator is multiplicative along the classical flow, the corresponding quasiclassical zeta function is entire for nice hyperbolic flows, and its eigenvalue spectrum contains the spectrum of the semiclassical zeta function. The advantage of the quasiclassical zeta function is that it has a larger analyticity domain than the original semiclassical zeta function; the disadvantage is that it contains eigenvalues extraneous to the quantum problem. Numerical investigations indicate that the presence of these extraneous eigenvalues renders the original Gutzwiller-Voros semiclassical zeta function preferable in practice to the quasiclassical zeta function presented here. The cumulant expansion of the exact quantum mechanical scattering kernel and the cycle expansion of the corresponding semiclassical zeta function part ways at a threshold given by the topological entropy; beyond this threshold quantum mechanics cannot resolve fine details of the classical chaotic dynamics.

1 Introduction

What we shall describe here is very much in the spirit of early quantum mechanics, and had physicists of the period been as familiar with classical chaos as we are today, this theory would have been developed in 1920's. The main idea is this: in the Bohr–de Broglie visualization of quantization, one places a wave instead of a particle on a Keplerian orbit around the hydrogen nucleus. The quantization condition is that allowed orbits are only those for which such a wave is stationary; from this follows the Balmer spectrum, the old quantum theory, and the more sophisticated theory of Schrödinger and others. Today we are very aware of the fact that integrable systems are exceptional and that chaos is the rule. So, can the Bohr quantization be generalized to chaotic systems? The answer was provided by Gutzwiller in 1971; the trace of the quantum evolution operator for a chaotic system in a semiclassical approximation is given by the Gutzwiller trace formula, an oscillating sum over all periodic orbits of the system.

There is however a hidden intellectual challenge in Gutzwiller's derivation: the derivation is based on the semiclassical Van Vleck approximation $K(x,x',t)$ to the quantum propagator which does not satisfy the semigroup property

$$\int \mathrm{d}x''\, K(x,x'',t_1)K(x'',x',t_2) \neq K(x,x',t_1+t_2)\ . \tag{1}$$

In the literature this problem is usually sidestepped by saying that an equality holds if the integral is carried out by the saddle point method. Here we offer an alternative "quasiclassical" quantization scheme based on a quasiclassical evolution operator which is multiplicative along the flow. Our main result is the *quasiclassical trace formula* for the quantization of a Hamiltonian dynamical system. For a system of 2 degrees of freedom the quasiclassical trace formula takes the form

$$\operatorname{tr}\mathcal{L}^t(E) = \sum_p T_p \sum_{r=1}^{\infty} \frac{\delta(t-rT_p)\,\mathrm{e}^{\frac{\mathrm{i}}{\hbar}(S_p-ET_p)r-\mathrm{i}\pi\frac{m_p}{2}r}}{|\Lambda_p|^{r/2}(1-1/\Lambda_p^r)^2(1-1/\Lambda_p^{2r})}\ .$$

Throughout this paper we reserve the term "quasiclassical" to distinguish this class of formulae from the original Gutzwiller formulae which we shall refer to as "semiclassical".

Search for the above formula was motivated by the classical periodic orbit theory, where convergence of cycle expansions is under much firmer control than in the semiclassical quantizations. One of the main lessons of the classical theory is that the "exponential proliferation of orbits" in itself is not the problem; what limits the convergence of cycle expansions for generic flows is the proliferation of the grammar rules, or the "algorithmic complexity". Indeed, for nice hyperbolic flows a theorem of H. H. Rugh (1992) asserts that the appropriate spectral determinants are entire and that their cycle expansions converge superexponentially.

On the basis of close analogy between the classical and the quantum zeta functions, it has been hoped (Cvitanović 1992) that for nice hyperbolic systems the semiclassical zeta functions of Gutzwiller (1988) and Voros (1988) should also be entire. This hope was dashed by Eckhardt and Russberg (1992) who established that the semiclassical zeta functions for the 3-disk repeller have poles. Their result had in turn motivated guesses for spectral determinants with improved convergence properties by Cvitanović and Rosenqvist (1993) and Cvitanović et al. (1993), which eventually lead to the first derivation of the above trace formula by Cvitanović and Vattay (1993). In this paper we offer a different derivation and interpretation of this formula.

Improved analyticity has been very useful in sorting out the relative importance of the semiclassical, diffraction (Wirzba (1992), Wirzba (1993), Vattay, Wirzba and Rosenqvist (1994)) and quantum contributions (Gaspard and Alonso Ramirez (1992), Vattay (1996), Vattay (1994), Vattay and Rosenqvist (1996)). One had also hoped that improved analyticity would yield cycle expansions that would converge faster with the maximal cycle length truncation than the Gutzwiller-Voros type zeta functions. As is shown here, this is not the case. Improved analyticity comes at a cost; the quasiclassical zeta functions predict extraneous eigenvalues which are purely classical and do not belong to the quantum spectrum, but their presence degrades significantly the convergence of the cycle expansions. Furthermore, the investigation of Wirzba (1996) has clarified the relationship between the cumulant expansion of the exact quantum mechanical scattering kernel and the cycle expansion of the semiclassical zeta function; the order of expansion at which the two part their ways is determined by the value of the topological entropy, and beyond this threshold quantum mechanics fails to resolve the arbitrarily fine details of the classical chaotic dynamics.

The paper is organized as follows: in Sect. 2 through Sect. 4 we develop the quasiclassical evolution operator formalism for a semiclassical approximation to the Schrödinger equation, and in Sect. 5 we derive the trace and zeta function formulae for quasiclassical quantization. In Sect. 6 we confront in numerical experiments the cycle expansions of the quasiclassical zeta functions with the cycle expansions of the more standard semiclassical zeta functions and dynamical zeta functions, as well as with the exact quantum mechanical results, and in Sect. 7 we explain the distinction between the asymptotic nature of quantum mechanical cumulant expansions and the convergence of semiclassical cycle expansions. Appendices contain some technical details as well as a discussion of the relation of the quasiclassical quantization to the Selberg zeta function.

2 Quantum Mechanics in Hydrodynamical Form

The Schrödinger equation for a particle in a d-dimensional potential V is

$$\left(\mathrm{i}\hbar\frac{\partial}{\partial t}+\frac{\hbar^2}{2}\Delta - V(q)\right)\psi(q,t)=0\ , \tag{2}$$

where $\psi(q,t)$ is the wave function, and we set the particle mass $m = 1$ throughout. The ansatz

$$\psi = \varphi \mathrm{e}^{\mathrm{i}S/\hbar} \tag{3}$$

is as old as quantum mechanics itself. Schrödinger's first wave mechanics paper was submitted 27 January 1926. Submission date for Madelung (1926) "quantum theory in hydrodynamical form" paper, where this ansatz is interpreted as a fluid flow, was 25 October 1926.

Substituting the ansatz into (2), differentiating, and separating the result into the real and imaginary parts (under assumption that both φ and S are real functions) yields

$$\frac{\partial S}{\partial t}+\frac{1}{2}(\nabla S)^2+V(q)-\frac{\hbar^2}{2}\frac{\Delta\varphi}{\varphi}=0 \tag{4}$$

$$\frac{\partial\varphi}{\partial t}+\nabla S\nabla\varphi+\frac{1}{2}\Delta S\varphi=0\ . \tag{5}$$

The $\hbar^2$ term has many names and is called the "quantum potential" by Bohm (1952), "enthalpy" by Spiegel (1995), by fluid dynamics analogy, or "quantum pressure" by Feynman (1972). While Schrödinger in his 21 June 1926 paper noted that $\rho = \varphi\varphi^*$ satisfies the continuity equation, it was Born who (in a footnote of his 24 June 1926 paper) identified ρ as the probability density. Interpretations of quantum mechanics bifurcate here; keeping the $\hbar$ term in the potential (4) leads to the Madelung "fluid" theory. Shifting the $\hbar$ term into the second equation enforces that S satisfies the classical Hamilton-Jacobi equation

$$\frac{\partial S}{\partial t} + \frac{1}{2}(\nabla S)^2 + V(q) = 0 \ , \tag{6}$$

while the "diffusive" $\hbar$ term in the equation for the amplitude

$$\frac{\partial \varphi}{\partial t} + \nabla S \nabla \varphi + \frac{1}{2}\Delta S \varphi = \frac{i\hbar}{2}\Delta\varphi \ , \tag{7}$$

motivates the "stochastic" interpretation of Nelson (1985).

2.1 Semiclassical Approximation

Our goal here is to study the semiclassical approximation of quantum mechanics, with $\hbar$ small, and concentrate on the leading order expressions. This can be achieved by setting $\hbar$ formally zero in either the "hydrodynamic" or the "stochastic" picture. Either way we get

$$\frac{\partial S}{\partial t} + \frac{1}{2}(\nabla S)^2 + V(q) = 0 \tag{8}$$

$$\frac{\partial \varphi}{\partial t} + \nabla S \nabla \varphi + \frac{1}{2}\Delta S \varphi = 0 \ . \tag{9}$$

As long as we concentrate on the leading semiclassical contribution, we can steer clear of the passions aroused by the differences between different interpretations of quantum mechanics, and follow the original Gutzwiller derivation of the semiclassical trace formula via Van-Vleck approximation to the quantum propagator, Gutzwiller (1971), Gutzwiller (1990).

Nevertheless, the procedure is unsatisfactory in the sense that in order to get an operator with the semigroup property we need to impose the saddle point condition. In order to overcome this problem we have to learn more about the technical details of the semiclassical dynamics first. This analysis will show that the semiclassical wave function evolution can be described as an evolution over an extended dynamical space.

3 Semiclassical Evolution as a Set of ODE's

We now examine the semiclassical approximation to the quantum wave evolution (a linear partial differential equation) and show that it can be reformulated in terms of a finite number of ordinary differential equations. We start by traversing a well trodden ground: Hamilton's 1823 formulation of wave mechanics.

3.1 Hamilton's Wave Mechanics

In the wave equation (2) q is *not* a variable; the variable is the wave function ψ that evolves with time, and one can think of ψ as an (infinite dimensional) vector where q plays a role of an index. $S(q,t)$ plotted as a function of the position q for two different times looks something like Fig. 1(a). A smooth

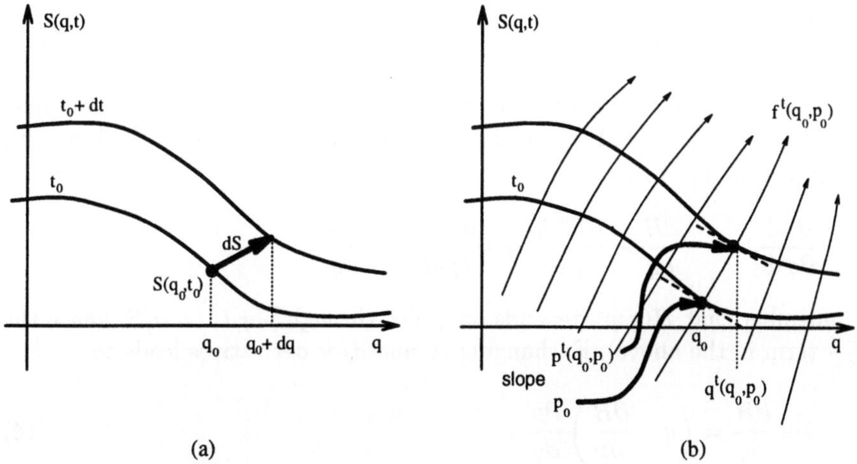

Fig. 1. (a) A wavefront $S(q,t)$ plotted as a function of the position q for two different times. (b) The phase of the wavefront $S(q,t)$ transported by a swarm of "particles"; Hamilton's equations (15) construct $S(q,t)$ by transporting $q_0 \to q(t)$ and p_0, the slope of $S(q_0,t_0)$, to $p_0 \to p(t)$.

"wavefront" $S(q,t_0)$ deforms smoothly with time into the "wavefront" $S(q,t)$ at time t. At this point one can ask: could we think of this front as a swarm of particles that move in such a way that if we know $S(q,t)$ and its slope $\partial S/\partial q$ at q at initial time $t=t_0$, we can construct a corresponding piece of $S(q,t)$ and its slope at time t, Fig. 1(b)? For notational convenience, define

$$p_i = p_i(q,t) := \frac{\partial S}{\partial q_i}, \quad i = 1,2,\ldots,d \ . \tag{10}$$

In the semiclassical approximation (4) reduces to the *Hamilton-Jacobi* equation

$$\frac{\partial S}{\partial t} + H\left(q, \frac{\partial S}{\partial q}\right) = 0 \ , \tag{11}$$

where $H(q,p)$ is the Hamiltonian, in this case

$$H = p^2/2m + V(q) \ . \tag{12}$$

For sake of simplicity we set $m = 1$ throughout. We shall also assume that the Hamiltonian is time independent (energy is conserved) and separable into a sum of kinetic and potential parts.

Infinitesimal variation of $S(q,t)$, Fig. 1(a), is given by

$$\mathrm{d}S = \mathrm{d}t\frac{\partial S}{\partial t} + \mathrm{d}q\frac{\partial S}{\partial q} \ .$$

Dividing through by $\mathrm{d}t$ and substituting (11) we obtain

$$\frac{\mathrm{d}S}{\mathrm{d}t} = -H(q,p) + \dot{q}p \ . \tag{13}$$

The "velocity" $\dot{q}$ is arbitrary, and now comes Hamilton's idea: can we adjust $\dot{q}$ so that p is promoted to a variable *independent* of q? Take a $\frac{\partial}{\partial q}$ derivative of both sides of (13):

$$\frac{\partial}{\partial q}\frac{\mathrm{d}}{\mathrm{d}t}S = -\frac{\partial H}{\partial q} - \frac{\partial H}{\partial p}\frac{\partial p}{\partial q} + p\frac{\partial}{\partial q}\frac{\mathrm{d}}{\mathrm{d}t}q + \dot{q}\frac{\partial p}{\partial q} \ .$$

(remember that $H(q,p)$ depends on q also through $p(q,t) := \partial_q S$, hence the $\frac{\partial H}{\partial p}$ term in the above). Exchanging ∂_q and d/dt derivatives leads to

$$\dot{p} + \frac{\partial H}{\partial q} = \left(\dot{q} - \frac{\partial H}{\partial p}\right)\frac{\partial p}{\partial q} \ . \tag{14}$$

Now we use the freedom of choosing $\dot{q}$, and trade the $\frac{\partial p}{\partial q}$ dependence for a set of ordinary differential equations, the Hamilton's equations

$$\dot{q} = \frac{\partial H(q,p)}{\partial p} \ , \qquad \dot{p} = -\frac{\partial H(q,p)}{\partial q} \tag{15}$$

with the "wavefront" $S(q,t)$ replaced by the action increment $S^t(q_0,p_0)$, the integral of (13) evaluated along the phase space flow $(q_0,p_0) \to (q(t),p(t))$:

$$S^t(q_0,p_0) = \int_{t_0}^{t} \mathrm{d}\tau \ \{\dot{q}(\tau)\cdot p(\tau) - H(q(\tau),p(\tau))\} \ . \tag{16}$$

If the energy is conserved, $H(q(\tau),p(\tau)) = E$, and the second term is simply $(t_0 - t)E$.

To summarize: the Hamilton-Jacobi *partial* differential equation (11) for the evolution of a wave front can be reformulated as a finite number of *ordinary* differential equations of motion which increment the initial action $S(q_0,t_0)$ by the integral (16) along the phase space trajectory $(q(\tau),p(\tau))$. In order to obtain the full quasiclassical evolution we also have to deal with the amplitude evolution (9).

3.2 Amplitude Evolution

The amplitude evolution (9) now takes place in the velocity field given by

$$v(q,t) = \nabla S(q,t) \ . \tag{17}$$

We can define $q(t) = f^t(q)$ as a solution of the differential equation

$$\dot{q} = v(q,t) \tag{18}$$

with initial condition $q(0) = q$ at time $t = 0$. This solution will coincide with $q^t(q, \nabla S(q,0))$, which is the q solution of the Hamilton's equations with initial conditions $q' = q$ and $p' = \nabla S(q',0)$. We introduce the notation $\kappa(q,t) = \Delta S(q,t)$ and write (9) as

$$\left\{\frac{\partial}{\partial t} + v(q,t)\cdot\nabla + \frac{1}{2}\kappa(q,t)\right\}\varphi(q,t) = 0 \ . \tag{19}$$

This is a linear equation in φ, so its solution can be written in terms of its Green's function as

$$\varphi(q,t) = \int \mathrm{d}q' \, \tilde{L}^t(q,q')\varphi(q',0) \tag{20}$$

where the kernel $\tilde{L}^t(q,q')$ is the special solution of (19) with initial condition $\tilde{L}^0(q,q') = \delta(q-q')$. It is easily checked by direct substitution into (20) and (19) that this Green's function is given by

$$\tilde{L}^t(q,q') = \exp\left\{\frac{1}{2}\int_0^t \kappa(f^\tau(q'),\tau)\,\mathrm{d}\tau\right\}\delta(q - f^t(q')) \ , \tag{21}$$

where an extra negative contribution to (19) results from $v(q,t)\nabla\delta(q-f^t(q')) = -(\nabla v(q,t))\delta(q - f^t(q'))$ and $\nabla v(q,t) = \kappa(q,t)$.

3.3 Quasiclassical Evolution

The whole quasiclassical evolution procedure can now be summarized. First we take our initial wave function $\psi(q,0)$. We pick a function $S(q,t)$, a solution of (8), and compute the initial amplitude $\varphi(q,0) = \mathrm{e}^{-\mathrm{i}S(q,0)/\hbar}\psi(q,0)$. We evolve this amplitude for time t and put back the phase:

$$\psi(q,t) = \mathrm{e}^{\mathrm{i}S(q,t)/\hbar}\int \mathrm{d}q' \, \tilde{L}^t(q,q')\mathrm{e}^{-\mathrm{i}S(q',0)/\hbar}\psi(q',0) \ . \tag{22}$$

The whole evolution can be cast into the semiclassical evolution operator

$$\psi(q,t) = \int \mathrm{d}q' \, L^t(q,q',S)\psi(q',0) \ , \tag{23}$$

where

$$L^t(q,q',S) = \tag{24}$$
$$\exp\left\{\frac{\mathrm{i}}{\hbar}(S(q,t)-S(q',0))+\frac{1}{2}\int_0^t \kappa(f^\tau(q'),\tau)\,\mathrm{d}\tau\right\}\delta(q-f^t(q')) \ .$$

The functional dependence on $S(q',0)$ sounds somewhat discouraging; we have to see it in an explicit form in order to understand the machinery of this operator.

The most complicated looking object here is the function

$$\lambda(q',t)=\int_0^t \kappa(f^\tau(q'),\tau)\,\mathrm{d}\tau \ .$$

We do not need the full information about $S(q,t)$ in order to compute this integral of $\Delta S(q,t)$ along the trajectory; as we shall see, an ODE suffices to evaluate this function. Consider the curvature matrix

$$\mathbf{M}_{ij}=\frac{\partial^2 S(q,t)}{\partial q_i \partial q_j} \ . \tag{25}$$

The time evolution equation for this matrix is obtained by taking the second derivatives of (8):

$$\frac{\partial \mathbf{M}}{\partial t}+v(q,t)\cdot\nabla\mathbf{M}+\mathbf{M}^2+\mathbf{D}^2V=0 \ , \tag{26}$$

where $\mathbf{D}^2V$ is the second derivative matrix of the potential. The first two terms combine to the full time derivative, and the evolution of $\mathbf{M}$ along a trajectory is given by

$$\dot{\mathbf{M}}=-\mathbf{M}^2-\mathbf{D}^2V \ . \tag{27}$$

So in the extended dynamical space we do not only keep track of q and slope of S at q, but also the curvature of S at q, see Fig. 2. Let us denote the solution of this ODE along a trajectory with starting point (q,p) and an initial matrix $\mathbf{M}$ by $\mathbf{M}^t(q,p,\mathbf{M})$. The function $\lambda(q,t)$ now can be expressed as

$$\lambda(q,t)=\int_0^t \mathrm{d}\tau\,\mathrm{tr}\,\mathbf{M}^\tau(q,p,\mathbf{M}) \tag{28}$$

with p initialized as $p=\nabla S(q,0)$.

Another point where the "functional dependence" can be simplified is the phase term. We can make the replacement

$$S(q,t)-S(q',0)=S^t(q',p') \tag{29}$$

in the kernel (25), where S^t is the integral (16) with initial point $(q',p'=\nabla S(q',0))$ and $t_0=0$.

With these observations the kernel (25) can be written as

$$L^t(q,q',S)=\int \mathrm{d}p'\,\mathrm{d}\mathbf{M}'\,\mathrm{e}^{\mathrm{i}S^t(q',p')/\hbar+\frac{1}{2}\int_0^t \mathrm{d}\tau\,\mathrm{tr}\,\mathbf{M}^\tau(q',p',\mathbf{M}')}\times$$
$$\delta(q-q^t(q',p'))\,\delta(p'-\nabla S(q',0))\,\delta(\mathbf{M}'-\mathbf{D}^2S(q',0)) \ , \tag{30}$$

where we have made the functional dependence explicit.

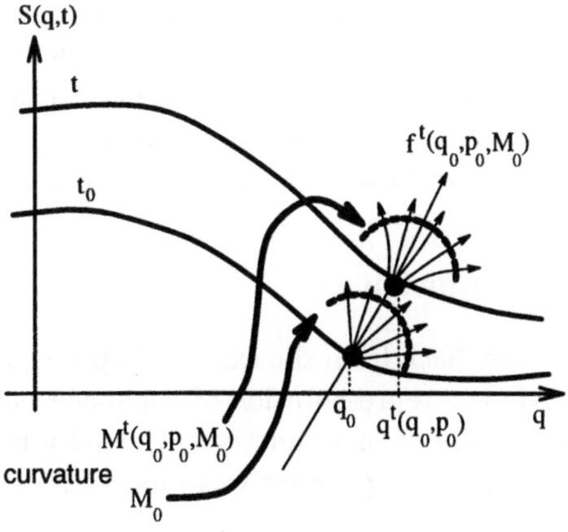

Fig. 2. The evolution of the curvature matrix M_{ij} of the wavefront $S(q,t)$ along the trajectory $[q(t), p(t), M(t)]$ in the extended dynamical space.

4 Quasiclassical Evolution Operator

If we write the time evolution of a wave function we get

$$\begin{aligned}\psi(q,t) = \int \mathrm{d}q'\,\mathrm{d}p'\,\mathrm{d}\mathbf{M}'\,W^t(q',p',\mathbf{M}')\,\delta(q-q^t(q',p'))\\ \times\psi(q',0)\,\delta(p'-\nabla S(q',0))\,\delta(\mathbf{M}'-\mathbf{D}^2S(q',0))\ ,\end{aligned} \tag{31}$$

where $W^t(q,p,\mathbf{M})$ is a short hand notation for the exponential in (30). We now make a new proposal: let us regard the last deltas $\delta(p'-\nabla S(q',0))\,\delta(\mathbf{M}'-\mathbf{D}^2S(q',0))$ as a part of the wave function. In other words, we think of $\Psi(q',p',\mathbf{M}') = \psi(q',0)\,\delta(p'-\nabla S(q',0))\,\delta(\mathbf{M}'-\mathbf{D}^2S(q',0))$ as a function defined on the $(q,p,\mathbf{M})$ space. We can multiply (31) by $\delta(p-\nabla S(q,t))\,\delta(\mathbf{M}-\mathbf{D}^2S(q,t))$ and write the evolved function in the extended space as

$$\Psi(q,p,\mathbf{M}) = \int \mathrm{d}q'\,\mathrm{d}p'\,\mathrm{d}\mathbf{M}'\,\mathcal{L}^t(q,p,\mathbf{M}|q',p',\mathbf{M}')\,\Psi(q',p',\mathbf{M}')\ ,$$

where the kernel of this integral operator shall be referred to as the *quasiclassical evolution operator*

$$\begin{aligned}\mathcal{L}^t(q,p,\mathbf{M}|q',p',\mathbf{M}') = \mathrm{e}^{\mathrm{i}S^t(q',p')/\hbar+\frac{1}{2}\int_0^t \mathrm{d}\tau\,\mathrm{tr}\,\mathbf{M}^\tau(q',p',\mathbf{M}')}\\ \times\delta(q-q^t(q',p'))\,\delta(p-p^t(q',p'))\,\delta(\mathbf{M}-\mathbf{M}^t(q',p',\mathbf{M}'))\ .\end{aligned} \tag{32}$$

Here the quantities $\nabla S(q,t)$ and $\mathbf{D}^2S(q,t)$ are computed from their initial values and replaced with $p^t(q',p')$ and $\mathbf{M}^t(q',p',\mathbf{M}')$ using (29).

So, what does this mean? We have constructed an evolution operator which acts on functions of the $(q,p,\mathbf{M})$ space. Because of the three delta functions the evolution operator has the semigroup property. However, there will be a price to pay: while a wave function can be embedded into the enlarged space, not all the functions living in the enlarged space represent functions in the old space. The spectrum of the quasiclassical operator will contain the semiclassical spectrum, but as we shall see in Sect. 6, it will also contain extraneous eigenvalues without quantum mechanical counterpart.

4.1 Wave Packet Evolution

There is also an easy way back from the extended space to the original one. If the function $\Psi(q,p,\mathbf{M})$ is a representation of a q space wave function or represents a linear combination of such functions, the delta function dependence on p and $\mathbf{M}$ ensures that a q dependent wave function can be recovered by

$$\psi(q,t) = \int \mathrm{d}p\,\mathrm{d}\mathbf{M}\,\Psi^t(q,p,\mathbf{M}) \ . \tag{33}$$

The quasiclassical evolution introduced here is closely related to the Gaussian wave packet evolution theories of Heller (1975), Heller, Tomsovic and Sepúlveda (1992). There a wave packet

$$\psi(q,0) = A_0 \mathrm{e}^{\mathrm{i}p_0(q-q_0)/\hbar + \frac{1}{2\hbar}(q-q_0)\mathbf{M}_0(q-q_0)} \tag{34}$$

is "launched" at $t = 0$, with the parameters $(q_0, p_0, \mathbf{M}_0)$ evolving in time according to the equations we have for q, p and $\mathbf{M}$, and with the amplitude evolving as

$$A^t = A_0 \mathrm{e}^{\mathrm{i}S^t(q_0,p_0)/\hbar - \frac{1}{2}\int_0^t \mathrm{d}\tau\,\mathrm{tr}\,\mathbf{M}^\tau(q_0,p_0,\mathbf{M}_0)} \ . \tag{35}$$

Initial wave functions can be decomposed into a linear combination of wave packets and the pieces can be evolved separately. Each packet is characterized by a phase point in the $(q,p,\mathbf{M})$ phase space and evolves according to (15) and (27), with clouds of points representing initial wave packets evolving as in the Heller, Tomsovic and Sepúlveda (1992) picture.

4.2 A Classical Motivation for the Extended Dynamical Space

The above discussion might lead the reader to believe that the extended dynamical phase space is a peculiarity of quantum quasiclassics. However, what we have done is an example of a much more general procedure for constructing multiplicative evolution operators in settings where the multiplicative property seems to have been lost.

The problem can be illustrated by the Ruelle (1987) "thermodynamic" evolution operator of form

$$\mathcal{L}^t(x,x') = \mathrm{e}^{h^t(x')}\,\delta(x - f^t(x'))\,\frac{1}{|\Lambda^t(x')|^{\beta-1}} \ ,$$

with $\Lambda^t(x)$ an eigenvalue of the Jacobi matrix $\mathbf{J}^t(x)$ (see Appendix A) and $h^t(x)$ is a weight additive along the trajectory $f^t(x)$. For one-dimensional maps this operator is multiplicative, but *not* so for flows with two or more transverse dimensions, for the simple reason that the eigenvalues of successive stability matrices are in general not multiplicative,

$$\Lambda_{ab} \neq \Lambda_a \Lambda_b \ .$$

Here $\mathbf{J}_{ab} = \mathbf{J}_b \mathbf{J}_a$ is the Jacobian matrix of the trajectory consisting of consecutive segments a and b, $\mathbf{J}_a$ and $\mathbf{J}_b$ are the stability matrices for these segments separately, and Λ's are their leading eigenvalues. It was this lack of multiplicative property for Λ's that had for long time frustrated attempts to construct evolution operators whose spectrum contains the semiclassical Gutzwiller spectrum, until the method presented here was developed.

The main idea, extending the dynamical system to the tangent space of the flow, is suggested by one of the standard numerical methods for evaluation of Lyapunov exponents; instead of computing eigenvalues of linearized stability matrices, one monitors the growth rate of separation between nearby trajectories, i.e. one adjoins the d-dimensional tangent space $\xi \in TU_x$ to the d-dimensional dynamical evolution space $x \in U \subset \mathbb{R}^d$. The dynamics in the $(x, \xi) \in U \times TU_x$ space is governed by the system of equations of variations, Arnold (1978):

$$\dot{x} = \mathbf{v}(x) \, , \quad \dot{\xi} = \mathbf{D}\mathbf{v}(x)\xi \ .$$

Here $\mathbf{D}\mathbf{v}(x)$ is the derivative matrix of the flow. We write the solution as

$$x(t) = f^t(x_0) \ , \quad \xi(t) = \mathbf{J}^t(x_0) \cdot \xi_0$$

with the tangent space vector ξ transported by the transverse stability matrix $\mathbf{J}^t(x_0) = \partial x(t)/\partial x_0$. Multiplicative evolution operators and the corresponding trace and determinant formulae for such flows are given in Cvitanović and Vattay (1993) and Pollner and Vattay (1996).

5 Quasiclassical Trace and Determinant Formulae

Determination of the approximate eigenvalues of the Schrödinger operator (2) is now reduced to the computation of the eigenvalues of the quasiclassical evolution operator (32). But before we do this, a warning is in order. The spectrum of the new operator *contains* the semiclassical spectrum, i.e. we might find eigenvalues beyond those found in quantum mechanics. Optimally these extraneous eigenvalues should be filtered out, but at present we know of no practical technique for doing this, other than comparison with the exact quantum mechanical spectrum.

We shall determine the eigenvalues of our operator by first deriving the classical trace formula (Cvitanović and Eckhardt (1991), Cvitanović et al.

(1996)), and then determining the zeros of the associated Fredholm determinant, in this context called the quasiclassical zeta function. The (p,q) integrations can be carried out first, and yield a weighted sum over primitive periodic orbits p and their repetitions r

$$\operatorname{tr} \mathcal{L}^t(E) = \sum_p T_p \sum_{r=1}^{\infty} \frac{\delta(t - rT_p)\, \mathrm{e}^{\frac{\mathrm{i}}{\hbar}(S_p - ET_p)r}}{|\det(1 - \mathbf{J}_p^r)|} \Delta_{p,r} \ . \tag{36}$$

By the periodicity condition $\delta(t - rT_p)$ the $\mathbf{M}$ trace is restricted to a transverse Poincaré section of the flow, evaluated at a prime cycle completion $t = T_p$, or its r-th repeat

$$\Delta_{p,r} = \int \mathrm{d}\mathbf{M}\, \delta\big(\mathbf{M} - \mathbf{M}^{rT_p}(q,p,M)\big)\, \mathrm{e}^{\frac{r}{2}\int_0^{T_p} \mathrm{d}\tau \operatorname{tr} \mathbf{M}^\tau(q,p,\mathbf{M})} \ . \tag{37}$$

The integration of this part requires some skill and it is left for Appendix A. It turns out that this last integral can also be expressed in terms of the eigenvalues of the full phase space Jacobian matrix $\Lambda_1, \Lambda_2, \cdots, \Lambda_{d+1} = 1/\Lambda_1, \cdots, \Lambda_{2d} = 1/\Lambda_d$. Putting all ingredients together we get the *quasiclassical trace formula* for the quantization of a Hamiltonian dynamical system in $(d+1)$ configuration dimensions, i.e. restricted to the fixed energy shell in the $2(d+1)$ phase space:

$$\operatorname{tr} \mathcal{L}^t(E) = \sum_p T_p \sum_{r=1}^{\infty} \prod_{i=1}^{d} \frac{\delta(t - rT_p)\, \mathrm{e}^{(\frac{\mathrm{i}}{\hbar}(S_p - ET_p) - \mathrm{i}\pi m_p/2)r}}{|\Lambda_{p,i}|^{r/2}(1 - 1/\Lambda_{p,i}^r)^2(1 - 1/\Lambda_{p,i}^{2r})} \ . \tag{38}$$

Here $T_p(E) = \oint dt$ is the p-cycle period, $S_p(E) = \oint pdq$ the cycle action evaluated along the periodic orbit on the energy surface $H = E$, m_p the Maslov index, and $\Lambda_{p,1}, \Lambda_{p,2}, \cdots, \Lambda_{p,d}$ are the d expanding eigenvalues of the transverse Jacobian matrix of the flow belonging to the p-cycle. The period is related to the action through $T_p(E) = \frac{\partial}{\partial E} S_p(E)$. The associated quasiclassical zeta function is given by

$$Z_{\mathrm{qc}}(E) = \exp\left\{ -\sum_{p,r} \frac{1}{r} \prod_{i=1}^{d} \frac{|\Lambda_{p,i}|^{-r/2} \mathrm{e}^{\frac{\mathrm{i}}{\hbar} S_p(E) r - \mathrm{i}\pi \frac{m_p}{2} r}}{(1 - 1/\Lambda_{p,i}^r)^2 (1 - 1/\Lambda_{p,i}^{2r})} \right\} \tag{39}$$

(see e.g. Cvitanović et al. (1996) for the trace ↔ zeta functions relationship). This *quasiclassical zeta function* is our main result. The zeros of $Z_{\mathrm{qc}}(E)$ yield the spectrum of the "quasiclassical" evolution operator.

5.1 The Semiclassical Zeta Function

The formulae derived above differ from those of the semiclassical periodic orbit theory for hyperbolic flows as originally developed by Gutzwiller (1971) in terms of traces of the Van Vleck semiclassical Green's functions. The *semiclassical Gutzwiller trace formula* has topologically the same structure as the quasiclassical trace formula (38):

$$\operatorname{tr} G(E) = \overline{g}(E) + \frac{1}{\mathrm{i}\hbar} \sum_p T_p \sum_{r=1}^{\infty} \frac{\mathrm{e}^{\frac{\mathrm{i}}{\hbar} S_p(E) r - \mathrm{i}\pi \frac{m_p}{2} r}}{\left|\det\left(\mathbf{1} - \mathbf{J}_p^r\right)\right|^{\frac{1}{2}}} . \tag{40}$$

The Gutzwiller trace formula differs from the quasiclassical trace formula in two aspects. One is the volume term $\overline{g}(E)$ in (40) which is a missing from our version of the classical trace formula. While an overall pre-factor does not affect the location of zeros of the determinants, it plays a role in relations such as the zeta function functional equations of Berry and Keating (1990). The other difference is that the quantum kernel leads to a square root of the cycle Jacobian $1/\sqrt{\det(1 - \mathbf{J}_p)}$, a reflection of the relation probability = (amplitude)^2. This difference does not effect the leading eigenvalues (which coincide for the semi- and quasiclassical quantizations), but has a dramatic effect on the convergence of respective zeta functions.

The precise relation between the semiclassical zeta functions and the quasiclassical zeta functions is given in Appendix C.

In the remainder of the paper we shall investigate the relative merits of the quasiclassical quantization compared to the Gutzwiller semiclassics and the exact quantum mechanics.

6 Numerical Convergence of Cycle Expansions and Extraneous Eigenvalues

A 3-disk repeller is one of the simplest classically completely chaotic scattering systems and provides a convenient numerical laboratory for testing both the ideas about chaotic dynamics and for computing exact quantum mechanical spectra, see Eckhardt (1987), Gaspard and Rice (1989a)-(1989c), Cvitanović and Eckhardt (1989). The 3-disk repeller consists of a free point particle moving in the two-dimensional plane and scattering specularly off three identical disks of radius a centered at the corners of an equilateral triangle of side length R. The discrete C_{3v} symmetry reduces the dynamics to motion in a fundamental domain, and the spectroscopy to irreducible subspaces A_1, A_2 and E. All our computations are performed for the fully symmetric subspace A_1.

In this section we address the following question: which of the three approximate quantization zeta functions is the best in predicting the exact quantum mechanical scattering resonances

(a) the semiclassical zeta function of Gutzwiller (1988) and Voros (1988)

$$Z_{\mathrm{sc}}(z;k) = \exp\left\{ - \sum_p \sum_{r=1}^{\infty} \frac{1}{r} \frac{z^{r n_p} t_p^r}{1 - 1/\Lambda_p^r} \right\} = \prod_p \prod_{j=0}^{\infty} \left(1 - \frac{z^{n_p} t_p}{\Lambda_p^j} \right) \tag{41}$$

(b) the dynamical zeta function of Ruelle (1987), the $j = 0$ part of the semiclassical zeta function

$$\zeta^{-1}(z;k) = \exp\left\{-\sum_p \sum_{r=1}^{\infty} \frac{1}{r} z^{rn_p} t_p^r \right\} = \prod_p (1 - z^{n_p} t_p) \tag{42}$$

(c) or the quasiclassical zeta function (39)

$$Z_{\rm qc}(z;k) = \exp\left\{-\sum_p \sum_{r=1}^{\infty} \frac{1}{r} \frac{z^{rn_p} t_p^r}{\left(1 - 1/\Lambda_p^r\right)^2 \left(1 - 1/\Lambda_p^{2r}\right)}\right\}$$
$$= \prod_p \prod_{j=0}^{\infty} \prod_{l=0}^{\infty} \left(1 - \frac{z^{n_p} t_p}{\Lambda_p^{j+2l}}\right)^{j+1} \quad ? \tag{43}$$

Here

$$t_p = \mathrm{e}^{\mathrm{i}kL_p - \mathrm{i}m_p\pi/2} / |\Lambda_p|^{\frac{1}{2}} \tag{44}$$

is the weight of the $p^{\rm th}$ prime cycle, n_p its topological length and z a book-keeping variable for keeping track of the topological order in cycle expansions — the above zeta functions are Taylor-expanded in z around $z = 0$ up to a given cycle expansion order and only then z is set to $z = 1$ (see also (51) below). L_p is the length of the $p^{\rm th}$ cycle, m_p its Maslov index together with the group theoretical weight of the studied C_{3v} representation, and Λ_p its stability (the expanding eigenvalue of the $p^{\rm th}$ Jacobian matrix).

The results of comparing finite cycle expansion truncations of the above zeta functions with each other and with the exact quantum mechanical results computed with the methods outlined in Sect. 7 are summarized in Figs. 3 and 4. Resonances are plotted as the real part of the resonance wavenumber (resonance "energy") vs. the imaginary part of the wavenumber (resonance "width"). We have computed several thousands of exact quantum mechanical as well as approximate A_1 resonances for the 3-disk repeller with center-to-center separation $R = 6a$. Further and considerably more detailed numerical results are available from Wirzba and Henseler (1995).

Some of the features of the resonance spectra have immediate interpretation. The mean spacing of the resonances is approximately $2\pi/\bar{L}$, where $\bar{L}$ is the average of the lengths L_0 and L_1 of the two shortest cycles of topological length one. The data also exhibit various beating patterns resulting from the interference of cycles of nearly equal lengths; e.g. the leading beating pattern is of order $2\pi/\Delta L$, where ΔL is the difference of the lengths L_1 and L_0.

In Fig. 3 the cycle expansion includes cycles up to topological length four. Already at this order the four leading resonance bands are well approximated by the semiclassical zeta function (41) (in fact, for $\mathrm{Re}\, k \lesssim 75/a$ already cycles up to length two suffice to describe the first two leading resonance bands). Neither the quasiclassical zeta function (43) nor the dynamical zeta function (42) perform quite as well. The reason is that the quasiclassical as well as the dynamical zeta function predict extra resonances which are absent in the exact quantum mechanical calculation. The accessible resonances close to the

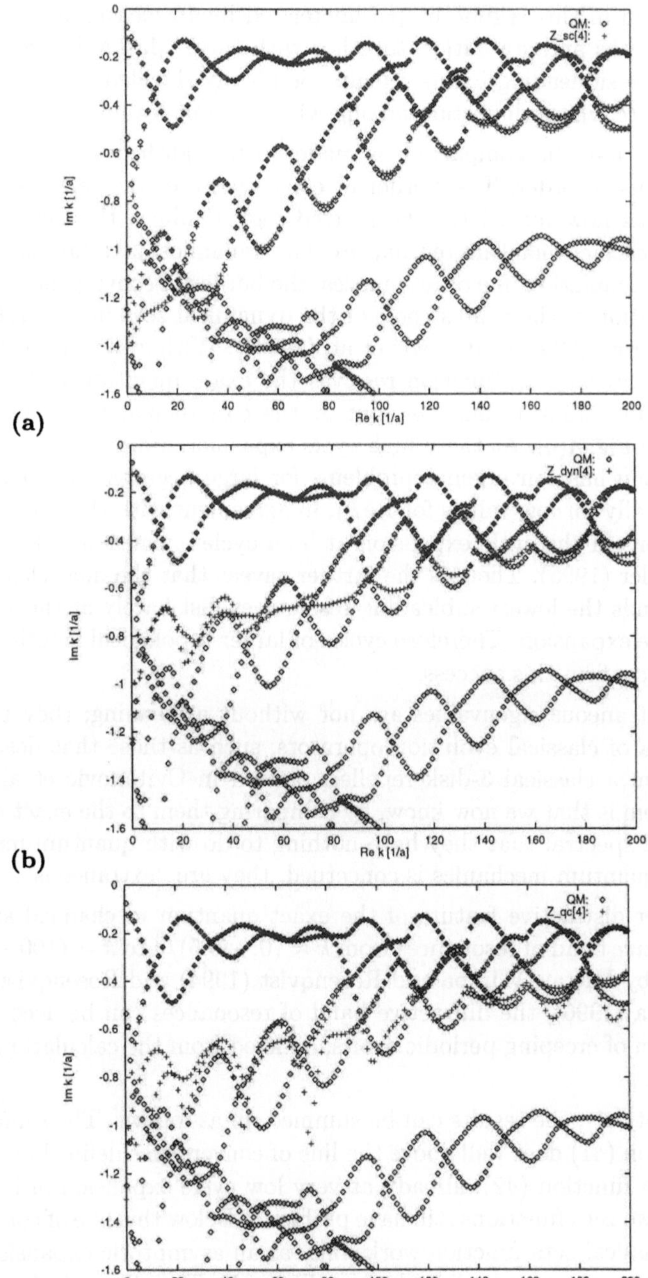

Fig. 3. The A_1 resonances of the 3-disk repeller with $R = 6a$. The exact quantum mechanical data are denoted by diamonds. The semiclassical ones are calculated up to 4th order in the cycle expansion and are denoted by crosses. (a) semiclassical zeta function (41), (b) dynamical zeta function (42), (c) quasiclassical zeta function (43).

real axis can in this regime be parameterized by 16 measured numbers, i.e. 8 cycle lengths and stabilities, together with the 8 Maslov indices. It turns out that the subleading bands remain completely shielded all the way up to $\text{Re}\, k \approx 950/a$ where they start mixing with the four leading ones.

In Figs. 4a-c the comparison is made up to eighth, respectively twelfth cycle expansion order. The border of convergence of the semiclassical zeta function has now moved (in the plotted region) above the fifth and sixth band of the exact quantum resonances. The dynamical zeta function exhibits a sharp accumulation line of resonances, the border of convergence controlled by the location of the nearest pole of the dynamical zeta function (Eckhardt and Russberg (1993), Cvitanović et al. (1993)). With cycles up to length 12 the quasiclassical zeta function resolves the exact quantum fifth and sixth bands of subleading resonances, but at the cost of many extraneous resonances, see Fig. 4(c). At these high cycle expansion orders the quasiclassical zeta function has convergence problems for large negative imaginary k values (especially for low values for $\text{Re}\, k$), in agreement with the expected large cancellations in the cycle expansion at high cycle expansion orders, Wirzba and Henseler (1995). There is the further caveat that the quasiclassical zeta function finds the lowest subleading resonances just barely at the 12th order in the cycle expansion. Therefore cycles of larger topological length would be needed to confirm this success.

The extraneous eigenvalues are not without a meaning; they belong to the spectra of classical evolution operators, such as those that describe the escape from a classical 3-disk repeller, plotted in Cvitanović et al. (1993). The problem is that we now know, by comparing them to the exact quantum mechanical spectra, that they have nothing to do with quantum mechanics. As far as quantum mechanics is concerned, they are "extraneous".

Another distinctive feature of the exact quantum mechanical spectra is the *diffractive* band of resonances from $k \approx (0.-\text{i}0.5)/a$ to $k \approx (100.-\text{i}1.6)/a$. As shown by Vattay, Wirzba and Rosenqvist (1994) and Rosenqvist, Vattay and Wirzba (1996), the diffractive band of resonances can be accounted for by inclusion of creeping periodic orbits, omitted from the calculations undertaken here.

Qualitatively, the results can be summed up as follows. The semiclassical zeta function (41) does well above the line of convergence defined by the dynamical zeta function (42), already at very low cycle expansion orders where the other two zeta functions still have problems. Below this line of convergence the semiclassical zeta function works only as an asymptotic expansion; when it works, it works very well and very efficiently. The dynamical zeta function does eventually as well for the leading resonances as the semiclassical one. As experimentally these are the only resonances accessible, one can – for practical purposes – limit the calculation just to this zeta function. The quasiclassical zeta function finds all known subleading quantum resonances, but at a high expense: the rate of convergence is poor compared to the semiclassical

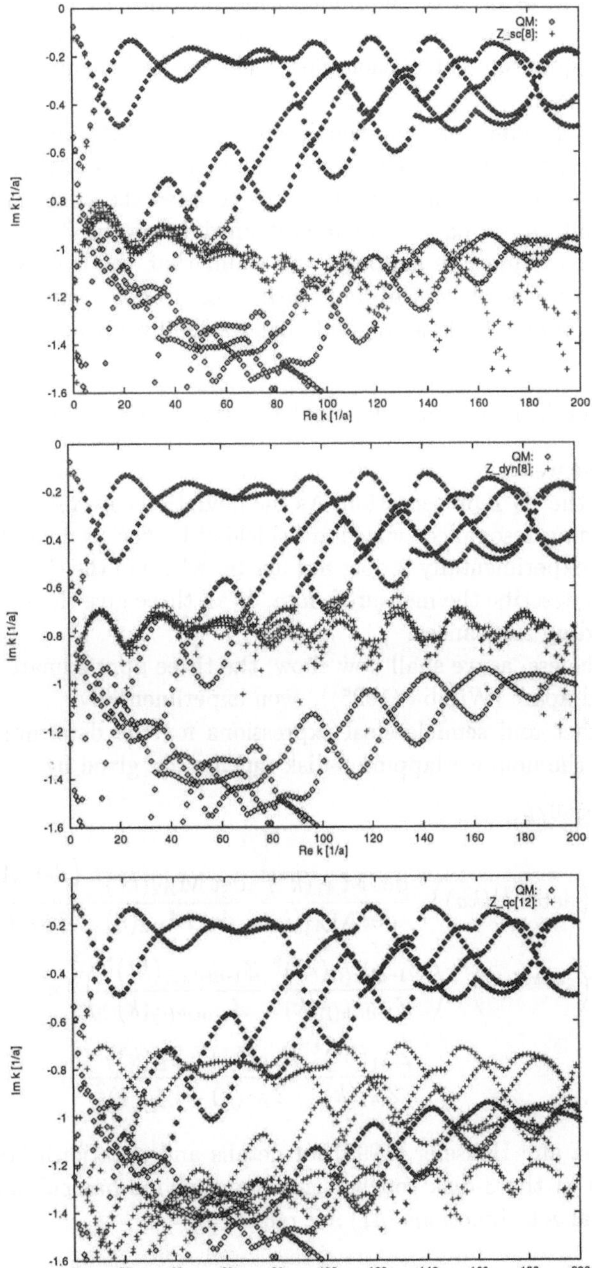

(a)

(b)

(c)

Fig. 4. The A_1 resonances of the 3-disk repeller with $R = 6a$. The exact quantum mechanical data are denoted by diamonds, the semiclassical ones by crosses. **(a)** semiclassical zeta function (41) up to 8th order in the cycle expansion, **(b)** dynamical zeta function (42) up to 8th, **(c)** quasiclassical zeta function (43) up to 12th order.

zeta function, as most of the information provided by longer cycles is used to determine the extraneous resonance bands, with no quantum counterpart. Without a quantum calculation, one could not tell the extraneous from the real resonances.

As a by-product of this calculation we can state an empirical rule of thumb: Each new cycle expansion or cumulant order is connected with a new line of subleading resonances. This rule relates the cycle expansion truncations limit, $n \to \infty$ (where n is defined below in (51)), and the limit $\mathrm{Im}\, k \to -\infty$. Numerics supports the claim that the cycle expansion limit $n \to \infty$ and the semiclassical limit $\mathrm{Re}\, k \to \infty$ do not commute deep down in the lower complex k plane, a point that we shall return to in Sect. 7.

6.1 Exact Versus Semiclassical Cluster Phase Shifts

In the above we compared the exact and semiclassical resonances of the 3-disk repeller in the A_1 representation. As the deviations are most pronounced for the subleading resonances which are shielded by the leading ones, one could argue that experimentally it does not matter which of the three zeta functions are used to describe the measured data, as all three give the same predictions for the leading resonances.

Nevertheless, as we shall now show, the three approximate quantizations can be told apart (Wirzba (1995)), even experimentally.

The exact and semiclassical expressions for the determinant of the S-matrix for the non-overlapping 3-disk repeller are given by

$$
\begin{aligned}
\det\ & \mathbf{S}^{(3)}(k) \\
= \ & \left(\det \mathbf{S}^{(1)}(ka)\right)^3 \frac{\det \mathbf{M}_{A1}(k^*)^\dagger}{\det \mathbf{M}_{A1}(k)} \frac{\det \mathbf{M}_{A2}(k^*)^\dagger}{\det \mathbf{M}_{A2}(k)} \frac{\left(\det \mathbf{M}_E(k^*)^\dagger\right)^2}{\left(\det \mathbf{M}_E(k)\right)^2} \\
\xrightarrow{\text{s.c.}} \ & \left(\mathrm{e}^{-\mathrm{i}\pi N(k)}\right)^6 \left(\frac{Z_{\text{1-disk(l)}}(k^*)^*}{Z_{\text{1-disk(l)}}(k)} \frac{Z_{\text{1-disk(r)}}(k^*)^*}{Z_{\text{1-disk(r)}}(k)}\right)^3 \times \\
& \times \frac{Z_{A1}(k^*)^*}{Z_{A1}(k)} \frac{Z_{A2}(k^*)^*}{Z_{A2}(k)} \frac{Z_E(k^*)^{*2}}{Z_E(k)^2} \ .
\end{aligned} \tag{45}
$$

(See Wirzba and Henseler (1995) for details and notation.) For the A_1 representation of the 3-disk repeller the quantum mechanical kernels and the semiclassical zeta functions (41) are related by

$$
\frac{\det \mathbf{M}_{A1}(k^*)^\dagger}{\det \mathbf{M}_{A1}(k)} \xrightarrow{\text{s.c.}} \frac{Z_{A1}(k^*)^*}{Z_{A1}(k)} \tag{46}
$$

Both sides of (45) and (46) respect unitarity, and if the wave number k is real, both sides can be written as $\exp\{\mathrm{i}2\eta(k)\}$ with a real *phase shift* $\eta(k)$. We define the total phase shift for the coherent part of the 3-disk scattering problem (here always understood in the A_1 representation) for the exact quantum mechanics as well as for the three approximate quantizations by:

$$
\begin{aligned}
e^{2i\eta_{\mathrm{qm}}(k)} &:= \frac{\det \mathbf{M}(k^*)^\dagger}{\det \mathbf{M}(k)} & e^{2i\eta_{\mathrm{sc}}(k)} &:= \frac{Z_{\mathrm{sc}}(k^*)^*}{Z_{\mathrm{sc}}(k)} \\
e^{2i\eta_{\mathrm{dyn}}(k)} &:= \frac{\zeta^{-1}(k^*)^*}{\zeta^{-1}(k)} & e^{2i\eta_{\mathrm{qc}}(k)} &:= \frac{Z_{\mathrm{qc}}(k^*)^*}{Z_{\mathrm{qc}}(k)} \,.
\end{aligned}
\tag{47}
$$

This phase shift definition should be compared with the cluster phase shift given in section 4 of Lloyd and Smith (1972). The important point here is that the coherent or cluster phase shift of $\det \mathbf{S}(k)$ is in principle experimentally accessible: one just has to construct the elastic scattering amplitude from the measured cross sections, and subtract the single disk contributions.

So, $\eta_{\mathrm{qm}}(k)$ is a "measurable" quantity, useful to us as a different method for discriminating between the various zeta functions. An example is given in Fig. 5 where the zeta functions in the numerators as well as in the denominators in (47) have been expanded up to cycles of topological length 12. The phase shifts are compared in the window $104/a \le k \le 109/a$, a typical window sufficiently narrow to resolve the rapid oscillations, with k sufficiently big that the diffraction effects are unimportant. The performance of the original semiclassical zeta function is again the best. We stress that in contrast to the subleading resonances studied in Sect. 6 (which are completely shielded from experimental detection by the leading resonances), phase shifts are *hard* data, in principle extractable from measured cross sections.

In conclusion: One can tell the three candidate zeta functions apart even experimentally. We have again confirmed that the semiclassical zeta function is the best.

7 Semiclassics Versus Asymptotic $\hbar$ Expansion

So far we have tested various approximate quantization proposals against each other and against exact quantum mechanics. Now we turn to a deeper question: how seriously should we take these cycle expansions in the first place? We will show here, following Wirzba (1996), that the semiclassical zeta function is approximating its quantum mechanical counterpart, the "characteristic KKR determinant" (Kohn and Rostoker (1954), Lloyd and Smith (1972), Berry (1981)) as an asymptotic series and therefore makes sense only as a truncated series.

Let $\det \mathbf{M}(k) = \det(\mathbf{1} + \mathbf{A}(k))$ be the characteristic KKR determinant of the 3-disk repeller in the A_1 representation, where the pertinent kernel $\mathbf{A}(k)$ expressed in the angular momentum basis relative to the half-disk in the fundamental domain reads (see Gaspard and Rice (1989c))

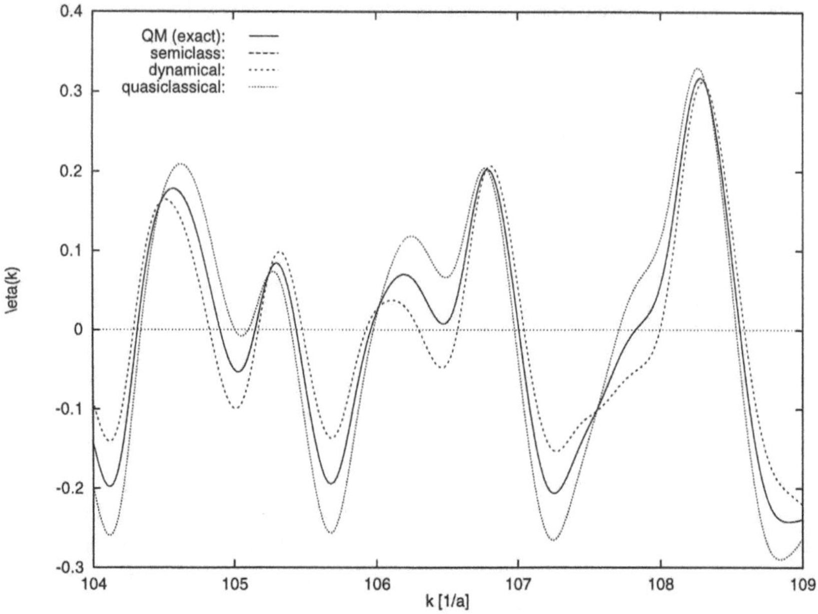

Fig. 5. The coherent cluster phase shifts of the 3-disk scattering system in the A_1 representation with $R = 6a$. The exact quantum mechanical data compared to the predictions of the semiclassical zeta function (41), the dynamical zeta function (42) and the quasiclassical zeta function (43) calculated up to 12th order in the cycle expansion. The semiclassical zeta function and the exact quantum mechanical data coincide within the resolution of the plot.

$$\begin{aligned}\mathbf{A}(k)_{m,m'} = d(m)d(m')\frac{J_m(ka)}{H^{(1)}_{m'}(ka)}&\left\{\cos\left(\frac{\pi}{6}(5m-m')\right)H^{(1)}_{m-m'}(kR)\right.\\&\left.+(-1)^{m'}\cos\left(\frac{\pi}{6}(5m+m')\right)H^{(1)}_{m+m'}(kR)\right\}\end{aligned} \tag{48}$$

with $0 \le m, m' < \infty$ and

$$d(m) := \begin{cases}\sqrt{2} \text{ for } & m > 0\\ 1 \text{ for } & m = 0\ .\end{cases}$$

Let $Q_m(k)$ denote the mth cumulant of $\det \mathbf{M}(k)$, i.e. the coefficient of z^m in the Taylor expansion of $\det(\mathbf{1} + z\mathbf{A}(k))$. $Q_m(k)$ satisfies the Plemelj-Smithies recursion relation (Wirzba and Henseler (1995))

$$\begin{aligned}Q_m(k) &= \frac{1}{m}\sum_{j=1}^{m}(-1)^{j+1}Q_{m-j}(k)\,\mathrm{Tr}(\mathbf{A}^j(k)) \quad \text{for } m \ge 1\\ Q_0(k) &\equiv 1\ ,\end{aligned}$$

where $\mathrm{Tr}(\mathbf{A}^j(k))$ is the trace of the jth power of the kernel $\mathbf{A}(k)_{m,m'}$ evaluated in the angular momentum basis, $\{|m\rangle\}$, relative to the half-disk in the fundamental domain.

The semiclassical analog of the characteristic determinant $\det(\mathbf{1}+z\mathbf{A}(k))$ is the semiclassical zeta function (41). More precisely, the cycle expansion of the semiclassical zeta function truncated at the topological order n is the semiclassical analog of the quantum cumulant expansion of $\det(\mathbf{1}+z\mathbf{A}(k))$ truncated at the same order. Thus $c_m(k)$, the corresponding semiclassical m^{th} order cycle expansion term of $Z_{\text{sc}}(k)$, is constructed from the semiclassical equivalent of the Plemelj-Smithies recursion relation:

$$c_m(k) = -\frac{1}{m}\sum_{j=1}^{m} c_{m-j}(k)\sum_{p,r} n_p \frac{\delta_{n_p r,j}\, {t_p}^r}{1-1/\Lambda_p^r} \qquad \text{for } m \geq 1 \tag{49}$$

$$c_0(k) \equiv 1 \ ,$$

with t_p defined in (43). The cycle expansion (Cvitanović (1988)) follows from the semiclassical limit

$$\mathrm{Tr}(\mathbf{A}^j(k)) \overset{\text{s.c.}}{\longrightarrow} (-1)^j \sum_{p,r} n_p \frac{\delta_{n_p r,j}\, {t_p}^r}{1-1/\Lambda_p^r} + \text{diffractive creeping orbits} \ . \tag{50}$$

In summary, the nth order truncated cumulant and cycle expansions are given by

$$\det \mathbf{M}(k)|_n = \sum_{m=0}^{n} Q_m(k)\,, \qquad Z_{\text{sc}}(k)|_n = \sum_{m=0}^{n} c_m(k) \tag{51}$$

where the notation $\cdots|_n$ indicates that the corresponding determinant or zeta function has been truncated at cumulant/cycle expansion order n. The following facts are known:

1. The cumulant sum

$$\lim_{n\to\infty} \det \mathbf{M}(k)|_n = \lim_{n\to\infty} \sum_{m=0}^{n} Q_m(k) = \det \mathbf{M}(k)$$

is absolutely convergent, $\sum |Q_m(k)| < \infty$, because of the trace class property of $\mathbf{A}(k) \equiv \mathbf{M}(k) - \mathbf{1}$ for non-overlapping, non-touching n-disk repellers (Wirzba and Henseler (1995)).

2. The semiclassical cycle expansion sum converges above an accumulation line (which runs below and approximately parallel to the real wave number axis, see Fig. 4(a)) given by the leading poles of the leading dynamical zeta function, $\zeta^{-1}(k)$, or the leading zeros of the subleading zeta function, $\zeta_1^{-1}(k)$ (Eckhardt and Russberg (1993), Cvitanović et al. (1993), Cvitanović and Vattay (1993)).

3. The *truncated* semiclassical cycle expansion sum $Z_{\mathrm{sc}}(k)|_n$ can approximate the quantum mechanical result as an asymptotic series even below the semiclassical zeta function border of convergence, Wirzba and Henseler (1995).

We have checked numerically that the following formulae relate the mth cumulants and cycle expansion terms on the real k-axis with the corresponding quantities inside the complex k plane — at least as long as the condition $|\mathrm{Im}\, k| \ll |\mathrm{Re}\, k|$ is satisfied: for the quantum mechanical cumulants of order m we have the approximate leading order relation (under assumption that the diffraction effects are negligible)

$$Q_m(\mathrm{Re}\, k + \mathrm{i}\mathrm{Im}\, k) \sim Q_m(\mathrm{Re}\, k)\, \mathrm{e}^{-m\bar{L}\mathrm{Im}\, k} \; . \tag{52}$$

$\bar{L} \approx R - 2a$ is the average length of the cycles of topological length one. We have also checked numerically that the corresponding relation for the semiclassical cycle expansion terms of order m is also approximately valid:

$$c_m(\mathrm{Re}\, k + \mathrm{i}\mathrm{Im}\, k) \sim c_m(\mathrm{Re}\, k)\, \mathrm{e}^{-m\bar{L}\mathrm{Im}\, k} \; . \tag{53}$$

Furthermore, on the basis of Fig. 6 we conjecture that for arbitrary values of the center-to-center separation R of the non-overlapping 3-disk repeller ($R > 2a$) the following relations hold on the real wave number axis (k real):

$$c_m(k) \approx Q_m(k) \quad \text{with} \quad 1 \gg |c_m(k)| \qquad \text{if} \;\; ka \gtrsim 2^{m-1}\frac{\bar{L}}{a} \; , \tag{54}$$

$$1 \gg |c_m(k)| \gg |Q_m(k)| \qquad \text{if} \;\; ka \lesssim 2^{m-1}\frac{\bar{L}}{a} \; . \tag{55}$$

7.1 The Meaning of It All

Where does the boundary $ka \approx 2^{m-1}\bar{L}/a$ come from?

This boundary follows from a combination of the uncertainty principle with ray optics and the non-vanishing value for the topological entropy of the 3-disk repeller. When the wave number k is fixed, quantum mechanics can only resolve the classical repelling set up to the critical topological order n given by (54). The quantum wave packet which explores the repelling set has to disentangle 2^n different sections of size $d \sim a/2^n$ on the "visible" part of the disk surface (which is of order a) between any two successive disk collisions. Successive collisions are separated spatially by the mean flight length $\bar{L}$, and the flux spreads with a factor $\bar{L}/a$. In other words, the uncertainty principle bounds the maximal sensible truncation in the cycle expansion order by the highest quantum resolution attainable for a given wavenumber k.

The upper limit n for which $c_m(k)$ with $m \leq n$ approximates $Q_m(k)$ increases with increasing $\mathrm{Re}\, k$. For $n > m(\mathrm{Re}\, ka)$, defined in (55), the cycle expansion terms and cumulant terms deviate so much from each other, that beyond this order the contributions of longer cycle expansions have nothing to

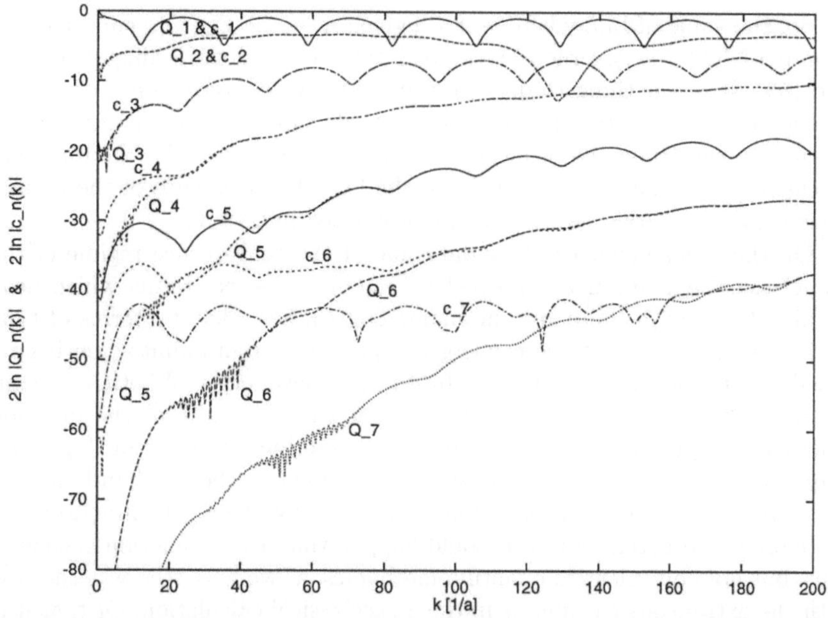

Fig. 6. Comparison of the absolute values of the first seven quantum mechanical cumulant terms, $|Q_n(k)|^2$, with the corresponding semiclassical cycle expansion terms, $|c_n(k)|^2$, of the semiclassical zeta function (41) evaluated on the real wave number axis k. Note that the deviations between quantum mechanics and semiclassics decrease with increasing $\mathrm{Re}\, k$, but increase with increasing cycle expansion order n. The value of $\mathrm{Re}\, k$ where the quantum mechanical and semiclassical curves join is approximately given by $\mathrm{Re}\, ka \sim 2^{n+1}$ where n is the order of the cumulant/cycle expansion term and a is the radius of the disk. The data are for the A_1 subspace of the 3-disk repeller with center-to-center separation $R = 6a$.

do with quantum mechanics. The fact that $Z_{\mathrm{sc}}(k)|_n$ – even in its convergence regime – is a good approximation to quantum mechanics *only* up to a finite n is usually not noticed, as the terms in (55) are exponentially small on or close to the real axis and sum therefore to a tiny quantity. In other words, for $n > m(\mathrm{Re}\, ka)$ and close to the real k axis, the absolute error $|c_n(k) - Q_n(k)|$ is still small, the relative error $|c_n(k)/Q_n(k)|$ on the other hand is tremendous. With increasing negative $\mathrm{Im}\, k$, however, using the scaling rules (52) and (53), the deviations (55) are blown up, such that the relative errors $|c_n(k)/Q_n(k)|$ eventually become visible as absolute errors $|c_n(k) - Q_n(k)|$ (see e.g. the resonance calculation of Wirzba and Henseler (1995)). For $\mathrm{Im}\, k$ above the boundary of convergence these errors still sum up to a finite quantity which might, however, not be negligible any longer. Below the convergence line these errors sum up to infinity.

So, the value of Imk where — for a given n — the $Z_{sc}(k)|_n$ sum deviates from $\det \mathbf{M}(k)|_n$ is governed by the real part of k and the scaling rules (52) and (53). It has nothing to do with the boundary of convergence of $Z_{sc}(k)$, as a good approximation is given by the *finite* sum of terms satisfying (54). Therefore, the truncated semiclassical expansion can describe the quantum mechanical resonance data even *below* the line of convergence of the infinite cycle expansion series, as we have already noted in Sect. 6.

On the other hand, the boundary line of the convergence regime of the semiclassical expansion is governed by $c_m(k)$, $m \to \infty$, terms which have nothing to do with the quantum analog $Q_m(k)$, i.e. solely by terms of type (55). The reason is that the convergence property of an infinite sum is governed by the *infinite* tail and not by the first few terms. Whether a semiclassical expansion converges or not is a separate issue from the question whether the quantum mechanical data are described well or not. The *convergence property* of a semiclassical zeta functions on the one hand and the *approximate description of quantum mechanics* by these zeta functions are therefore two different issues. It could happen that a zeta function is convergent, but not equivalent to quantum mechanics, as we have seen was the case with the extraneous resonances in the quasiclassical calculation. Or that it is not convergent in general, but its finite truncations nevertheless approximate well quantum mechanics, as is the case for the Gutzwiller-Voros semiclassical zeta function(41).

We conclude that the exponential rise of the number of cycles with increasing cycle expansion order n is the physical reason for the breakdown of the cycle expansion of the semiclassical zeta function(41) with respect to the exact quantum mechanical cumulant expansion.

8 Summary and Conclusions

In conclusion, we have constructed a classical evolution operator for the quasiclassical wave function evolution, and derived the corresponding trace and determinant formulae for periodic orbit quasiclassical quantization of chaotic dynamical systems.

Improved analyticity has been very useful in sorting out the relative importance of the semiclassical, diffraction and quantum contributions. However, one hope for consequence of the superexponential convergence of the cycle expansions of the new Fredholm determinant was that they would converge faster with the maximal cycle length truncation than the more familiar Gutzwiller-Voros and Ruelle type zeta functions. As is shown here, this is not the case. Improved analyticity comes at cost; extraneous eigenvalues are purely classical and do not belong to the quantum spectrum, but their presence degrades significantly the convergence of the cycle expansions.

The analysis sheds new light on the differences between the classical and semiclassical spectra; in particular, we have made explicit for the case of

n-disk repellers the quantum limitations on the phase space resolution by classical orbits, in the spirit of Bogomolny's analysis (Bogomolny (1992)) of the finite resolution of phase space for the bound systems.

In spite of its laggard performance as a putative competitor to the semiclassical quantization, the mere fact that there exists an alternative "quasiclassical" quantization that follows directly from the Schrödinger equation without recourse to path integrals and saddle points is of intellectual interest. It is still possible that a more ingeniously constructed "classical" evolution operator would also perform better than the semiclassical zeta function in practice.

Acknowledgements. It was Dieter Wintgen's success in applying cycle expansions to atomic physics that has motivated the above "in depth" study of various semiclassical quantizations. We are grateful to Per E. Rosenqvist for a long and fruitful collaboration, and in particular for the 3-disk repeller data set used in the numerical experiments of this paper. We have striven to meet Gregor Tanner's exacting standards. G.V. thanks the grant of the Hungarian Ministry of Culture and Education (MKM 337) and of the OTKA (F019266/F17166/T17493). G.V. and A.W. acknowledge the support of the International Relations Offices of Germany and Hungary. A.W. is grateful to Michael Henseler for discussions and would like to thank the Center for Chaos and Turbulence Studies at the Niels Bohr Institute for hospitality during his visits to Copenhagen.

Appendices

A Calculation of Trace M

In this appendix we calculate the trace (37). The equations of motion for a time independent Hamiltonian (15) can be written as

$$\dot{x}_m = \omega_{mn} \frac{\partial H}{\partial x_n} , \qquad \omega = \begin{bmatrix} 0 & \mathbf{I} \\ -\mathbf{I} & 0 \end{bmatrix} , \qquad m, n = 1, 2, \ldots, 2d , \tag{56}$$

where $x = [q, p]$ is a phase space vector, $\mathbf{I} = [d \times d]$ the unit matrix, and ω the $[2d \times 2d]$ symplectic form $\omega_{mn} = -\omega_{nm}$, $\omega^2 = -1$. The linearized motion in the vicinity of a phase space trajectory $x(t) = [q(t), p(t)]$ is given by the Jacobian matrix

$$\delta x(t) = \mathbf{J}^t(x)\, \delta x(0) , \qquad J^t(\xi)_{ij} = \frac{\partial x_i(t)}{\partial \xi_j} , \qquad \xi = x(0) .$$

The equations of motion of $\mathbf{J}$ follow from (56)

$$\frac{\mathrm{d}}{\mathrm{d}t}\mathbf{J}^t(x) = \mathbf{L}(x,t)\,\mathbf{J}^t(x)\ , \qquad \text{with} \qquad \mathbf{L}(x,t)_{mn} = \omega_{mk}H_{kn}(x)|_{x(t)}\ . \tag{57}$$

where $H_{kn} = \partial_k\partial_n H$ is the matrix of second derivatives of the Hamiltonian. $\mathbf{L}$ is infinitesimal generator of symplectic (or canonical) transformations which leaves ω invariant

$$\mathbf{L}^T\omega + \omega\mathbf{L} = 0\ . \tag{58}$$

$\mathbf{J}$ is a symplectic matrix, as it preserves the symplectic bilinear invariant ω:

$$\mathbf{J}^T\omega\mathbf{J} = \omega\ . \tag{59}$$

From this follows that $\det\mathbf{J} = 1$, and that the transpose $\mathbf{J}^T$ and the inverse $\mathbf{J}^{-1}$ are also symplectic; $\mathbf{J}^{-1} = -\omega\mathbf{J}^T\omega$. Hence if Λ is an eigenvalue of $\mathbf{J}$, so are $1/\Lambda$, Λ^* and $1/\Lambda^*$.

Let $\mathbf{j}$ be the *configuration* space Jacobian matrix

$$\mathbf{j}^t_{ij}(x) := \frac{\mathrm{d}q_i(t)}{\mathrm{d}q_j(0)}\ , \qquad j^t(x) := \det\mathbf{j}^t(x)\ , \tag{60}$$

and j the *configuration* space Jacobian evaluated on the q-space projection of the phase-space trajectory $x(t)$ passing through the $t = 0$ initial point $x = (q,p)$. The curvature matrix (25) is related to the configuration space Jacobian matrix (60) by

$$M_{ij}(x,t) = \frac{\partial v_i}{\partial q_j} = \frac{\partial q_k(0)}{\partial q_j(t)}\frac{\mathrm{d}}{\mathrm{d}t}\frac{\partial q_i(t)}{\partial q_k(0)} = \left(\frac{1}{\mathbf{j}^t}\right)_{kj}\left(\frac{\mathrm{d}}{\mathrm{d}t}\mathbf{j}^t\right)_{ik}\ ,$$

so the configuration space Jacobian matrix satisfies

$$\frac{d}{dt}\mathbf{j}^t = \mathbf{M}\,\mathbf{j}^t \tag{61}$$

and is given by the exponentiated time-ordered integral of the trace of $\mathbf{M}$

$$\det\mathbf{j}^t(x) = \mathrm{T}e^{\int_0^t \mathrm{d}\tau\,\mathrm{tr}\,\mathbf{M}^\tau}\ . \tag{62}$$

The *full phase space* Jacobian matrix $\mathbf{J}$ is given by

$$\begin{bmatrix}\delta\mathbf{q}'\\ \delta\mathbf{p}'\end{bmatrix} = \mathbf{J}\begin{bmatrix}\delta\mathbf{q}\\ \delta\mathbf{p}\end{bmatrix} = \begin{bmatrix}\mathbf{J}_{qq} & \mathbf{J}_{qp}\\ \mathbf{J}_{pq} & \mathbf{J}_{pp}\end{bmatrix}\begin{bmatrix}\delta\mathbf{q}\\ \delta\mathbf{p}\end{bmatrix}\ , \tag{63}$$

where $\delta\mathbf{q}$, $\delta\mathbf{p}$ are d-dimensional infinitesimal tangent space vectors, and $\mathbf{J}_{qq}$, $\mathbf{J}_{qp}$, $\mathbf{J}_{pq}$ and $\mathbf{J}_{pp}$ are the $[d \times d]$ submatrices of the full $[2d \times 2d]$ Jacobian matrix. (To save paper, we suppress the t, q, p dependence for the time being). Take a derivative $\partial/\partial\delta q_i$ of both sides of (63), keeping terms to linear order in δq. This expresses the configuration Jacobian matrix $\mathbf{j}$ and the curvature matrix (25) $\mathbf{M}'$ in terms of the $\mathbf{J}$ and the initial $\mathbf{M}$

$$\begin{bmatrix} \mathbf{j} \\ \mathbf{M}'\mathbf{j} \end{bmatrix} = \mathbf{J} \begin{bmatrix} \mathbf{I} \\ \mathbf{M} \end{bmatrix} . \tag{64}$$

Using (61) we see that $\mathbf{J}$ evolves the configuration Jacobian matrix and its time derivative

$$\begin{bmatrix} \mathbf{j}^t \\ \frac{d}{dt}\mathbf{j}^t \end{bmatrix} = \mathbf{J} \begin{bmatrix} \mathbf{j}^0 \\ \frac{d}{dt}\mathbf{j}^0 \end{bmatrix} ,$$

where the initial condition is $\mathbf{j}^0 = \mathbf{1}$ for $t = 0$.

To spell it out: for a given initial set of $\delta\mathbf{q}$'s and $\delta\mathbf{p}$'s, the projection of the phase space volume onto the configuration space is given by the configuration space Jacobian matrix $\mathbf{j}$

$$\mathbf{j} = \mathbf{j}^t(q, p, \mathbf{M}) := \mathbf{J}_{qq} + \mathbf{J}_{qp}\mathbf{M} , \tag{65}$$

and the matrix of curvatures $\mathbf{M}'$ is evolved recursively by

$$\mathbf{M}' = \mathbf{M}^t(q_0, p_0, \mathbf{M}_0) := (\mathbf{J}_{pq} + \mathbf{J}_{pp}\mathbf{M}) \frac{1}{\mathbf{J}_{qq} + \mathbf{J}_{qp}\mathbf{M}} , \tag{66}$$

where the q, p, t dependence is hidden in $\mathbf{J}$. We also note that transposing (64), multiplying from the right by $\omega\mathbf{J}$, and using the symplectic invariance (59) yields an alternative formula for the configuration space Jacobian matrix

$$\left(\frac{1}{\mathbf{j}}\right)^T = \mathbf{J}_{pp} - \mathbf{M}'\mathbf{J}_{qp} . \tag{67}$$

Evaluation of the trace (37) requires a first variation in all of the dynamical space coordinates X, including $\delta\mathbf{M}'$. From (66) together with (67) we obtain

$$\delta\mathbf{M}' = \mathbf{J}_{pp}\,\delta\mathbf{M}\frac{1}{\mathbf{j}} - \mathbf{M}'\,\mathbf{J}_{qp}\,\delta\mathbf{M}\frac{1}{\mathbf{j}} = \left(\frac{1}{\mathbf{j}}\right)^T \delta\mathbf{M}\,\frac{1}{\mathbf{j}} , \tag{68}$$

so the trace (37) is simply reinstated

$$\Delta_{p,r} = \sum \frac{(\det \mathbf{j}_p)^{r/2}}{\left|\det\left(\mathbf{1} - \frac{\partial}{\partial \mathbf{M}}\mathbf{M}^{T_p r}(M)\right)\right|} = \sum \frac{(\det \mathbf{j}_p)^{r/2}}{\left|\det\left(\mathbf{1} - \mathbf{j}_p^{-r} \otimes \mathbf{j}_p^{-r}\right)\right|} . \tag{69}$$

The sum is over all $\mathbf{M}$ that satisfy the fixed point condition

$$\mathbf{M}^{T_p}(q, p, M) = \mathbf{M} . \tag{70}$$

Consider now $\mathbf{j}$ for a periodic orbit p; $\mathbf{j}$ is a $[d \times d]$ matrix with eigenvalues and eigenvectors

$$\mathbf{j}\,\mathbf{e}_i = \Lambda_i \mathbf{e}_i , \qquad i = 1, 2, \cdots, d .$$

Multiply (64) from the right by the $2d$-dimensional vector $[\mathbf{e}_i, \mathbf{e}_i]$; we see that an eigenvalue of $\mathbf{j}$ is also an eigenvalue of the $[2d \times 2d]$ phase space Jacobian matrix:

$$\Lambda_i \begin{bmatrix} \mathbf{e}_i \\ \mathbf{M}\mathbf{e}_i \end{bmatrix} = \mathbf{J} \begin{bmatrix} \mathbf{e}_i \\ \mathbf{M}\mathbf{e}_i \end{bmatrix} ,$$

Furthermore, transposing this equation, multiplying it from the right by $\Lambda_i^{-1}\omega\mathbf{J}$, and using the symplectic condition (59) yields the associated left eigenvector with eigenvalue $1/\Lambda_i$,

$$[\mathbf{e}_i^T, \mathbf{e}_i^T\mathbf{M}]\,\omega\Lambda_i^{-1} = [\mathbf{e}_i^T, \mathbf{e}_i^T\mathbf{M}]\,\omega\mathbf{J} \ .$$

In this way the $(\Lambda_i, 1/\Lambda_i)$ pairs of eigenvalues of the $[2d \times 2d]$-dimensional phase space Jacobian matrix correspond to the d eigenvalues of the d-dimensional $\mathbf{j}$. As the d eigenvalues of $\mathbf{j}$ generate the d pairs of eigenvalues of $\mathbf{J}$, the sum (69) gets 2^d contributions $\Lambda_1^{\pm 1}\Lambda_2^{\pm 1}\cdots\Lambda_d^{\pm 1}$. Each of these is expanding on its own $\mathbf{M}$ subspace, and the dominant one is the most expanding one, so we keep from (69) only the modulus of the leading term (the phase will be treated in the next section)

$$\left|\widetilde{\Delta}_{p,r}\right| = \prod_{i=1}^{d} \frac{|\Lambda_{p,i}|^{r/2}}{1 - 1/\Lambda_{p,i}^{2r}} \ . \tag{71}$$

The dynamics in the tangent space can be restricted to a unit eigenvector neighborhood corresponding to the largest eigenvalue of the Jacobian matrix. On this neighborhood the largest eigenvalue of the Jacobian matrix is the only fixed point, and the quasiclassical zeta function obtained by keeping only the largest term in the $\Delta_{p,r}$ sum in (69) is also entire, Cvitanović and Vattay (1993).

So, (very pleasantly) as Λ_i are also eigenvalues of the configuration space Jacobian matrix $\mathbf{j}$, the extra trace over $\mathbf{M}$ comes for free; we have *already* computed the eigenvalue set $\{\Lambda_1, 1/\Lambda_1, \cdots, \Lambda_d, 1/\Lambda_d\}$ for every full (q,p) phase space cycle p.

B Maslov Indices

The square root of the configuration space Jacobian (62) is also a time ordered integral

$$\left(\det \mathbf{j}^t(x)\right)^{\frac{1}{2}} = \mathrm{T}\exp\left\{\frac{1}{2}\int_0^t d\tau\, \mathrm{tr}\,(\mathbf{M}^\tau)\right\} \ . \tag{72}$$

$\mathbf{M}$ diverges at caustics; for example, for $d = 1$ Poincaré sections (such as for billiards) $\mathbf{M} = \partial p/\partial q$ diverges whenever a trajectory points in the p-axis direction. Close to a singularity, where

$$\mathbf{M}(t \to t^c) = \infty \ ,$$

we can neglect the non-leading terms from (27) and use the solution of

$$\dot{\mathbf{M}} = -\mathbf{M}^2 \ , \tag{73}$$

after the symmetric matrix $\mathbf{M}$ has been transformed into a diagonal form. The time ordered integral close to the singularity is dominated by

$$\left(\det\left\{\frac{\mathrm{d}q_i(t^c_+)}{\mathrm{d}q_j(t^c_-)}\right\}\right)^{1/2} = \exp\left(\frac{1}{2}\int_{t^c_-}^{t^c_+}\frac{R}{\tau + \mathrm{i}\epsilon - t_c}\mathrm{d}\tau\right) \ ,$$

where $t^c_\pm = t^c \pm \eta$ are infinitesimally close to t^c and the integration variable τ is shifted to $\tau + \mathrm{i}\epsilon$, because the corresponding wave packet should start out with a positive phase before it encounters the first singularity. This integral can be computed by taking the limit $\epsilon \to 0$,

$$\left(\det\left\{\frac{\mathrm{d}q_i(t^c_+)}{\mathrm{d}q_j(t^c_-)}\right\}\right)^{1/2} = \exp(-\mathrm{i}\pi(R/2))\left|\det\left\{\frac{\mathrm{d}q_i(t^c_+)}{\mathrm{d}q_j(t^c_-)}\right\}\right|^{\frac{1}{2}} \ . \tag{74}$$

Note that the phase only results from the delta function part of the integrand, whereas the principle value contributes just to the modulus which has been already calculated in (71). Between two singular points the time ordered integral is positive and gives the absolute value of the volume ratio. R counts the number of rank reductions of the matrix $\mathbf{M}$ along the classical path, and it is a function of the initial condition $\mathbf{M}_0$; for a periodic orbit it is an invariant property of the cycle.

C Gutzwiller Trace Formula vs. Quasiclassics

Consider a generalization of the quasiclassical zeta function (39), weighted by extra powers of $\Lambda_{p,i}$:

$$F_n(k) = \exp\left(-\sum_p\sum_{r=1}^{\infty}\frac{1}{r}\prod_{i=1}^{d}\frac{|\Lambda_{p,i}|^{-r/2}\,\mathrm{e}^{\frac{\mathrm{i}}{\hbar}S_p(k)r - \mathrm{i}\pi\frac{m_p}{2}r}}{\Lambda^{nr}_{p,i}(1-1/\Lambda^r_{p,i})^2(1-1/\Lambda^{2r}_{p,i})}\right) \ . \tag{75}$$

The weight $1/(1-x)$, $x = 1/\Lambda^r_{p,i}$ of the p-th term in the exponent of the semiclassical zeta function (41) can be related to the quasiclassical zeta function cycle weight $1/(1-x)^2(1-x^2)$ in (75) by multiplying it by

$$1 = \frac{1}{(1-x)(1-x^2)} - \frac{x}{(1-x)(1-x^2)} - \frac{x^2}{(1-x)(1-x^2)} + \frac{x^3}{(1-x)(1-x^2)} \ .$$

From this it follows that the semiclassical zeta function function (41) for Axiom A flows is meromorphic in the complex k plane, as it can be written as a ratio of entire functions; for 2-dimensional Hamiltonian systems

$$Z_{sc}(k) = \frac{F_0(k)F_3(k)}{F_1(k)F_2(k)} , \tag{76}$$

where $F_n(k)$ includes only (71), the first term in the $\Delta_{p,r}$ sum (69). The zeros of the semiclassical zeta function coincide with the ones obtained from $F_0(k) = Z_{qc}(k)$, and the leading poles should arise from $F_1(k)$. In two dimensions, i.e. $d = 1$, (75) can be resummed as

$$F_n(k) = \prod_p \prod_{j=0}^{\infty} \prod_{l=0}^{\infty} \left(1 - \frac{t_p}{\Lambda_p^{n+j+2l}}\right)^{j+1} , \tag{77}$$

where t_p is defined in (44).

D Selberg Zeta Function

The question that arises naturally in discussing semiclassical quantization is following: if the usual semiclassical evolution is not multiplicative, why does it anyway yield the *exact* quantization in the case of the Selberg trace formula? And what does the quasiclassical quantization yield for flows on surfaces of constant negative curvature?

The Selberg (1956) zeta function for geodesic flows on surfaces of constant negative curvature is exceptional: in this very special case the multiplicativity is guaranteed by the Bowen-Series (1979) map, which reduces the two-dimensional flow to a direct product of 1-dimensional maps, and makes it possible to construct the associated transfer operators in terms of one variable, Mayer (1990).

The essence of the construction is the following: In the Poincaré halfplane representation the dynamics is described by the free Hamiltonian

$$\hat{H} = \frac{1}{2y^2}(p_x^2 + p_y^2) \tag{78}$$

whose classical trajectories are circle segments. The centers of the circles always lie on the $y = 0$ axis, and any free trajectory can be characterized by x_f and x_b, the forward and backward intersection points of its circle with the $y = 0$ axis. The polygonal billiards in the x, y plane are defined in terms of walls which themselves are geodesics, hence also characterized by their footpoints x_1, x_1', x_2, x_2' ... (see Fig. 7). A reflection off a wall changes the direction of the particle, with the new trajectory characterized by a new pair of footpoints x_f' and x_b'. The new forward footpoint will be the image of the old footpoint with respect to an inversion transformation on the circle of the wall. For example, a reflection off the wall x_n, x_n' of radius $R_n = |x_n - x_n'|/2$ and center $x_n^c = (x_n + x_n')/2$ is described by

$$x_f' = f_n(x_f) = x_n^c + R_n^2/(x_f - x_n^c) \ .$$

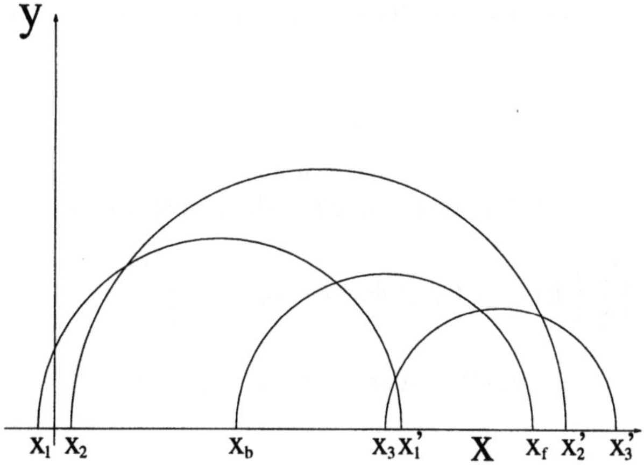

Fig. 7. A typical arrangement on the Poincaré halfplane. The half circles with footpoints (x_i, x_i'), $i = 1, 2, 3$ are the billiard walls. The forward and backward footpoints (x_b, x_f) represent a trajectory.

The forward footpoint and the index of the wall determine uniquely the next forward footpoint. The footpoint of a periodic orbit reflected off walls $\epsilon_1\epsilon_2...\epsilon_{n_p}$ respectively is determined by the equation

$$x_p = F_{\epsilon_1\epsilon_2...\epsilon_{n_p}}(x_p) = f_{\epsilon_{n_p}}(f_{\epsilon_{n_p-1}}(...f_{\epsilon_1}(x_p)...)) \ . \tag{79}$$

The hyperbolic length of this periodic orbit is $l_{\epsilon_1\epsilon_2...\epsilon_{n_p}} = \log|F'_{\epsilon_1\epsilon_2...\epsilon_{n_p}}(x_p)|$, and its stability eigenvalue is also given by the derivative $F'_{\epsilon_1\epsilon_2...\epsilon_{n_p}}(x_p)$. The stability factor is the product of derivatives evaluated along the orbit

$$F'_{\epsilon_1\epsilon_2...\epsilon_{n_p}} = \prod_{i=1}^{n_p} F'_{\epsilon_i} \ ,$$

and is multiplicative without any need for further manipulations. This property makes the polygonal billiards on surfaces of constant negative curvature unique and atypical.

The Fredholm determinant of the 1-dimensional Perron-Frobenius operator

$$\mathcal{L}(y, x, k) = |f'(x)|^{1/2+ik}\delta(y - f(x)) \ ,$$

where f is the appropriate footpoint mapping and $k = \sqrt{E - 1/4}$ is the wave number, is precisely the Gutzwiller-Voros semiclassical zeta function for this problem, $Z_{sc}(E) = \det(1 - \mathcal{L}_0)$. Unlike the generic situation discussed in this paper, the semiclassical zeta function is in this case an entire function. However, the spectrum of the quasiclassical zeta function $Z_{qc}(E)$ defined in this paper contains spurious zeroes in the complex plane in addition to the true zeroes on the real k axis. These spurious zeroes are the eigenvalues of weighted operators of type

$$\mathcal{L}_m(y,x,k) = |f'(x)|^{1/2+m+ik}\delta(y - f(x)) \ , \tag{80}$$

where m is an integer number. Since the $F_n(k)$'s (see (77)) can be expressed in terms of the Fredholm determinants of these operators as

$$F_n(k) = \prod_{l=0}^{\infty} \left\{ \det\left(1 - \mathcal{L}_{n+l}\right) \det\left(1 - \mathcal{L}_{n+l+1}\right) \right\}^{l+1} \ , \tag{81}$$

$Z_{\rm sc}(k) = \det(1 - \mathcal{L}_0)$ results under the relation (76), too.

References

Arnold V. I. (1978): *Ordinary Differential Equations* (MIT Press, Cambridge, Mass.)

Artuso R., Aurell E., Cvitanović P. (1990): Recycling of strange sets: I. cycle expansions. Nonlinearity **3**, 325–359

Berry M. V. (1981): Quantizing a Classically Ergodic System: Sinai's Billiard and the KKR Method. Ann. Phys. (N.Y.) **131**, 163–216

Berry M. V., Keating J. P. (1990): A rule for quantizing chaos? J. Phys. **A 23**, 4839–4849

Bogomolny E. B. (1992): On dynamical zeta function. CHAOS **2**, 5–14

Bohm D. (1952): A suggested Interpretation of the Quantum Theory in Terms of "Hidden" Variables. I. Phys. Rev. **85**, 166

Born M. (1926): Z. f. Physik **37**, 863

Bowen R., Series C. (1979): Publ. Math. Inst. Hautes Etud. Sci. **50**, 153–170

Cvitanović P. (1988): Invariant measurement of strange sets in terms of cycles. Phys. Rev. Lett. **61**, 2729–2733

Cvitanović P., ed. (1992): *Periodic Orbit Theory - theme issue*, CHAOS **2**, 1–158

Cvitanović P., Rosenqvist P. E., Rugh H. H., Vattay G. (1993): A Fredholm determinant for semi-classical quantization. CHAOS **3**, 619–636

Cvitanović P., et al., (1996): *Classical and Quantum Chaos: A Cyclist Treatise*, http://www.nbi.dk/~predrag/QCcourse/

Cvitanović P., Eckhardt B. (1989): Periodic-Orbit Quantization of Chaotic Systems. Phys. Rev. Lett. **63**, 823–826

Cvitanović P., Eckhardt B. (1991): Periodic orbit expansions for classical smooth flows. J. Phys. **A 24**, L237–L241

Cvitanović P., Rosenqvist P. E. (1993): A new determinant for quantum chaos, in *From Classical to Quantum Chaos*, Dell'Antonio G. F., Fantoni S., Manfredi V. R., eds., Soc. Italiana di Fisica Conf. Proceed. **41** (Ed. Compositori, Bologna), 57–64

Cvitanović P., Vattay G. (1993): Entire Fredholm determinants for evaluation of semi-classical and thermodynamical spectra. Phys. Rev. Lett. **71**, 4138–4141

Eckhardt B. (1987): Fractal properties of scattering singularities. J. Phys. **A**: Math. Gen. **20**, 5971–5979

Eckhardt B., Russberg G. (1993): Resummations of classical and semi-classical periodic orbit expressions. Phys. Rev. **E 47**, 1578–1588

Eckhardt B., Russberg G., Cvitanović P., Rosenqvist P. E., Scherer P. (1995): *Pinball Scattering*, in *Quantum chaos between order and disorder*, Casati G. and Chirikov B. eds. (Cambridge University Press, Cambridge), 405–433

Feynman R. P. (1972): *Statistical Mechanics - a set of lectures*, (W. A. Benjamin, Mass.)

Gaspard P., Alonso Ramirez D. (1992): Ruelle classical resonances and dynamical chaos: The three- and four-disk scatterers. Phys. Rev. **A 45**, 8383–8397

Gaspard P., Rice S.A. (1989a): Scattering from a classically chaotic repellor. J. Chem. Phys. **90**, 2225-2241

Gaspard P., Rice S.A. (1989b): Semiclassical quantization of the scattering from a classically chaotic repellor. J. Chem. Phys. **90**, 2242–2254

Gaspard P., Rice S.A. (1989c) Exact quantization of the scattering from a classically chaotic repellor. J. Chem. Phys. **90**, 2255–2262

Gutzwiller M.C. (1971): Periodic Orbits and Classical Quantization Conditions. J. Math. Phys. **12**, 343–358

Gutzwiller M.C. (1988): J. Phys. Chem. **92**, 3154

Gutzwiller M.C. (1990): *Chaos in Classical and Quantum Mechanics* (Springer, New York)

Heller E. J. (1975): Time-dependent approach to semiclassical dynamics. J. Chem. Phys. **62**, 1544–1555

Heller E. J., Tomsovic S., Sepúlveda A. (1992): Time domain approach to semiclassical dynamics: Breaking the log time barrier. CHAOS **2**, 105–116

Holland P. R. (1993): *The quantum theory of motion, - An account of the de Broglie-Bohm causal interpretation of quantum mechanics* (Cambridge Univ. Press, Cambridge)

Kohn W., Rostoker N. (1954): Solution of the Schrödinger equation in periodic lattices with an application to methalic lithium. Phys. Rev. **94**, 1111–1120

Lloyd P., Smith P.V. (1972): Multiple-scattering theory in condensed materials. Adv. Phys. **21**, 69–142

Madelung E. (1926): Z. f. Physik **40**, 332

Nelson E (1985): *Quantum Fluctuations* (Princeton Univ. Press, Princeton)

Mayer D. H. (1990): On the Thermodynamic Formalism for the Gauss Map. Comm. Math. Phys. **130**, 311–333; Mayer D. H. (1992): in *From Phase Transitions to Chaos*, Györgyi G., et al., eds., 521–529

Pollner P., Vattay G. (1996): New Method for Computing Topological Pressure. Phys. Rev. Lett. **76**, 4155–4158

Rosenqvist P. E., Vattay G., Wirzba A. (1996): Application of the Diffraction Trace Formula to the Three Disk Scattering System. J. Stat. Phys. **83**, 243–257.

Ruelle D. (1987): *Statistical Mechanics, Thermodynamical Formalism* (Addison-Wesley, Reading Mass.)

Rugh H. H. (1992): The Correlation Spectrum for Hyperbolic Analytic Maps. Nonlinearity **5**, 1237–1263; see also ref. Cvitanović et al. (1993) for a discussion of the applicability of the theorem

Schrödinger E. (1926): Annalen der Physik **79**, 361, 489, **80**, 437, **81**, 109

Selberg A. (1956): J. Ind. Math. Soc.**20**, 47

Spiegel E.A. (1995): private communication

Tanner G. (1995): private communication

Vattay G. (1994): Differential equations to compute $\hbar$ corrections of the trace formula. chao-dyn/9406005, unpublished

Vattay G. (1996): Bohr-Sommerfeld Quantization of Periodic Orbits. Phys. Rev. Lett. **76**, 1059–1062

Vattay G, Rosenqvist P. E. (1996): Periodic Orbit Quantization beyond Semiclassical Approximation. Phys. Rev. Lett. **76**, 335–339

Vattay G., Wirzba A., Rosenqvist P. E. (1994): Periodic Orbit Theory of Diffraction. Phys. Rev. Lett. **73**, 2304–2307

Vattay G., Wirzba A., Rosenqvist P. E. (1995): Inclusion of Diffraction Effects in the Gutzwiller Trace Formula. Proc. Int. Conf. *Dynamical Systems and Chaos*, Tokyo, May 1994, chao-dyn/9408005, (World Scientific, Singapore) Vol. **2**, 463–466

Voros A. (1988): Unstable periodic orbits and semiclassical quantisation. J. Phys. **A**: Math. Gen. **21**, 685–692

Wirzba A. (1992): Validity of the semiclassical periodic orbit approximation in the 2-and 3-disk problems. CHAOS **2**, 77–83

Wirzba A. (1993): Test of the Periodic Orbit Approximation in n-Disk Systems. Nucl. Phys. **A 560**, 136–150

Wirzba A. (1995): Zeta functions in scattering problems evaluated on the real wave number axis. http://crunch.ikp.physik.th-darmstadt.de/~wirzba/, unpublished

Wirzba A. (1996): Comparison of the quantum mechanical cumulant and the semi-classical curvature expansion for classically chaotic scattering systems. http://crunch.ikp.physik.th-darmstadt.de/~wirzba/, unpublished

Wirzba A., Henseler M. (1995): The missing-link between the quantum mechanical and semi-classical determination of scattering resonance poles. http://crunch.ikp.physik.th-darmstadt.de/~wirzba/, unpublished

Semiclassical Cross Sections

Bruno Eckhardt and Bruno Hüpper

FB Physik und Institut für Chemie und Biologie des Meeres,
C.v. Ossietzky Universität, Postfach 25 03, D-26111 Oldenburg, Germany

Abstract. We discuss the semiclassical formalism for calculating cross sections in photodissociation of small molecules. The general procedures for calculating the smooth background in the classical limit and the leading order quantum corrections, as well as for the semiclassical periodic orbit and recurrent orbit contributions are outlined. As a specific example, we discuss an application to the photodissociation spectrum of H_2O.

1 Introduction

Spectra of highly excited atoms and molecules confront us with a very complex and wildly fluctuating set of transitions (Gay 1994). Their interpretation and calculation in terms of a complete set of eigenstates of the Hamiltonian is prohibitively complicated and, as we now know, not particularly rewarding. Arguments used long ago in nuclear physics (Porter 1965) and supported by semiclassical calculations suggest that the state to state fluctuations are nearly uncorrelated and can be described by statistical ensembles constrained only by a few symmetry properties (Haake 1991, Bohigas 1991). It is only for larger differences in energy that system specific correlations show up in the transition strength (Wintgen 1987, Main et al. 1994). This suggests that the system specific features can be extracted from a suitable smoothing in energy and indeed that has been used successfully for a variety of systems (e.g. Ezra et al. 1991, Stein & Stöckmann 1992)

A very useful alternative to the full quantum mechanics focusses on the time evolution of states and operators rather than matrix elements directly (Heller 1991). Since time and energy domain are related by a Fourier transform, a smoothed spectrum in energy can be related to the behaviour of the propagator in a finite time interval, perhaps sufficiently short so that simple trajectory and interference based methods become available. In particular, the smoothed cross sections in energy can be obtained from a Fourier transform of the classical correlation function of the transition operator. In certain cases this provides already most of the information that can be extracted (Gomez Llorente & Pollak 1992).

The aim of this article is to sketch the semiclassical arguments behind the above description, to derive the main formulas and to discuss their consequences, both in models and in concrete applications. The principle method is a stationary phase approximation of various quantum propagators and traces.

In leading order in $\hbar$ Gutzwiller's (1967 et seq.) semiclassical trace formula is obtained, but there are important improvements both in the smooth background term and in the periodic orbit weights for matrix elements which are needed for quantitative agreement in the extreme quantum regime, as applies e.g. for H_2O.

We close this introduction with a description of the main objects and an outline of the sections. Section 2 contains a brief survey of the main techniques. In section 3 we discuss the calculation of the smooth background term within the Wigner-Weyl formalism. Section 4 deals with the periodic orbit contribution including a discussion of zeta functions and the like. In section 5 we discuss an application to the photodissociation of water on the $\tilde{A}B_1$ surface. We conclude with a few remarks on fluctuation properties in section 6.

2 Preliminaries

Without loss of generality we may assume that the systems can be described by a kinetic energy quadratic in momenta and a potential energy V, so that the Hamiltonian is

$$H = \frac{\mathbf{p}^2}{2m} + V(\mathbf{x}) \,. \tag{1}$$

We furthermore assume the potential to go to infinity as the distance grows, so that the spectrum is discrete. Open systems may be closed by introducing walls far away from the region of interest; this will not have a significant effect on the spectra smoothed over an interval ΔE if the time needed to run from the interaction zone to the boundary and back is larger than the smoothing time $T_E = 2\pi/\Delta E$. The bounded Hamiltonian then has an infinite number of eigenenergies E_n and eigenstates $|n\rangle$. The scattering resonances in open systems are reflected in a clustering of levels.

By Fermi's golden rule the strengths of the observed transitions are proportional to the matrix elements $M_{i,n} = \langle n|\mu|i\rangle\langle i|\mu|n\rangle$, where $|i\rangle$ describes the initial state and μ the dipole operator. All these matrix elements can be obtained from the matrix element weighted density of states

$$\rho_\Pi(E) = \sum_n \langle n|\Pi|n\rangle \delta(E - E_n) \,, \tag{2}$$

with the projection onto the initial state

$$\Pi = \mu|i\rangle\langle i|\mu \,. \tag{3}$$

There are two approaches to obtain information on these matrix elements. The first one employs a Laplace transform,

$$\Theta_\Pi(\beta) = \int_0^\infty e^{-\beta E} \rho_\Pi(E) dE \tag{4}$$

$$= \sum_n \langle n|\Pi|n\rangle e^{-\beta E_n}$$

$$= \mathrm{Tr}\, \Pi e^{-\beta H}\,, \tag{5}$$

and results in a relation to the canonical partition function in statistical mechanics (Wigner 1932). The high-temperature limit $\beta \to 0$ probes the large energy limit $E \to \infty$ and provides information on the average behaviour of the density of states and the transition strengths (Simon 1979). The leading order term in an expansion for small β (high temperature expansion) yields the phase space average of the classical observable. Higher order corrections can be calculated e.g. by use of Wigner's expansion for the statistical operator, as described in section 3.

For the second approach, a Fourier transformation

$$\Theta_\Pi(t) = \int_0^\infty e^{itE/\hbar} \rho_\Pi(E) dE = \sum_n \langle n|\Pi|n\rangle e^{iE_n t/\hbar}$$

$$= \mathrm{Tr}\, \Pi e^{iHt/\hbar}\,, \tag{6}$$

relates the density of states to the quantum mechanical propagator. The semiclassical approximation to the propagator then reveals the significant contributions from periodic orbits, as discussed further in section 4.

Quantitative calculations require both next to leading order terms in the mean part and efficient resummation techniques for the periodic orbit contributions. Particularly useful formulae based on dynamical zeta functions arise if there is a simple symbolic coding for the periodic orbits and if the dynamics is sufficiently unstable so that only a few orbits dominate (Eckhardt et al. 1995a). This is the case for open systems like the dissociation of water as studied in section 5. In other cases, in particular in closed systems, the formulae presented here can be used to calculate smoothed spectra, where the contributions are again limited to a few orbits.

3 Wigner-Weyl Transforms and Phase Space Averages

The smooth background follows from the high temperature expansion of the statistical operator (Wigner 1932, Heller 1978). To relate it to classical phase space averages, we use the Wigner-Weyl transform which associates with any quantum mechanical operator A a classical phase space function

$$A_W(\mathbf{P}, \mathbf{Q}) = \int d^N \mathbf{s}\, \langle \mathbf{Q} - \mathbf{s}/2|A|\mathbf{Q} + \mathbf{s}/2\rangle e^{i\mathbf{P}\cdot\mathbf{s}/\hbar}\,. \tag{7}$$

Dimensions are chosen such that

$$\mathrm{Tr}\, A = \int \frac{d^N \mathbf{P} d^N \mathbf{Q}}{(2\pi\hbar)^N} A_W(\mathbf{P}, \mathbf{Q})\,. \tag{8}$$

In particular, for a Gaussian initial state of width σ,

$$\langle x|i\rangle = \frac{1}{\pi^{1/4}\sigma^{1/2}} e^{-(x-x_0)^2/2\sigma^2}\,, \tag{9}$$

the Wigner transform of $|i\rangle\langle i|$ is Gaussian in position and momentum,

$$W(P,Q) = 2\exp\left\{-\frac{(Q-x_0)^2}{\sigma^2} - \frac{\sigma^2 P^2}{\hbar^2}\right\}. \tag{10}$$

Note that the width in momentum is $\hbar/\sigma$ and vanishes as $\hbar \to 0$. This has consequences for the semiclassical expansion of the traces, as we will point out below.

The trace of a product of two operators equals the classical trace of the product of the Weyl transforms,

$$\operatorname{Tr} AB = \int \frac{d^N\mathbf{P}d^N\mathbf{Q}}{(2\pi\hbar)^N} A_W B_W\,, \tag{11}$$

but this does not extend to the full product, i.e.

$$(AB)_W = A_W B_W + \text{terms of higher order in } \hbar\,. \tag{12}$$

Therefore, the Wigner transform of the quantum propagator becomes

$$\left(e^{-\beta H}\right)_W = e^{-\beta H}(1 + O(\hbar)). \tag{13}$$

To calculate the higher order corrections we turn to the defining equation for the statistical operator,

$$-\frac{\partial}{\partial\beta}P = HP = \frac{1}{2}(HP + PH)\,. \tag{14}$$

The symmetrized version on the right hand side is particularly well suited for application of the Weyl transformation. The corresponding equation for P_W becomes

$$-\frac{\partial}{\partial\beta}P_W = H_W \cos\left\{\frac{\hbar}{2}\Lambda\right\} P_W \tag{15}$$

with the differential operator

$$\Lambda = \overleftarrow{\nabla}_{\mathbf{P}}\overrightarrow{\nabla}_{\mathbf{Q}} - \overleftarrow{\nabla}_{\mathbf{Q}}\overrightarrow{\nabla}_{\mathbf{P}} \tag{16}$$

where the differentiations act to the left or to the right as indicated. An obvious consequence of the above equation is that the corrections contain only even powers of $\hbar$.

The calculation of these corrections is rather cumbersome and can be aided by formal manipulation programs like MAPLE or MATHEMATICA or REDUCE. The most important leading order terms can be obtained rather directly by substitution of the ansatz

$$P_W = \exp\{-\beta H + \hbar^2 c_2 + \hbar^4 c_4 + \ldots\} \tag{17}$$

in the equations of motion (15), expansion of all terms in powers of $\hbar$ and comparison of coefficients. This gives differential recursion relations for the coefficients involving lower order coefficients and derivatives of the Hamiltonian. Specifically, the first term becomes (e.g. Wigner 1932, Heller 1978)

$$c_2 = -\beta^2 f_2 + \beta^3 f_3 \tag{18}$$

with

$$f_2 = \frac{1}{8m}\Delta V \quad \text{and} \quad f_3 = \frac{1}{24m}(\nabla V)^2 + \frac{1}{24m^2}(\mathbf{P}\nabla)^2 V\,. \tag{19}$$

With this information we can now obtain the leading order terms of the smooth cross section,

$$\sigma_0(E) = \mathrm{Tr}\, A_W \delta(E-H) - \hbar^2 \frac{\partial^2}{\partial E^2}\, \mathrm{Tr}\, A_W\, f_2\, \delta(E-H) + \hbar^2 \frac{\partial^3}{\partial E^3}\, \mathrm{Tr}\, A_W\, f_3\, \delta(E-H)\,, \tag{20}$$

where the trace is the classical one. Alternatively, keeping the terms in the exponent, one finds an Airy function approximation to the delta function,

$$\sigma_0(E) = \int \frac{d^N\mathbf{P} d^N\mathbf{Q}}{(2\pi\hbar)^N} e^{-(H-E)f_2/3f_3 - 2\hbar^2 f_2^3/27 f_3^3} \times \frac{1}{(3\hbar^2 f_3)^{1/3}} \mathrm{Ai}\left(\frac{H-E+\hbar^2 f_2^2/f_3}{(3\hbar^2 f_3)^{1/3}}\right) A_W\,. \tag{21}$$

Both expressions have their merits.

In order to demonstrate the necessity of $\hbar$ corrections, Fig. 1 compares the Weyl term with the Gaussian projector matrix element weighted density of states for the harmonic oscillator as a function of Planck's constant. The parameter controlling the approach to the semiclassical limit is the ratio between the width σ of the wave packet and the characteristic length scale of the oscillator, $\ell = \sqrt{\hbar/m\omega}$,

$$r = \sigma/\ell = \sigma\sqrt{\frac{m\omega}{\hbar}}\,. \tag{22}$$

The Wigner transform of the initial state has a maximum in phase space near $x = x_0$ and $p = 0$. The leading order classical phase space average then yields a cross section which increases up to $E_{max} \approx m\omega^2 x_0^2/2$ and decreases for higher energy. The quantum calculations show that for large $\hbar$ and small r the quantum cross section is shifted towards higher energies. However, already for r near 0.5, this effect is barely noticeable.

In general, such expansions are asymptotic in $\hbar$ and not convergent. The structure of such asymptotic expansions as well as the specific properties of

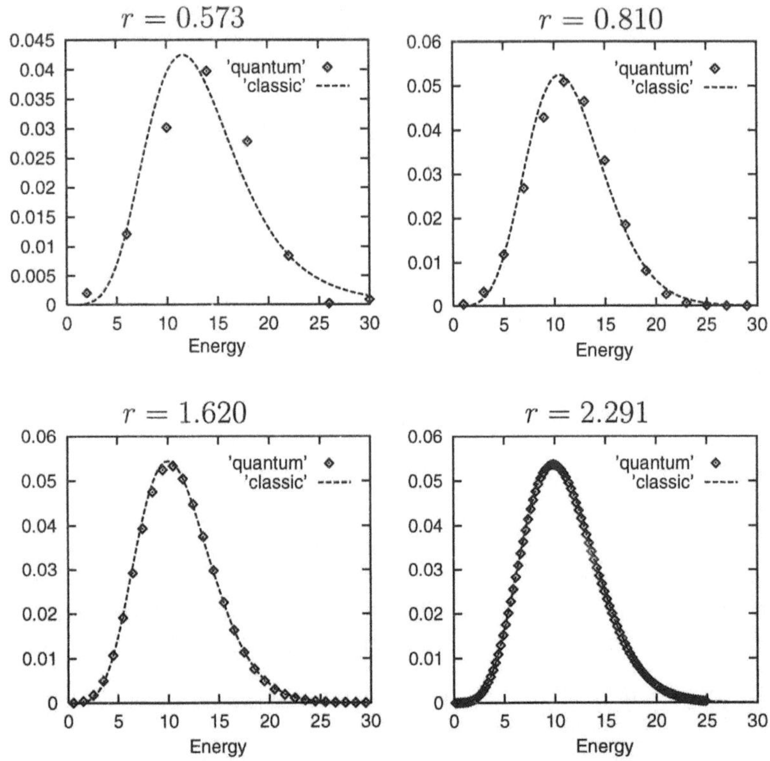

Fig. 1. A comparison between the n-th harmonic oscillator eigenstate matrix elements $\langle n|\sigma\rangle\langle\sigma|n\rangle$ of the projector onto a Gaussian $\sigma(q)$ and the corresponding classical phase space average of the Wigner transform $(2\pi\hbar)^{-1}\int dqdpW_\sigma(q,p)\delta(E-H(q,p))$ manifestly shows deviations for small $r=\sigma\sqrt{m\omega/\hbar}$ (here $m=\omega=1, \sigma=\sqrt{21/16}$)

the Wigner-Weyl expansion for the density of states in billiards have been studied by Berry and Howls (1995). The calculation of higher order terms is very often complicated by the lack of explicit formulae or efficient computational schemes. In order to illustrate the typical behaviour for photodissociation cross sections, we turn to a simple model: a Gaussian initial state excited onto a linear potential (Heller 1978, Eckhardt & Hüpper to be published).

The wave functions on the upper, linear potential surface $V(x) = -ax$ are given by Airy functions, normalized in the energy, and the initial state is given by a Gaussian, eq. 10. The exact transition probability is then given by

$$\sigma_{\text{exact}}(E) = \frac{1}{a}\left(\frac{2ma}{\hbar^2}\right)^{\frac{2}{3}} \left| \int_{-\infty}^{\infty} \Psi_{\text{g}}(x)\, \text{Ai}\left(-\left(\frac{2ma}{\hbar^2}\right)^{\frac{1}{3}}\left(x+\frac{E}{a}\right)\right) dx \right|^2 \tag{23}$$

Rescaling the length $\tilde{x} = x/\lambda$ in the above expression in units of $\lambda =$

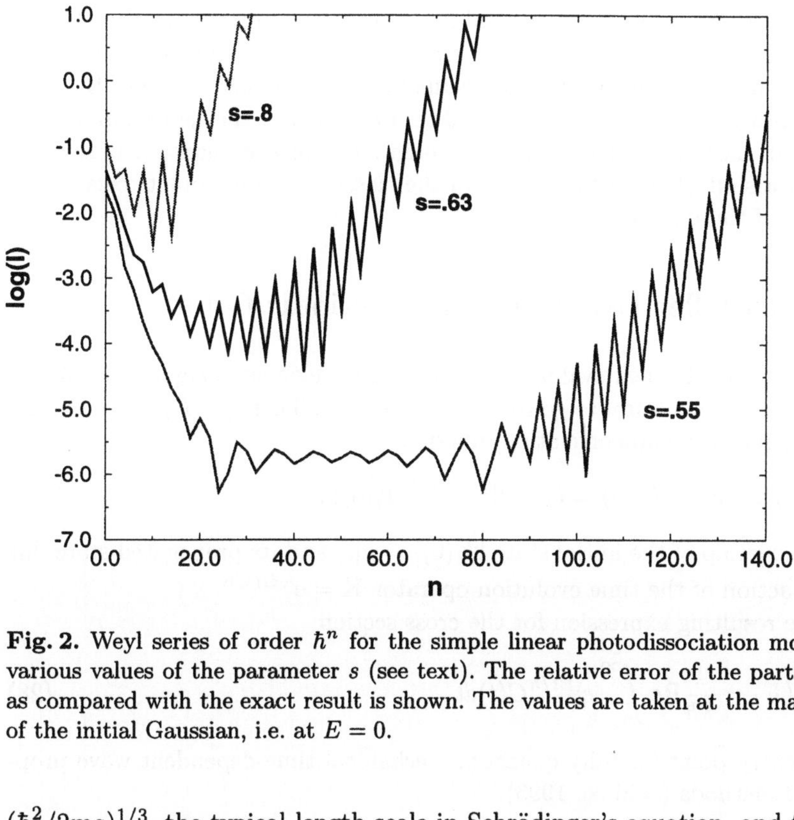

Fig. 2. Weyl series of order $\hbar^n$ for the simple linear photodissociation model for various values of the parameter s (see text). The relative error of the partial sum as compared with the exact result is shown. The values are taken at the maximum of the initial Gaussian, i.e. at $E = 0$.

$(\hbar^2/2ma)^{1/3}$, the typical length scale in Schrödinger's equation, and the energy by $\varepsilon = E/a\lambda$, leads to:

$$\sigma_{\text{exact}}(\varepsilon) = \frac{1}{a}\left|\int_{-\infty}^{\infty} \Psi_g(\tilde{x}) \operatorname{Ai}(-(\tilde{x}+\varepsilon))d\tilde{x}\right|^2 . \tag{24}$$

The Weyl series in terms of the appropriate expansion parameter

$$s = \frac{\lambda}{\sigma} \tag{25}$$

can be derived with the help of a purely algebraic procedure (Grammaticos & Voros 1979, Eckhardt & Hüpper to be published):

$$\sigma_{\text{asymp}}(\varepsilon) = \frac{2}{\pi a \lambda}\sum_{n=0}^{\infty}\frac{s^{3n}}{12^n n!}\int_0^{\infty} e^{-(y^2-\varepsilon)^2 s^2 - y^2/s^2} H_{3n}((y^2-\varepsilon)s)dy \quad ,(26)$$

where H_n are the Hermite polynomials of order n. (The Gaussian wave packet is located without loss of generality at $x_0 = 0$.) Typically, the asymptotic series develops a minimum before diverging (Fig. 2). The position of the minimum is controlled by the expansion parameter s: for small s, i.e. in the semiclassical regime of small $\hbar$ or broad initial state, the minimum occurs later and then takes on smaller values. Calculating corrections can then be useful. However, for large s, divergence sets in early and no quantitative results can be obtained without resummation of the series. Here, the Airy representation (21) contains a partial resummation of the series and is more useful. In particular, it captures the main effect of the quantum behaviour, namely a shift of the maximum from the classical turning point towards the interior of the potential.

4 Gutzwiller Expansions and Periodic Orbits

Within a semiclassical evaluation of the matrix element weighted density of states eq. 2 it is natural to go to the time domain. Then, $\rho_\Pi(E)$ is the Fourier transform of the autocorrelation function

$$C(t) \equiv \langle \Psi_i(0) | \Psi_i(t) \rangle = \langle \Psi_i(0) | e^{-i\hat{H}t/\hbar} | \Psi_i(0) \rangle \,, \tag{27}$$

i.e. the overlap of the initial state $|\Psi_i(0)\rangle = \mu|i\rangle$ and its propagated state due to the action of the time evolution operator $\mathbf{K} = e^{-i\hat{H}t/\hbar}$.

The resulting expression for the cross section

$$\rho_\Pi(E) = \frac{1}{\pi\hbar} \mathrm{Re} \int_0^\infty e^{iEt/\hbar} C(t) dt \tag{28}$$

is the entry point for fully quantum mechanical time-dependent wave propagation methods (Schinke 1993).

4.1 Semiclassical Trace Formulas

In order to connect this quantum mechanical expression to classical mechanics we write, first of all, the autocorrelation function as a trace, viz.

$$C(t) = \mathrm{tr}(\mathbf{K}\Pi) \,. \tag{29}$$

Replacing $\mathbf{K}$ and Π by their Wigner transforms, C becomes by eq. 11:

$$C(t) = \int \frac{d^N\mathbf{P} d^N\mathbf{Q}}{(2\pi\hbar)^N} \mathbf{K}_W(t) \Pi_W \,. \tag{30}$$

For the semiclassical evaluation of the above expression one substitutes a semiclassical approximation for the propagator and evaluates the remaining integrals in stationary phase approximation (Eckhardt et al. 1992a,b). If Π_W is sufficiently smooth, as for position, momenta and energy observables, the calculation proceeds as in the case of the density of states and results in

$$\rho_{\Pi}(E) = \int \frac{d\mathbf{Q}d\mathbf{P}}{h^N} \Pi_W(\mathbf{Q},\mathbf{P})\delta(E - H(\mathbf{Q},\mathbf{P}))$$
$$+\mathrm{Re}\ \frac{1}{\pi\hbar}\sum_p \sum_{r=1}^{\infty} w_{p,r}\Pi_{W,p}e^{(iS_p(E)/\hbar - i\pi\mu/2)r} \tag{31}$$

where

$$\Pi_{W,p} = \int_0^{T_p} dt\, \Pi_W(\mathbf{Q}_p(t),\mathbf{P}_p(t)) \tag{32}$$

is the time average of the observable evaluated along the classical trajectory $(\mathbf{Q}(t),\mathbf{P}(t))$ over one period. The weight of the orbit is given by

$$w_{p,r} = |\det(M_p^r - 1)|^{-1/2}\,. \tag{33}$$

For bounded systems the sum extends over all periodic orbits. In the case of an open system (closed perhaps by the method mentioned at the beginning of section 2) the summation can be restricted to all orbits trapped in the main interaction zone. In this way, the contributions from scattering resonances are obtained. For the projector type observables typical of cross sections, the smoothness requirements are not completely met since the width in momenta is of order $\hbar/\sigma$ (viz eq. 10). An improved semiclassical expression takes this into account and keeps second order terms of the action as well. The results of this calculation for a Gaussian initial state yield a change in the weight of the orbits, (Zobay and Alber 1993).

$$w_{p,r} = |2 - \mathrm{Tr}M_p^r + 2i\hbar/\sigma^2|^{-1/2}\,. \tag{34}$$

An important feature of this expression is the fact that in contrast to the previous one it is no longer singular at a bifurcation. The bifurcation is a property of a very specific trajectory, whereas neighbouring ones do not feel it. By averaging over some neighbourhood of the periodic orbit as indicated by the projector this feature is smeared out.

4.2 Zeta Functions and Cycle Expansion

The oscillatory part in eq. 31 for the cross section is now rewritten as the real part of the logarithmic derivative of a suitable zeta function (Eckhardt et al. 1992).

Specifically, for the two degrees of freedom system we consider here, M_p is a 2×2 matrix of determinant one. As we consider only unstable orbits, the eigenvalues are Λ_p and $1/\Lambda_p$. The denominator can then be expanded in a geometric series and the summation on repetitions can be performed. This gives

$$\rho_{\mathrm{osc}} = \mathrm{Re}\frac{1}{\pi\hbar}\sum_p \sum_{j=0}^{\infty} \frac{\Pi_{W,p}\tilde{t}_{p,j}}{1-\tilde{t}_{p,j}}\,, \tag{35}$$

where $\tilde{t}_{p,j} = e^{iS_p/\hbar - i\mu_p\pi/2}|\Lambda_p|^{-1/2}\Lambda_p^{-j}$. To arrive at zeta functions one has to introduce an auxiliary variable γ and the extended quantity $t_p^j = e^{-\gamma\Pi_{W,p}}\tilde{t}_{p,j}$. Because of the relations $\tilde{t}_{p,j} = t_p^j\big|_{\gamma=0}$ and $\frac{\partial t_p^j}{\partial\gamma}\Big|_{\gamma=0} = -\Pi_{\mathrm{W},p}\tilde{t}_{p,j}$, eq. 35 may be written

$$\rho_{\mathrm{osc}} = \mathrm{Re}\frac{1}{\pi\hbar}\sum_p\sum_{j=0}^{\infty}\frac{\partial}{\partial\gamma}\ln(1-t_p^j)\big|_{\gamma=0} , \tag{36}$$

so that finally

$$\begin{aligned}\rho_{\mathrm{osc}} &= \mathrm{Re}\frac{1}{\pi\hbar}\frac{\partial}{\partial\gamma}\ln Z(\gamma,E)|_{\gamma=0} \\ &= \mathrm{Re}\frac{1}{\pi\hbar}\frac{Z'(\gamma,E)}{Z(\gamma,E)}\bigg|_{\gamma=0}\end{aligned} \tag{37}$$

with the Selberg-type product (Voros 1988)

$$Z(E) = \prod_p\prod_{j=0}^{\infty}(1-t_p^j) \quad . \tag{38}$$

This expression can be evaluated using the cycle expansion (Artuso, Aurell & Cvitanović 1990a,b, Cvitanović & Eckhardt 1989).

Let us specify to the case of a hyperbolic system with two degrees of freedom and a complete ternary symbolic coding for the classical orbits (as e.g. hydrogen in a magnetic field of sufficient strength, Eckhardt & Wintgen 1990, Hansen 1995). Labelling the contributions from the periodic orbits by their symbolic code, we have to consider products of the form

$$Z(\gamma,E) = (1-t_0)(1-t_1)(1-t_2)(1-t_{01})(1-t_{02})(1-t_{12})\cdots , \tag{39}$$

which expand into

$$Z(\gamma,E) = 1 - t_0 - t_1 - t_2 - (t_{01} - t_0t_1 + t_{02} - t_0t_2 + t_{12} - t_1t_2)\cdots . \tag{40}$$

Upon taking the derivative, we have

$$\begin{aligned}\frac{\partial Z(\gamma,E)}{\partial\gamma} = \;&\Pi_{W,0}t_0 + \Pi_{W,1}t_1 + \Pi_{W,2}t_2 \\ &+ [\Pi_{W,01}t_{01} - (\Pi_{W,0}+\Pi_{W,1})t_0t_1] \\ &+ [\Pi_{W,02}t_{02} - (\Pi_{W,0}+\Pi_{W,1})t_0t_2] \\ &+ [\Pi_{W,02}t_{12} - (\Pi_{W,1}+\Pi_{W,2})t_1t_2]\cdots .\end{aligned} \tag{41}$$

The suggestive argument for convergence (Eckhardt and Russberg 1993) is that a long orbit t_{ab} can be decomposed into shorter ones (labelled a and b), which shadow it; therefore, the term $t_{ab} - t_at_b$ is small. However, the same can be expected of the observable: if the short orbits a and b are close to ab, then $\Pi_{W,ab}$ should also be close to $\Pi_{W,a} + \Pi_{W,b}$ and the convergence properties of the above cycle expansion can be expected to be the same as the ones for the density of states.

Thus, if only the primitive orbits dominate, the cross section can be expressed as

$$\rho_{\text{osc}}(E) = \text{Re}\frac{1}{\pi\hbar}\frac{\Pi_{w,0}t_0 + \Pi_{w,1}t_1 + \Pi_{w,2}t_2}{t_0 + t_1 + t_2} . \tag{42}$$

At first sight, this expression looks like a truncation of Gutzwiller's summation. However, it has to be noted that it includes the exponential proliferation of periodic orbits in a well defined approximate manner through the cycle expansion of the zeta function. This is noticed most prominently in the positions of the resonances, which are closer to the real axis than estimates based on single orbits would suggest. Thus it includes the stabilizing effect of the hyperbolic tangle. The local Lyapunov exponent of an orbit measures the escape from that orbit which partially is compensated by the large number of orbits forming the tangle.

5 Photodissociation of H_2O

In this section we will discuss the application of the above formalism to photodissociation of water in the first absorption band. Compared with bounded potentials, the scattering system has several advantages which make it particularly well suited for a semiclassical investigation: i) The structure of the periodic orbits is simple and can be captured by a good symbolic coding.
ii) The system is rather open, so that a few periodic orbits dominate.
iii) It has only two degrees of freedom, so that one can expect the systematic errors of the semiclassical expansion to remain small. We will summarize the semiclassical analysis, more details will be given elsewhere.

5.1 General Aspects

The photodissociation of water in the first absorption band

$$\text{H}_2\text{O}(\tilde{X}^1A_1) + \hbar\omega \rightarrow \text{H}_2\text{O}(\tilde{A}^1B_1) \rightarrow \text{H}(^2S) + \text{OH}(X^2\Pi, nj),$$

is known to be a prototype system for a fast and direct photodissociation process (Engel et al. 1992). The main features are summarized in Fig. 3.

The potential energy surface (PES) of the $\tilde{A}^1B_1$ state is repulsive, and therefore the H-OH bond breaks almost immediately when the parent molecule is excited. This behaviour is reflected in a broad and quite structurless absorption cross section. Superimposed on this broad background are modulations which decrease with increasing excitation energy (see Fig. 3.). The ground state $|\Psi(0)\rangle$, which is assumed to be of Gaussian shape, is transferred on the excited PES. There, the wave-packet $|\Psi(t)\rangle$ propagates and spreads. Parts of

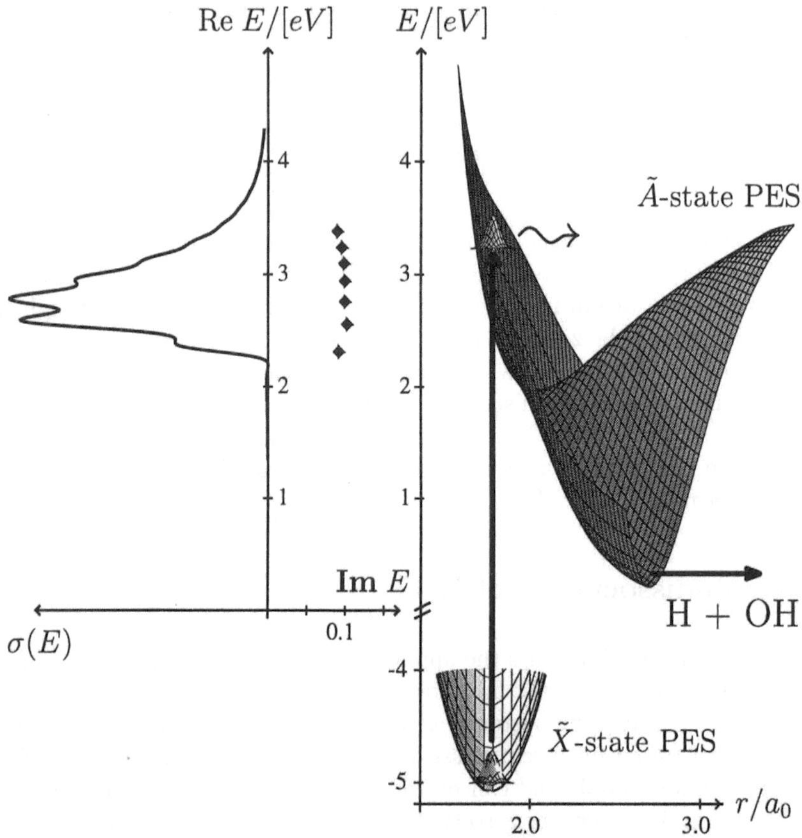

Fig. 3. The photoabsorption cross section $\sigma(E)$ (left part) of the water molecule on the $\tilde{A}B_1$ surface consists of a broad structure and small modulations due to resonances. The resonances are determined semiclassically from the zeros of the zeta function in the complex energy plane (diamonds). In the time dependent picture these resonances are recurrences of a Gaussian wavepacket initially located on the ground $\tilde{X}$-state PES and then, after excitation, propagating on the $\tilde{A}$-state PES (right hand side). The portrayal of the excited PES is a projection onto the symmetric stretch plane. In front, as well as symmetrically in the back, extend the exit channels for photodissociation. (see also Fig. 4.)

the wave-packet return to the initial excitation region leading to recurrences in the autocorrelation function $C(t) = \langle \Psi(0) | \Psi(t) \rangle$.

It has been suggested that the symmetric stretch motion is the dominant one and suffices to explain the oscillatory structures in the photodissociation of water (Engel et al. 1988, Henriksen et al. 1988). We confirm these findings in a more sophisticated semiclassical treatment in which we include systematically further periodic orbit contributions, the asymmetric stretch motion and the fixed stretch motion as well as combinations of these.

5.2 Classical Description

We confine our classical and semiclassical analysis to the two degrees of freedom corresponding to the two OH separations as experimental observations (Andresen et al. 1984) confirm that the OH($^2\Pi$) fragment is created in low rotational states. This is consistent with the observation that at short distances, the PES has a minimum at roughly the same bending angle as the PES of the ground state $\tilde{X}^1 A_1$, and therefore the anisotropy is locally very weak in the Franck-Condon regime. And, as the molecule dissociates from an initial $\nu_2 = 0$ ground bending state, no considerable torque is exerted on the OH molecule during dissociation.

The classical dynamics is determined by the Hamiltonian for the three nuclei on the PES, where we assume a bending angle of $\gamma = 104.52°$. With R_1 and R_2 the two OH separations and p_1, p_2 the conjugate momenta, the Hamiltonian takes the form

$$H = \frac{1}{2\mu}\left(p_1^2 + p_2^2\right) + \frac{x}{\mu} p_1 p_2 \cos\gamma + V(R_1, R_2)\,. \tag{43}$$

The potential energy surface (PES) $V(R_1, R_2, \gamma = 104.52°)$ is taken from Engel et al. (1988). The reduced two-body mass of OH is $\mu = m_{\text{O}} m_{\text{H}}/(m_{\text{O}} + m_{\text{H}}) = 16/17$ and the dimensionless factor $x = m_{\text{H}}/(m_{\text{O}} + m_{\text{H}}) = 1/17$.

The PES is symmetric in the two OH separations. Along the symmetric stretch $R_1 = R_2$ it has a minimum at $R_1 = 1.09$ Å, which is a saddle point in two dimensions separating the two dissociation channels. The zero point for the energy is located at the two-body breakup, i.e. as R_1 goes to infinity and $R_2 = 0.971$ Å. The three-body breakup with $R_1, R_2 \to \infty$ then has the energy $E = 4.621$ eV.

A small displacement of two initially nearby trajectories on the PES grows exponentially. All trajectories launched on the PES will run into one of the dissociation channels, except for an infinite number of periodic orbits which remain trapped in the PES. As we will show below, all periodic orbits on the excited PES are unstable from a little bit above the saddlepoint energy. Thus, we are confronted with a completely chaotic system, apart from a small energy domain on the top of the barrier. The symmetric stretch motion (Fig. 4a) exists from the saddle point energy $E \approx 1.977$ eV up to the three-body breakup. Slightly above the saddle point energy it undergoes a transition from unstable to stable at $E = 2.184154$ eV and then from stable to unstable at $E = 2.18516$ eV. At the lower critical point, the asymmetric stretch (Fig. 4b) and the fixed stretch (Fig. 4c) are born via a saddle-node bifurcation. While the fixed stretch is unstable for all energies the asymmetric stretch remains stable in a little interval of the energy before it tends to completely unstable behaviour. All orbits remain unstable up to the three-body breakup at $E = 4.621$ eV.

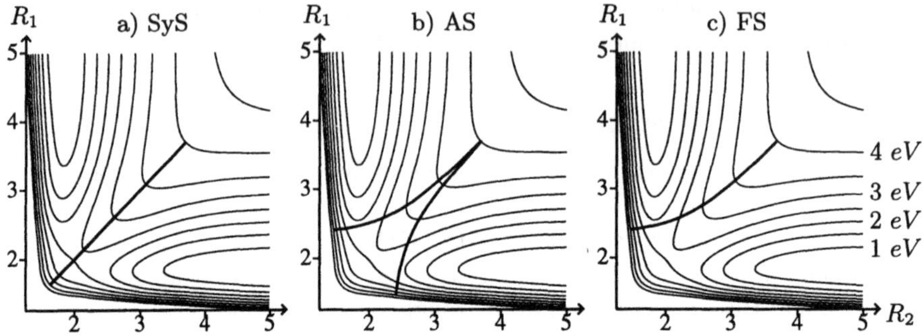

Fig. 4. Short periodic orbits on the water PES. R_1 and R_2 are the two OH separations in atomic units. a) Trajectory of the symmetric stretch motion. b) Trajectory of the asymmetric stretch motion and c) trajectory of the fixed stretch motion. It is very close to the asymmetric stretch motion and therefore only one orbit is shown.

In a completely chaotic system, like water on the very repulsive dissociative PES from $E \approx 2.2$ eV on, one may assume that there exists a periodic orbit for any symbol string, i.e. the symbolic dynamics is complete and cycle expansion methods as explained in the previous sections can be applied.

The symmetries are an important aspect of the problem. The PES has a shape similar to a quadrant of the 4-disk billiard that has been used to label the periodic orbits in the C_{4v} semi-parabolic potential for the quadratic Zeemann effect (Eckhardt and Wintgen 1990, Hansen 1995). As in that case we may introduce a ternary symbolic code labelling all periodic orbits. Following the notation of Cvitanović & Eckhardt (1993), the symmetric stretch motion is labelled with the symbol 2, the asymmetric stretch with 1, and the fixed stretch with symbol 0.

Another consequence of the C_{4v} symmetry is that the dynamics can be reduced to a fundamental domain, one octant of the billiard (Cvitanović & Eckhardt 1993). The advantage is that classical properties like actions, periods and stability properties of the full periodic orbit can be calculated in the fundamental domain with less computational effort.

The corresponding trajectories on the water PES are born below $E =$ 2.2 eV. Some run very close together, and the reduction to the fundamental domain becomes an essential simplification for the computation of stability properties and other classical quantities. Locating all periodic orbits of symbol length three turns out to be impossible with an accuracy of sixteen digits. But, as we will show below, corrections from longer orbits can be neglected for the calculation of the cross section.

5.3 Photodissociation Cross Section

As in previous investigations (Henriksen et al. 1988) we assume an initial state of Gaussian form,

$$\Psi_i(R_1, R_2) = \frac{1}{\sqrt{\pi}\sigma} \exp\left(-\frac{(R_1 - R_e)^2}{2\sigma^2} - \frac{(R_2 - R_e)^2}{2\sigma^2}\right) \tag{44}$$

with $R_e = 0.964$ Å and $\sigma = 0.0868$ Å.

Broad background For the calculation of the smooth background term the coordinate system is rotated,

$$u = \frac{R_1 + R_2}{\sqrt{2}} \quad , \quad v = \frac{R_2 - R_1}{\sqrt{2}} \quad . \tag{45}$$

The Wigner transform of the Gaussian eq. 44 then becomes

$$W_\Pi(u, v, p_u, p_v) = 4 \cdot \exp\left(-\frac{\sigma^2}{\hbar^2}(p_u^2 + p_v^2) - \frac{1}{\sigma^2}((u - u_e)^2 + v^2)\right) . \tag{46}$$

Inserting eq. 46 into eq. 20 and performing the integration over the momenta yields an integral expression for the background,

$$\rho_0(E) = \frac{2}{\pi\sqrt{kl}\hbar^2} \iint_{V(u,v) \leq E} du dv \; e^{-((u-u_e)^2+v^2)/\sigma^2} e^{-\alpha_+} I_0(\alpha_-) \tag{47}$$

with the mass dependent factors

$$k = \frac{1}{\mu}(1 + x\cos\gamma) \quad , \quad l = \frac{1}{\mu}(1 - x\cos\gamma) \tag{48}$$

and the arguments of the exponential and modified Bessel function

$$\alpha_\pm = (E - V(u, v))\left(\frac{1}{k} \pm \frac{1}{l}\right)\frac{\sigma^2}{\hbar^2} \quad . \tag{49}$$

This can be calculated by Simpson integration.

A qualitative physical picture of this first term is given by the reflection principle (Schinke 1993): The energy $E_{\max}$ of the maximum of $\rho_0(E)$ is roughly related to the center $\mathbf{Q_0}$ of the initial Gaussian via $V(\mathbf{Q_0}) = E_{\max}$. Fig. 5 shows the semiclassical background, as well as the background obtained from time-dependent wave-propagations. Although their global behaviour is in conformity, there is a deviation in the position of the maxima. This can be expected since the classical phase space density turns into a delta on the equi-potential line whereas wave mechanics shows a smoother behaviour, very much like an Airy function which will develop a maximum slightly away from the turning point. To illustrate this, we consider harmonic oscillator eigenfunctions along the symmetric stretch coordinate. In the semiclassical limit, i.e. high quantum number n, the eigenfunctions (Hermite polynomials) approach the classical behaviour. But, for low quantum numbers the eigenfunctions tend to show maxima towards the center of the potential. The ground state is even a Gaussian with a maximum at the center. To tackle this problem, the delta function has to be replaced by an Airy function, as

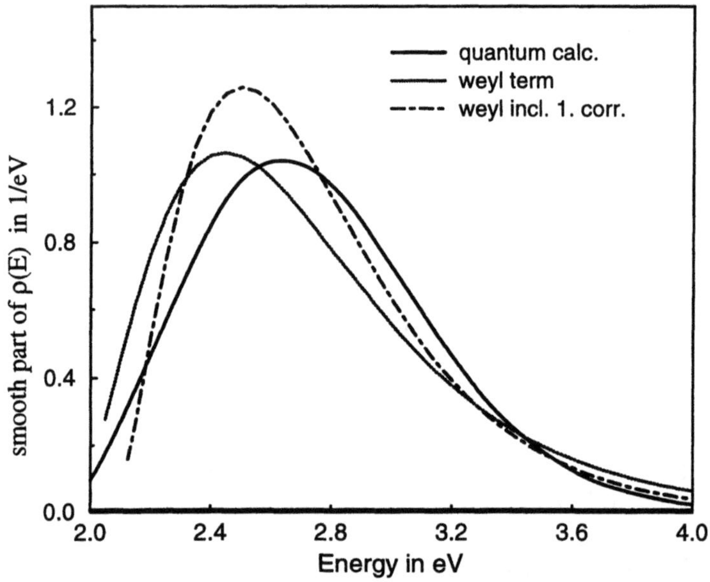

Fig. 5. Comparison of semiclassical and quantum background for the photodissociation of water. Note that the quantum mechanical curve is shifted towards higher energies, just as in the case of the harmonic oscillator, Fig. 1.

described before. As this function has to be multiplied with the initial Gaussian to obtain the smooth background, the maximum is now described by the first maximum of the Airy function and the main contribution to the mean part of $\rho_{\Pi}(E)$ is shifted towards higher energy. The first quantum correction including $O(\hbar^2)$ can be calculated numerically by integration over coordinate space following the procedure described in section 3.

It improves the lowest order Weyl term $\rho_0(E)$ properly for high energy but poorly for low energy, where the asymptotic expansion behaves like a divergent series already from its very first terms on (Eckhardt & Hüpper to be published).

Oscillatory part The position and the width of the resonances can be extracted from the zeros of the zeta function (40) for $\gamma = 0$, as they lead to poles in (37). It is sufficient to consider only those resonances that are situated close to the real energy axis. This allows us to confine the zeta function product to terms with $j = 0$. The action is expanded linearly for small complex energies,

$$S_p(E + i\varepsilon) = S_p(E) + i\varepsilon T_p(E) \; ; \tag{50}$$

the stability factors may be calculated at $\varepsilon = 0$, since the phase changes more rapidly with ε than the amplitude. The actions, periods and stability elements are calculated as functions of the energy and fitted by the method of least squares to polynomials to ease calculations in $Z(E,\varepsilon)$. If we consider only the symmetric stretch motion this yields simply:

$$Z_0(E,\varepsilon) = 1 - |\Lambda_0|^{-1/2} e^{iS_0/\hbar - \varepsilon T_0 - i3\pi/2} . \tag{51}$$

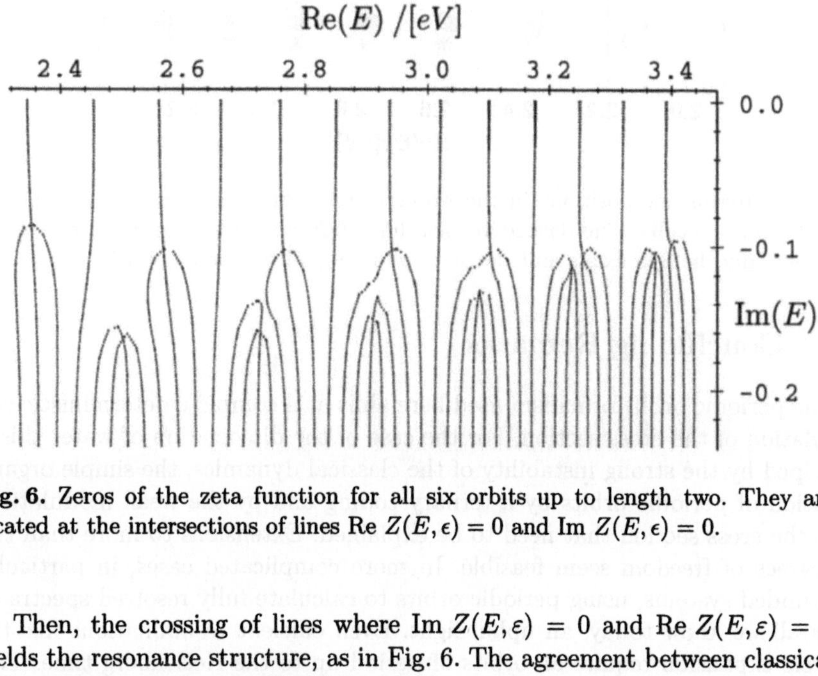

Fig. 6. Zeros of the zeta function for all six orbits up to length two. They are located at the intersections of lines Re $Z(E,\epsilon) = 0$ and Im $Z(E,\epsilon) = 0$.

Then, the crossing of lines where Im $Z(E,\varepsilon) = 0$ and Re $Z(E,\varepsilon) = 0$ yields the resonance structure, as in Fig. 6. The agreement between classical and quantum results (see Fig.7) is good.

The interesting part of the classical dynamics lies above the bifurcation energy of the asymmetric and fixed stretch. Then, in general, not only the symmetric stretch is of importance but all periodic orbits, according to their stability properties. The effect of taking into account also the other two fundamental orbits is twofold: at lower energies the real parts of the zeros are shifted a little bit downwards, whereas they are shifted upwards for higher energies, in accordance with the positions obtained from the quantum results.

But, as the weights Π_W of the asymmetric and fixed stretch are neglegibly small and the resonances are located sufficiently far away from the real energy axis, this shift does not appear in the photodissociation cross section. Secondly, the width becomes narrower compared with those obtained by the symmetric stretch alone and this tendency agrees well with the quantum width. Consequently, the positions and widths are not solely determined by the symmetric stretch alone and contain an interference effect of the three fundamental orbits.

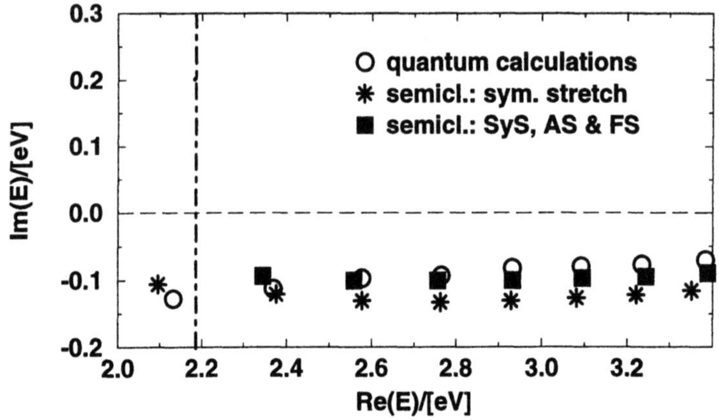

Fig. 7. Resonance positions in the water system obained quantum mechanically and semiclassically. The dashed vertical line at $E = 2.18$ eV separates the region where only the SyS exists and the one containing all fundamental orbits.

6 Concluding Remarks

The periodic orbit formalism used here aims at a complete deterministic calculation of the cross section. For the case of the dissociation of water this is helped by the strong instability of the classical dynamics, the simple organization of periodic orbits by a ternary coding and by the weak modulations in the cross section that need to be explained. Extensions to more than two degrees of freedom seem feasible. In more complicated cases, in particular bounded systems, using periodic orbits to calculate fully resolved spectra is, by all we know today, an uphill fight. Even efficient resummations like the cycle expansion require all orbits of period up to the Heisenberg time. This exponentially increasing number of contributions is in strong contrast to the algebraically increasing number of eigenvalues. But then the random matrix theory paradigm enters again: the information contained on the finest scale is universal and system independent. Currently, several efforts are under way to combine the periodic orbit formalism with random matrix theory. One example of such a hybrid approach might use short periodic orbits to calculate large scale features of the spectra and the statistical properties of classical trajectory segments of length the Heisenberg time to estimate the size and distribution of quantum fluctuations around the (modulated) background (Eckhardt et al. 1995b). For the spectral form factor this is within reach by the formalism of Bogomolny and Keating (1996).

References

Andresen P., Ondrey G.S., Titze B. and Rothe E.W. (1984): *J. Chem. Phys.* **80**, 2548

Artuso R., Aurell E. and Cvitanović P. (1990): *Nonlinearity* **3**, 325; *Nonlinearity* **3**, 361

Berry M.V. and Howls C.J. (1994): *Proc. R. Soc. Lond.* A **447**, 527-555

Bogomolny E.B. and Keating J. (1996): *Phys. Rev. Lett.* **77**, 1472

Bohigas O. (1991): in *Chaos and Quantum Physics, Proceedings of the NATO ASI Les Houches Summer School August 1989*, (ed. M.-J. Giannoni, A. Voros and J. Zinn-Justin), 87-199, Amsterdam: North Holland

Cvitanović P. and Eckhardt B. (1989): *Phys. Rev. Lett.* **63**, 823

Cvitanović P.and Eckhardt B. (1993): *Nonlinearity* **6**, 277

Eckhardt B. and Wintgen D. (1990): *J. Phys. B: At. Mol. Opt.* **23**, 355

Eckhardt B. (1992a) in *Quantum Chaos*, (ed. G. Casati, I. Guarneri and U. Smilansky), 77, Amsterdam: North Holland,

Eckhardt B., Fishman S., Müller K. and Wintgen D. (1992b): *Phys. Rev. A* **45**, 3531

Eckhardt B. and Russberg G. (1993): *Phys. Rev. E* **47**, 1578

Eckhardt B., Russberg G., Cvitanović P., Rosenquist P.E. and Scherer P. (1995a) in *Quantum Chaos between Order and Disorder*, (ed. G. Casati & B. Chirikov), 405-433, Cambridge University Press

Eckhardt B., Fishman S., Keating J., Agam O., Main J. and Müller K. (1995b): *Phys. Rev. E* **52**, 5893

Engel V., Schinke R. and Staemmler V. (1988): *J. Chem. Phys.* **88**, 129

Engel V., Staemmler V., Van der Wal R. L., Crim F. F., Sension R. J., Hudson B., Andresen P., Hennig S., Weide K. and Schinke R. (1992): *J. Phys. Chem.* **96**, 3201

Ezra G., Richter K., Tanner G. and Wintgen D. (1991): *J. Phys. B: At. Mol. Opt.* **24**, L413

Feynman R.P. and Hibbs A.R. (1965): *Quantum Mechanics and Path Integrals*, Mc Graw-Hill Book Company, Inc., New York

Gay J.C. (ed.) (1994) in *Irregular Atomic Systems and Quantum Chaos*, Gordon & Breach, Philadelphia

Gomez Llorente J. M. and Pollak E. (1992): *Annu. Rev. Phys. Chem.* **43**, 91-126

Grammaticos B. and Voros A. (1979), *Ann. Phys. (NY)* **123**, 359

Gutzwiller M.C.: *J. Math. Phys.* **8**, 1979 (1967); **10**, 1004 (1969); **11**, 1791 (1970); **12**, 343 (1971)

Haake F. (1991): *Quantum Signatures of Classical Chaos*, Springer, New York

Hansen K.T. (1995): *Phys. Rev. E* **51**, 1838

Heller E.J. (1978): *J. Chem. Phys.* **68**, 2066

Heller E.J. (1991) in *Chaos and Quantum Physics, Proceedings of the NATO ASI Les Houches Summer School August 1989*, (ed. M.-J. Giannoni , A. Voros and J. Zinn-Justin), 549, Amsterdam: North Holland

Henriksen N. E., Zhang J. and Imre D. G. (1988): *J. Chem. Phys.* **89**, 5607

Hillery M., O'Connell R.F., Scully M.O. and Wigner E.P. (1984): *Phys. Rep.* **106**, 121

Main J., Wiebusch G., Welge K., Shaw J. and Delos J. B. (1994): *Phys. Rev. A* **49**, 847

Porter C.E. (1965): *Statistical Theories of Spectra: Fluctuations.*, Academic Press

Schinke R. (1993): *Photodissociation Dynamics*, Cambridge University Press

Simon B. (1979): *Functional Integration and Quantum Physics*, Academic, New York
Stein J. and Stöckmann H.J. (1992): *Phys. Rev. Lett.* **68**, 2867
Voros A. (1988): *J. Phys. A: Math. Gen.* **21**, 685
Wigner E.P. (1932): *Phys. Rev.* **40**, 749
Wintgen D. (1987): *Phys. Rev. Lett.* **58**, 1589
Zobay O. and Alber G. (1994): *J. Phys. B: At. Mol. Opt.* **26**, L539

Resummation and the Turning Points of Zeta Functions

Jonathan P. Keating

School of Mathematics, University of Bristol, Bristol BS8 1TW UK
and
BRIMS, Hewlett-Packard-Laboratories, Filton Road,
Stoke Gifford, Bristol BS12 6QZ UK.

Abstract. We review various periodic orbit formulae for the zeta function whose zeros represent semiclassical approximations to the energy levels of chaotic systems. In particular, we focus on the Riemann-Siegel-resummed expression. The emphasis is on the ability of such formulae to reproduce the analytic properties of the spectral determinant, whose zeros are the exact quantum levels. As an example, the distribution of turning points is investigated and compared.

One of the central aims of quantum chaology is to calculate approximations to the energy levels of chaotic systems using classical trajectories. Most recent studies of this problem have been based on the semiclassical zeta function (Voros 1988), whose zeros represent the levels to leading order as $\hbar \to 0$. Using the trace formula (Gutzwiller 1971), this zeta-function can be expressed as a product over all periodic orbits of the underlying classical dynamics, and hence these orbits can, at the least in principle, be used to compute the zeros. The difficulty is, however, that the product does not converge in the region where the zeros lie. The goal has therefore been to find a formula that provides an analytic continuation outside its domain of convergence.

The two main candidates purporting to provide such a continuation are the cycle expansion (Cvitanovic & Eckhardt 1989, Eckhardt 1993) and Riemann-Siegel-type formulae (Berry & Keating 1990, Keating 1992, Berry & Keating 1992). Both are based on resummations of the series obtained by expanding the orbit product. In the case of the cycle expansion, this involves exploiting the shadowing property of the orbits in hyperbolic systems to generate a conditionally convergent re-ordering of the series. The motivation behind the approach is thus firmly rooted in classical mechanics, and its success relies on the system in question being sufficiently hyperbolic to possess a complete symbolic dynamics. The Riemann-Siegel method is almost entirely complementary in these respects. It is based on the analytical imposition of quantum hermiticity, in the form of a functional equation, onto the semiclassical series representing the zeta function. The result is an asymptotic expansion, each term of which is given by an absolutely convergent orbit sum that is effectively truncated at half the Heisenberg time T_H ($= 2\pi\hbar$ divided by the mean level separation). In this case, the motivation and the success of the approach relies upon the ability of leading-order semiclassical approximations to capture the essentially quantum structure underlying the zeta function. Put simply, the cycle expansion uses the chaoticity and Riemann-Siegel the quantumness of quantum chaology.

Both of these resummation methods have been applied successfully to calculate spectra. In particular, the cycle expansion provided the basis of perhaps the most important application to-date of semiclassical zeta-function techniques to a real physical system, namely the seminal work by Dieter Wintgen and his group on the helium atom (Ezra et al. 1991,). In addition, it has been used in several other studies involving, for example, the 3-disk scattering resonances (Eckhardt et al. 1995, Wirzba & Hensler 1995), and, in Riemann-Siegel form, the energy levels of bound systems (Tanner et al. 1991). Riemann-Siegel-type formulae themselves have been applied to various model systems (Szeredi & Goodings 1993, Keating & Sieber 1994) and, furthermore, constitute a central component of the closely related surface-of-section (Bogomolny 1992) and scattering matrix (Doron & Smilansky 1992) quantization techniques.

Impressive though the results of these energy-level calculations have been, we are probably still a long way from fully understanding the possibilities offered by resummation. It therefore seems worthwhile to pursue the approach further, especially in view of the fact that considerable (exponential) improvements in the efficiency of semiclassical methods based directly on periodic orbits are still needed to make them comparable with matrix-diagonalization techniques. Unfortunately, I can offer no new suggestions here. Instead, I would like to discuss an additional indication of the correctness of the basic resummation schemes already employed. This follows from the observation that most of the previous computations involve evaluating the semiclassical zeta function over various ranges of energy, but that the only data ultimately used are the positions of the zeros. The point is that the analytic structure of the function itself is often much more strongly dependent than the zeros on the convergence properties of the semicalssical representations employed, and that this then furnishes the possibility of making much more discriminating tests of the accuracy of the different methods of calculation.

As a simple example, I will here consider the distribution of the maxima and minima of a function that is equal to the semicalssical zeta function up to an overall phase, chosen to make it real on the line where the zeros lie. Specifically, I will prove that this function has exactly one maximum or, alternatively, one minimum between each pair of zeros, except possibly before the first zero, where it may have one of each. In a sense this is trivial, but not completely so. The result is certainly not surprising, and indeed could be guessed without much thought. Nevertheless, it is potentially very useful, since it turns out that this is precisely one of the properties that is satisfied by resummed but often not by unresummed semiclassical formulae, even if these both have almost exactly the same real zeros. This is one reason for giving a derivation. Another is that the first time I appreciated the relevance of this question was during a discussion with Dieter Wintgen in Copenhagen in 1991. It therefore seems not unfitting to set down the solution in a paper dedicated to his memory.

1 Semiclassical Zeta Functions and Resummation

For ease of presentation, I will consider the case of a 2-dimensional billard system corresponding to a domain Γ with boundary $\partial\Gamma$, although the results are much more general. The energy levels E_n are then the eigenvalues of the Laplacian in Γ with Dirichlet boundary conditions on $\partial\Gamma$; that is

$$\nabla^2\Psi_n = -E_n\Psi_n \tag{1}$$

in Γ, and $\Psi = 0$ on $\partial\Gamma$. These energy levels are, by construction, the only zeros of the spectral determinant

$$\Delta(E) = \prod_n \left(1 - \frac{E}{E_n}\right) \exp(E/E_n), \tag{2}$$

where the product converges and defines an entire function in the complex E- plane. Alternatively, and more usefully for our purposes here, Δ can also be written as a function of the momentum k, where $E = k^2$, and the corresponding momentum eigenvalues k_n, defined by $E_n = k_n^2$:

$$\Delta(k^2) = \prod_n \left(1 + \frac{k}{k_n}\right)\left(1 - \frac{k}{k_n}\right) \exp(k^2/k_n^2). \tag{3}$$

In the semiclassical limit, as $|k| \to \infty$, the density of momentum states

$$d_k(x) = \sum_n \{\delta(x - k_n) + \delta(x + k_n)\} \tag{4}$$

has a leading order asymptotic representation as a sum over all periodic orbits of the classical hamiltonian flow in Γ (Gutzwiller 1971). This in turn implies (Voros 1988, Keating & Sieber 1994) that in the same limit

$$\Delta(k^2) \approx B(k^2) \exp\left(-i\pi\bar{N}_k(k)\right) \frac{\zeta_s(k)}{\zeta_s(0)}, \tag{5}$$

where B is real and non-zero when k is real; $\bar{N}_k(x)$ is the mean of the momentum staircase N_k, defined by

$$N_k(x) = \int_0^x d_k(x')dx', \tag{6}$$

and so is an odd function of x (since d_k is even); and ζ_s is the semiclassical zeta function defined by

$$\zeta_s(k) = \prod_p \prod_{m=0}^{\infty} \left(1 - \sigma_p^m \exp\{-(\tfrac{1}{2}+m)\lambda_p l_p + ikl_p - \tfrac{1}{2}i\pi\mu_p\}\right) \tag{7}$$

where l_p is the length and λ_p the Liapunov exponent of the pth periodic orbit, μ_p is a Maslov-type index (Robbins 1991), and σ_p is the sign of the eigenvalues of the associated monodromy matrix. Explicit asymptotic formulae for B and $\bar{N}_k$ will be given in the next section. I will begin, however, by making some remarks about resummation.

The need for analytic continuation is immediately apparent from the fact that if the product (7) were to converge when k is real then the eigenvalues k_n would be the zeros of the individual factors. But these are neither real nor have the correct density. In fact, the product only converges provided $\mathrm{Im}\, k \geq \bar{\lambda}/2$, where $\bar{\lambda}$ is the mean Liapunov exponent. Hence in order to calculate energy levels using the periodic orbits in a chaotic system, some form of continuation from this region down to $\mathrm{Im}k = 0$ is required.

The first step in this direction is to expand the product as a sum over pseudo-orbits, that is, over all linear combinations of periodic orbits:

$$\zeta_s(k) = \sum_n C_n \exp(ikL_n), \tag{8}$$

where L_n denotes length of the nth pseudo-orbit, and the amplitude C_n is given in terms of the stabilities and Maslov indices of the contributing periodic orbits (Berry & Keating 1990). The point is that these amplitudes typically carry different signs, and so the possibility exists that the range of convergence of the sum may extend beyond that of the product (7). Any convergence in the extended region must, of course, be conditional, and so will depend of the ordering of the terms. Indeed, different orderings will give different results. The question is, is there any unique arrangement that is picked out by the physics of the problem?

In strongly hyperbolic systems the cycle expansion represents a natural candidate for such an ordering (Cvitanovic & Eckhardt 1989, Eckhardt 1993). It relies on the shadowing of long orbits by shorter ones, organised via the symbolic dynamics, to maximize the potential for cancellation in the pseudo-orbit sum. For a wide range of cases the resulting sum has been shown to lead to accurate approximations to the zeros of the zeta function, especially in systems that are not fully bound, for which the convergence problem is not so acute.

Riemann-Siegel resummation is based on a different approach. Rather than trying to find a preferred (conditionally) convergent ordering of the pseudo-orbits, it uses a quantum functional equation to re-express (8) in an asymptotic form in which the orbit sums involved converge in the absolute sense. This function equation is that Δ is an even function of the momentum k, because

$$\Delta\left(k^2\right) = \Delta\left((-k)^2\right). \tag{9}$$

Quantum mechanically, this is trivial and follows directly from the form of the spectral determinant (3), which itself is related to the fact that the energy levels E_n are necessarily real. However, its semiclassical implication, that to leading order in k^{-1}

$$\exp\left(-i\pi\bar{N}_k(k)\right)\zeta_s(k) = \exp\left(i\pi\bar{N}_k(k)\right)\zeta_s(-k), \tag{10}$$

is highly non-trivial, because it is manifestly satisfied neither by the orbit-product (7), nor by the pseudo-orbit sum (8). Put another way, the classical orbits have no intrinsic knowledge of quantum hermiticity. This is clearly related to the subtle convergence properties of the expansions involved, since, for example, if the orbit product (7) is substituted into (10) the two sides will not both converge simultaneously at any given point in the complex plane. It is in this observation that one sees the first hint that solving the hermiticity problem may also lead to a solution of the convergence problem.

The functional equation may be imposed upon the pseudo-orbit sum in the following way (for full details see Berry and Keating 1992). First

$$Z(k) = \exp\left(-i\pi\bar{N}_k(k)\right)\zeta_s(k) \tag{11}$$

is written in the form

$$Z(k) = \oint Z(k+s)\exp\left(-s^2\alpha^2/2|k|\right)\frac{ds}{2\pi i s}, \tag{12}$$

where the contour contains the origin, but no other singularities. This holds for any real α. Choosing the contour to be symmetric with respect to reflection through the origin, using the fact that the functional equation implies that $Z(-s) = Z(s)$, and assuming that Z is analytic in a strip about the real axis, we then have that

$$Z(k) = \int_{-\infty+i\varepsilon}^{\infty+i\varepsilon}\{Z(s+k)+Z(s-k)\}\exp\left(-s^2\alpha^2/2|k|\right)\frac{ds}{2\pi i s}. \tag{13}$$

Finally, lifting the contour up into the region where the pseudo-orbit sum converges (assuming this is possible) and integrating term-by-term leads to the main result, namely that to leading semiclassical order when k is real

$$Z(k) = Z_0(k,\alpha) + \sum_{m=3}^{\infty} Z_m(k,\alpha), \tag{14}$$

where if

$$\gamma_n(k) = L_n - \pi\bar{N}'_k(k), \tag{15}$$

$$Q^2(k,\alpha) = \alpha^2 + i\pi k\bar{N}''_k(k), \tag{16}$$

and $b_m(k)$ is defined by

$$\exp\left\{-i\pi\bar{N}_k(k+x)+i\pi\bar{N}_k(k)+i\pi x\bar{N}_k'(k)+\tfrac{1}{2}i\pi x^2\bar{N}_k''(k)\right\}$$
$$=1+\sum_{m=3}^{\infty}b_m(k)x^m, \tag{17}$$

then

$$Z_0(k,\alpha) \tag{18}$$
$$=2\,\mathrm{Re}\left[\exp\left(-i\pi\bar{N}_k(k)\right)\sum_n C_n\exp\left(ikL_n\right)\tfrac{1}{2}\,\mathrm{erfc}\left(\frac{\gamma_n(k)}{Q(k,\alpha)}\sqrt{\frac{k}{2}}\right)\right]$$

and, for $m \geq 3$,

$$Z_m(k,\alpha)=\frac{2}{\sqrt{\pi}}\left(\frac{k}{2}\right)^{m/2}\mathrm{Re}\left[\frac{i^m b_m(k)}{Q^m(k,\alpha)}\exp\left(-i\pi\bar{N}_k(k)\right)\right. \tag{19}$$
$$\left.\times\sum_n C_n\exp\left(ikL_n\right)\exp\left(\frac{-\gamma_n^2(k)k}{2Q^2(k,\alpha)}\right)H_{m-1}\left(\frac{\gamma_n(k)}{Q(k,\alpha)}\sqrt{\frac{k}{2}}\right)\right]$$

H_m being the mth Hermite polynomial.

The following observations suggest an interpretation for this formula. First, analytical estimates and numerical computations show that for the Riemann zeta-function the analogue of the m-sum (14) forms an initially rapidly converging asymptotic expansion (Berry & Keating 1992). The same has also been found numerically for the hyperbola billiard (Keating & Sieber 1994). We will thus assume that this is typically the case for all systems. Hence Z_0 represents the main term in the expansion, with later terms in the m-sum giving additional contributions that are relatively much smaller. That being said, these later terms still play an important role in the interpretation, because whilst each term on the right of (14) is a function of the parameter α, Z itself is not. Thus α-independence is a property of the whole sum and not of the individual terms.

Second, it is clear that the main term (18) is equal to twice the original pseudo-orbit sum, obtained by substituting (8) into (11), with each contribution multiplied by an erfc-function factor. The effect of this factor is straightforward: its value is close to 1 if $L_n < L_{RS}$, where $L_{RS}(k) = \pi\bar{N}_k'(k) = \pi\bar{d}_k(k)$, and close to 0 if $L_n > L_{RS}$. The transition between these two values is smooth and takes place over a length range whose size is proportional to $\alpha/\sqrt{k}$. In particular, it is sharp enough to ensure that the orbit sum is absolutely convergent. Indeed, if α is fixed, this transition becomes perfectly sharp in the semiclassical limit as $|k| \to \infty$. Z_0 is thus to a good approximation given by twice the real part of the original pseudo-orbit sum truncated when $L_n = L_{RS}$. This limit is precisely the Riemann-Siegel-lookalike formula conjectured in Berry & Keating (1990). For finite k the smoothing of the cut-off and the m-series in (14) are the additional features needed to make this semiclassically exact.

The fact that this method generates a resummation of the pseudo-orbit series is now clear, because it follows that the contributions from the orbits of length $L_n > L_{RS}$ have, approximately speaking, been replaced in the final result by the complex conjugate of the contributions from the orbits of length $L_n < L_{RS}$. This link between long and short orbits, first proposed in Berry (1986) and Berry & Keating (1990), and first derived in Keating (1992), is the central physical consequence of the mathematical analysis outlined above. In fact this property extends down to a more refined level: the contribution from the group of orbits with $L_n \approx L_{RS} + L$ is similarly related to that from the group with $L_n \approx L_{RS} - L$ (Keating 1991). It is this behaviour which makes the orbit sums converge and ensures that the functional equation is manifestly satisfied.

These surprising relations are a mathematical consequence of the functional equation and the pseudo-orbit sum (8). However, in most cases the zeta function is only a leading order semiclassical object, and so it is important to test the predictions explicitly in individual chaotic systems. One obvious way of doing this is to compare the contributions directly by summing up the corresponding sets of orbits. This has been done for the hyperbola billiard (Keating & Sieber 1994), and the results provide a detailed confirmation of the relationship. Another test is to use the resummed expression (14) to compute approximations to the energy levels. Here too the numerical support is strong. For example, in the case of the hyperbola billiard the first 46 levels were computed and the modulus of the error relative to the exact quantum values found to have an average of 3.4% and a standard deviation of 4.7% of the mean level spacing. Similar results have been reported for the wedge billard (Szeredi & Goodings 1993). However, as noted above, the ability to locate the positions of the zeros of the zeta function does not represent a particularly strict benchmark for the method of calculation employed. Indeed, good numerical results have also been obtained using unresummed formulae. It is for this reason that we now turn to the analytic structure of Z in order to provide a more discriminating test of the ideas associated with resummation.

2 Turning Points

The turning points of the quantum spectral determinant have a simple distribution: when E is real $\Delta(E)$ has a single maximum or, alternatively, a single minimum between each pair of neighbouring zeros. This fact may be proved as follows. Differentiating (2) gives

$$\frac{d}{dE}\left(\frac{\Delta'}{\Delta}\right) = -\sum_n \frac{1}{(E-E_n)^2} \ , \tag{20}$$

which in turn implies that

$$\frac{d}{dE}\left(\frac{\Delta'}{\Delta}\right) < 0 \qquad \text{for} \qquad E \neq E_n. \tag{21}$$

Thus between consecutive zeros of $\Delta, \Delta'/\Delta$ is a monotonically decreasing function and so Δ' can itself have at most one zero. Obviously there must be either a maximum or a minimum in the range, and hence the result follows.

The question is, to what extent is the analytical behaviour just described respected by the various semiclassical formulae discussed in the previous section? This issue is important because it follows from (20) that any additional maxima and minima must be associated with complex zeros. It is thus related to the fundamental problem of how far quantum hermiticity is preserved in leading-order semiclassical formulae, since complex zeros represent complex, and hence spurious or unphysical, energy levels. A key point now is the fact that semiclassically it is more natural to calculate $Z(k)$, definded by (11), rather than Δ. It thus follows from (5) that we are forced to investigate the distribution of the maxima and minima of

$$F(E) = \Delta(E)/B(E), \tag{22}$$

to which Z represents a leading-order approximation as $k \to \infty$.

The analytical structure of F clearly depends upon that of B. In certain special cases, such as the Riemann (see, for example, Keating 1993) and the Selberg (Voros 1987) zeta-functions, $B(E)$ has been calculated exactly. In general systems, however, only the leading-orders in its asymptotic expansion have been derived. These may be obtained from the terms in the corresponding (Weyl) expansion of the mean staircase $\bar{N}$. For example, for 2-dimensional planar billiards with Dirichlet boundary conditions

$$\bar{N}(E) = \frac{A}{4\pi}E - \frac{L}{4\pi}\sqrt{E} + C + \dots, \tag{23}$$

where A is the area, L the length of the perimeter, and C is given in terms of the smoothness and curvature of the boundary (Baltes & Hilf 1976). Then a simple calculation (Keating & Sieber 1994) leads to the result

$$B(E) = \exp\left\{\beta E + \frac{A}{4\pi}E(\log E - 1) + C\log E\right\}, \tag{24}$$

where

$$\beta = \lim_{X\to\infty}\left(\sum_{E_n \le X}\frac{1}{E_n} - \frac{A}{4\pi}\log X\right) \tag{25}$$

is a constant. In the case of the hyperbola billiard, for which the boundary $\partial\Gamma$ is the positive x- and y-axes and the line $y = 1/x$, and hence which is unbounded, the eigenvalues divide into two symmetry classes, being respectively even (+) and odd (−) about the line $y = x$. Then

$$\bar{N}(E) = \frac{1}{8\pi}E\log E + \frac{a}{8\pi}E + \frac{b^{\pm}}{8\pi}\sqrt{E} + c^{\pm} + \dots \tag{26}$$

where (Steiner & Trillenberg 1990) $a = 2(\gamma - \log 2\pi)$, γ being Euler's constant, $b^{\pm} = 4\pi^{3/2}/\Gamma^2\left(\frac{1}{4}\right) \pm 2\sqrt{2}$, and (Sieber 1991) $c^+ \approx -0.173$ and $c^- \approx 0.194$, from which it follows that (Keating & Sieber 1994)

$$B(E) \tag{27}$$
$$= \exp\left\{\beta' E + \frac{1}{16\pi} E \log^2 E + \frac{a}{8\pi} E(\log E - 1) - \frac{\pi}{16} E + c^{\pm} \log E\right\},$$

where the constant β' is given by

$$\beta' = \frac{\pi}{48} + \lim_{X\to\infty}\left(\sum_{E_n \le X} \frac{1}{E_n} - \frac{1}{16\pi}\log^2 X - \frac{a+1}{8\pi}\log X\right). \tag{28}$$

Armed with these expressions for B it is a simple matter to repeat for F the analysis performed for Δ at the beginning of this section. Consider first the case of a general bounded billard. Clearly

$$\frac{d}{dE}\left(\frac{F'}{F}\right) = -\sum_n \frac{1}{(E-E_n)^2} - \left(\frac{A}{4\pi E} - \frac{C}{E^2}\right), \tag{29}$$

and so F'/F is a monotonically decreasing function provided the term in the brackets is positive. This is obviously the case for $E > \tilde{E}$, where

$$\tilde{E} = \frac{4\pi C}{A}. \tag{30}$$

Thus for $E > \tilde{E}$, F behaves in the same way as Δ in that it has only a single maximum or minimum between adjacent zeros. Typically $\tilde{E}$ is smaller than the value of the first energy level E_0 because, roughly speaking, $\bar{N}(E_0) \approx 1/2 \Rightarrow E_0 \approx 2\pi/A$, and for billiards with smooth boundaries $C = 1/6$. For $E < \tilde{E}$ there is a competition between the two terms on the left-hand-side in (29). Clearly as $E \to 0$ the second term wins, because it is unbounded. Hence there is the possibility that before the first energy level there may be both a maximum and a minimum, since the derivative of F'/F may, and generally will, change sign.

For the hypberbola billard

$$\frac{d}{dE}\left(\frac{F'}{F}\right) = -\sum_n \frac{1}{(E-E_n)^2} - \left(\frac{\log E}{8\pi E} + \frac{a+1}{8\pi E} - \frac{c^{\pm}}{E^2}\right). \tag{31}$$

Again, when the term in the brackets is positive, F'/F is a monotonically decreasing function and so F has only one maximum or one minimum between consecutive zeros. This is clearly the case for $E > \tilde{E}$, where

$$\frac{1}{8\pi}\log\tilde{E} + \frac{a+1}{8\pi} = \frac{c^{\pm}}{\tilde{E}}. \tag{32}$$

In the even symmetry case there is no solution to this equation and so the behaviour just described persists down to $E = 0$. In the odd symmetry case there is one solution, which may be found numerically to be $\tilde{E} \approx 8.26$, that is, well below the first energy level $E_0 \approx 21.46$. Hence only below the first zero is the existence of two turning points possible. A detailed study of the balance between the terms in (31) shows that both a maximum and a minimum do indeed occur in this case.

3 Discussion

The distribution of the maxima and minima would not be of interest here if it did not distinguish between the resummed and unresummed formulae for the semiclassical approximation Z of F. Fortunately, it does. In particular, both functions have been studied in considerable detail for the hyperbola billiard (Keating & Sieber 1994). Numerical evaluations based on truncations of the divergent product (7) show a large number of additional turning points associated with spurious complex energy levels. This is not surprising: any truncation must vanish at the zeros of the individual factors, and these are clearly complex. (What is surprising is the accuracy with which the real zeros are approximated, but that is a different story).

More illuminating is the behaviour of the pseudo-orbit sum (8), because in this case the convergence properties are not immediately obvious. Again, without resummation this shows additional turning points whose number grows as the (arbitrary) truncation point is increased. Thus if the series does converge it is to a function whose analytic properties are fundamentally different from those of the spectral determinant. Crucially, this is not the case for the Riemann-Siegel resummed formula for Z, which, over the range of energies studied, exhibits exactly the same turning-point structure as F. Indeed, it provides a remarkably good approximation to F at all points, and not just at the zeros.

This study thus provides support both for the resummed formula (14) and, more importantly, for the ideas underlying its derivation, such as the orbit relations discussed in Sect. 1. Hence it would appear to indicate that the resummation programme, which at this stage is still largely mathematical, also has some interesting and non-trivial physical consequences that might repay further investigation in the atomic and molecular systems whose study is now generating some of the most exciting applications of quantum chaology.

Acknowledgements

It is a pleasure to acknowledge a helpful correspondece with Dr Martin Sieber on the ideas discussed in Sect. 2.

References

Baltes HP & Hilf ER (1976), *Spectra of finite systems*. Mannheim: Bibliographisches Institut.

Berry MV (1986), 'Riemann's zeta function: a model for quantum chaos?' In *Quantum chaos and statistical nuclear physics* (ed. T.H. Seligman & H. Nishioka). Springer Lecture Notes in Physics no. 263, 1-17.

Berry MV & Keating JP (1990), *J. Phys.* A**23**, 4839-4849.

Berry MV & Keating JP (1992), *Proc. R. Soc. Lond.* A**437**, 151-173.

Bogomolny EB (1992), *Nonlinearity* **5**, 805-866.

Doron E & Smilansky U (1992), *Nonlinearity* **5**, 1055-1084.

Cvitanovic P & Eckhardt B (1989), *Phys. Rev. Lett.* **63**, 823-826.

Eckhardt B (1993), 'Periodic orbit theory'. In *Quantum chaos* (ed. G. Casati, I. Guarneri & U. Smilansky), 77-111. Amsterdam: North-Holland.

Eckhardt B, Russberg G, Cvitanovic P, Rosenqvist PE & Scherer P (1995), in *Quantum Chaos between Order and Disorder* (ed. G. Casati & B. Chirikov), 405-433. Cambridge University Press.

Ezra GS, Richter K, Tanner G & Wintgen D (1991), *J. Phys.* B**24**, L413-L420.

Gutzwiller MC (1971), J. *Math. Phys.* **12**, 343-358.

Keating JP (1992), *Proc. R. Soc. Lond.* A**436**, 99-108.

Keating JP (1993), 'The Riemann zeta-function and quantum chaology'. In *Quantum chaos* (ed. G. Casati, I. Guarneri & U. Smilansky), 145-185. Amsterdam: North-Holland.

Keating JP & Sieber M (1994), *Proc. R. Soc. Lond.* A**447**, 413-437.

Robbins JM (1991), *Nonlinearity* **4**, 343-363.

Sieber M (1991), 'The hyperbola billiard: a model for the semiclassical quantization of chaotic systems'. Ph.D. thesis, University of Hamburg.

Steiner F & Trillenberg P (1990), *J. Math. Phys.* **31**, 1670.

Szeredi T & Goodings DA (1993), *Phys. Rev.* E**48**, 3529-3544.

Tanner G, Scherer P, Bogomolny EB, Eckhardt B & Wintgen D (1991), *Phys. Rev. Lett.* **67**, 2410-2413.

Voros A (1987), *Communs. Math. Phys.* **110**, 439-465.

Voros A (1988), *J. Phys.* **A21**, 685-692.

Wintgen D, Burgers A, Richter K & Tanner G (1994), *Progress of Theoretical Physics Supplement no. 116*, 121-142.

Wintgen D, Richter K & Tanner G (1992), *CHAOS* **2**, 19-33.

Wirzba A & Hensler M (1995), 'The missing link between the quantum mechanical and semi-classical determination of scattering resonance poles'. Preprint, Darmstadt.

Tunneling and the Semiclassical Propagator: A New Perspective

Neepa T. Maitra and Eric J. Heller

Department of Physics and
Harvard-Smithsonian Center for Astrophysics
Harvard University
Cambridge, MA 02138

Abstract. We consider quantum propagation in one dimension when there is an intervening barrier, from both the time domain and the energy domain viewpoints. We relate the issue of tunneling in the energy representation to flow across the barrier in real time. We find that time domain trajectories which have energies above the barrier do not correctly account for (energy domain) tunneling below the barrier when the time-energy Fourier transform is done exactly. The stationary phase transform (WKB result) cannot be thought of as an approximation to the Fourier integral, because of a branch cut in the complex-time plane. In fact, in the time domain, to obtain accurate semiclassical amplitudes for cross barrier flow at long times, below barrier tunneling trajectories need to be included. The Fourier transform from the time domain into the energy domain cannot be "improved" by performing it numerically rather than by stationary phase; instead, it is the time Green function kernel of the integral which needs correction (by the addition of tunneling paths) if one keeps time real.

1 Introduction

Despite the fact that the word "tunneling" hardly appeared in Dieter Wintgen's papers and never in his abstracts, the word was far from banned from his lexicon. It was known to his friends and associates that Dieter, the dominant force in the application of periodic orbit theory to atomic physics (or to anything else for that matter), had become pessimistic about making further breakthroughs using periodic orbits. It seems likely, however, that rather than abandon semiclassics altogether Dieter would have turned increasingly to the corrections and underpinnings of periodic orbit theory, as seen for example in the posthumously submitted Tanner and Wintgen (1995) paper.

Suppose we start at x_1 to the left of a barrier and travel to x_2 to the right of the barrier, in time T. Since to be at position x_1 with certainty at time 0 is to have completely uncertain momentum and therefore completely uncertain energy, there is always a supply of classical paths surmounting the barrier in just such a way as to arrive at x_2 at time T. At long times the energy of the arriving trajectories approaches the top of the barrier, since the slowing down near the barrier top is needed to effect the time delay necessary to arrange arrival at x_2 at long times. Thus the barrier problem in the time domain seemingly does not involve "tunneling" or classically forbidden paths. The semiclassical Green function

$$G^{sc}(x_2,T;x_1,0) = \frac{1}{\sqrt{2\pi i\hbar}} \left|\frac{\partial^2 S(x_2,x_1,T)}{\partial x_1 \partial x_2}\right|^{\frac{1}{2}} \exp\left[\frac{iS(x_2,x_1,T)}{\hbar}\right] \tag{1}$$

can thus apparently be constructed with ordinary "over the barrier" trajectories. We will return to this point shortly.

Assuming the semiclassical time Green function computed in this way is reasonable, then to obtain the semiclassical energy domain Green function, one performs the Fourier transform from the time domain into the energy domain *by stationary phase*:

$$G^{sc}(x_2,x_1;E) = \int_{sp} e^{iET/\hbar} G^{sc}(x_2,T;x_1,0)\, dT \tag{2}$$

The kernel of the stationary phase Fourier transform is the time domain semiclassical propagator, which is always nonvanishing because of the "over the barrier" trajectories. However for energies below the barrier top, no real time stationary phase point is found, forcing a search in the complex time plane for a saddle point. This search is rewarded with the well known classically forbidden imaginary (euclidean) time tunneling trajectories, which asymptotically give the correct fixed energy tunneling amplitudes (the WKB tunneling expression, see Sect. 3.2). Since $G^{sc}(x_2,T;x_1,0)$ is apparently under control it is tempting to think that if the original (real time) integral were performed exactly (rather than by stationary phase) one would get an even better answer, given only in terms of classically allowed over the barrier trajectories (since $G^{sc}(x_2,T;x_1,0)$ is). The smallness of the tunneling amplitude in this view would be due to the rapid oscillation of the integrand. This view was at the foundation of work by Keshavamurthy and Miller (1994), but was called into doubt by Grossman and Heller (1995). Here, we lay the question to rest by showing below-barrier tunneling trajectories are needed even in the time domain. Although it is not straightforward to quantify the time-domain tunneling contribution in general, an analysis is possible for large times and distant endpoints. We show explicitly that in this limit tunneling trajectories in fact dominate the contribution from over the barrier trajectories. Conventional semiclassics (above-barrier trajectories) breaks down at long times but this is where our correction works well. The corrected time domain semiclassical Green function could then be numerically Fourier transformed to give a good tunneling amplitude at fixed energy.

We find that the integral in (2), normally approximated by saddle point integration for energies below the barrier top, cannot generally give an improved transmission coefficient under exact (numerical) evaluation. Moreover the integral being approximated by complex saddle point integration in the derivation of the WKB tunneling expression is *not* the analytic continuation of the real time integral (2) due to an intervening branch point, which is another major point of this paper.

In Sect. 2 we begin with a brief review of the time domain semiclassical propagator. Then we present a quasiclassical picture of barrier transmission which suggests that in the generic case the semiclassics based on classically allowed trajectories is incomplete. We indicate how a correction may be made. At long times and distant endpoints we obtain a close approximation to the exact propagator where the conventional semiclassical propagator completely fails. In Sect. 3, we investigate the tunneling problem in the energy domain where we point out some problems associated with using the conventional semiclassical propagator and we find the insight gained from the quasiclassical approach in the time domain to be very useful.

2 The Time Domain Semiclassical Propagator

Consider a particle of mass m in one dimension interacting under the Hamiltonian $H = \frac{p^2}{2m} + V$, p being the momentum and V a time-independent potential. The probability amplitude for the particle, initially at x_1, to be found at x_2 after time T, is given by the propagator

$$G(x_2, T; x_1, 0) = \langle x_2|e^{-iHT/\hbar}|x_1\rangle = \int \mathcal{D}[x(t)]e^{\frac{i}{\hbar}S[x(t)]} \tag{3}$$

where $\mathcal{D}[x(t)]$ implies a summation over all paths $x(t)$ linking x_1 to x_2 in time T and $S[x(t)]$ is the classical action $\int L dt$ for the (not necessarily classical) path $x(t)$ linking x_1 to x_2 in time T. The formula on the right-hand side, due to Feynman (Feynman and Hibbs (1965)), provides a natural basis for a semiclassical evaluation of the propagator, the exact evaluation of which is, in general, very difficult. (See Grosche and Steiner (1995) for a recent review of some exactly solvable path integrals.) The reader is referred to the existing literature Holstein and Swift (1982), Rajaraman (1975), Schulman (1981) for details of the derivation of the semiclassical propagator. The essence of the technique lies in stationary phase evaluation of the path integral, and reduces the formidable sum over all paths to a sum over only classical paths:

$$\begin{aligned} &G^{sc}(x_2, T; x_1, 0) \\ &= \sum_{\substack{classical\\paths}} \frac{1}{\sqrt{2\pi i\hbar}} \left|\frac{\partial^2 S_{cl}(x_2, x_1, T)}{\partial x_2 \partial x_1}\right|^{1/2} \times \exp\left(\frac{i}{\hbar}S_{cl}(x_2, x_1, T) - \frac{i\nu\pi}{2}\right) \\ &= \sum \left|\frac{1}{2\pi i\hbar p_{cl}(x_1)p_{cl}(x_2)\int_{x_1}^{x_2}\frac{dx}{p_{cl}(x)^3}}\right|^{1/2} \\ &\times \exp\left(\frac{i}{\hbar}\int_{x_1}^{x_2} p_{cl}(x)dx - \frac{i}{\hbar}E_{cl}T - \frac{i\nu\pi}{2}\right) \end{aligned} \tag{4}$$

where $p_{cl} = \sqrt{2m(E_{cl} - V(x))}$ is the classical momentum at x of a particle with energy E_{cl}, and E_{cl} is determined by the classical relation

$$T = \int_{x_1}^{x_2} \frac{dx}{\sqrt{2m(E_{cl} - V(x))}}. \tag{5}$$

ν is the Maslov index, counting the number of conjugate points the trajectory encounters. In the special case of a harmonic or linear potential, which generates a quadratic action, this gives the exact propagator, due to the exactness of the stationary phase method for quadratic exponents. The contour taken for the integral in (5) is the same as for the action integral, and reflects the behavior of the associated classical trajectory.

In this paper we are concerned mostly with transmission across a barrier, with x_1 and x_2 being on opposite sides of a maximum of the potential. For any time T we can always find one, and only one, classical path, with $E_{cl} > V_{max}$, which would contribute in the above formula (4).

The question now arises: what role if any do tunneling paths play in this semiclassical approach? As this semiclassical recipe calls only for ordinary real time classical paths linking x_1 to x_2 in time T, the issue arises whether classically forbidden phenomena such as tunneling through or quantum reflection above a barrier (see Fig. 1) are somehow encoded in the classical trajectories, in the sense mentioned above: i.e. does *explicit* real time Fourier transform over the "classically allowed" Green function into the energy domain give good results? This indeed works for the inverted harmonic oscillator, exactly describing the below barrier tunneling in terms of ordinary trajectories going over the top of the barrier. This seems to support the view that the below barrier energy domain results are indeed encoded in the real time classical trajectories. We shall nonetheless show here that this is in fact false for generic potentials which vanish as $x \to \pm\infty$. Specifically, the above barrier trajectories provide an incomplete description not only in the energy domain, where tunneling has precise meaning, but also in the time domain, where more precisely we should use the term "classically forbidden path" instead of tunneling.

For definiteness, we consider propagation across the Eckart barrier $V(x) = V_o \text{sech}^2(\alpha x)$ with $V_o = \alpha = 1$ and mass $m = 1$. In a quasiclassical phase space picture, a state $|x\rangle$ may be represented in phase space by a vertical line through x. For our propagation problem we thus consider each point on a line at x_1 as an initial condition, and the overlap of the phase space distribution obtained after evolving these under the classical equations of motion for time T with a vertical line at x_2 corresponds to the classical analog of the quantum propagator. Equivalently, since $\langle x_2|x_1(t)\rangle = \langle x_2|e^{-iHt/\hbar}|x_1\rangle = \langle x_2|(e^{iHt/2\hbar})^\dagger e^{-iHt/2\hbar}|x_1\rangle = \langle x_2(-t/2)|x_1(t/2)\rangle$, we may take the overlap of the phase space distribution from a vertical line at x_1 evolved for time $T/2$ with that of a vertical line at x_2 evolved backward in time for $T/2$. Such a quasiclassical propagation reveals some interesting features: in Fig. 2 we launch an initial state at x_1 and focus on the case $x_1 = -x_2$.

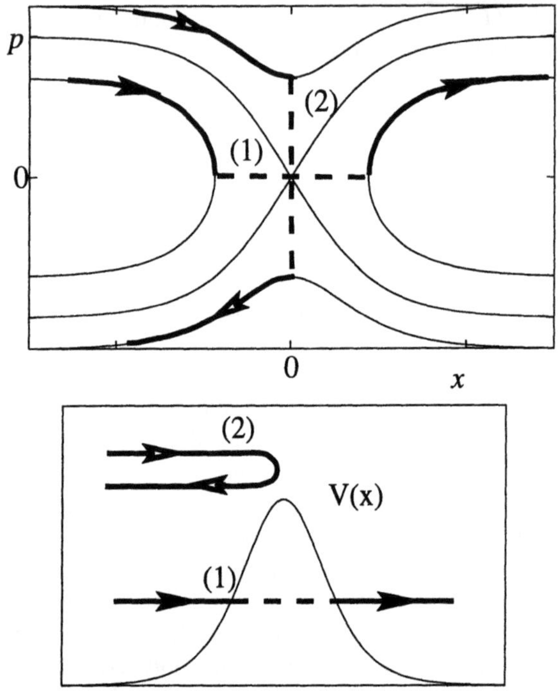

Fig. 1. Whether the semiclassical propagator, which involves classical paths only, can describe classically forbidden processes such as tunneling through a barrier (1) and above-barrier reflection (2), is the subject of this paper. The lower figure shows schematically the processes and the potential and the upper figure is the associated phase space.

For our purposes it suffices to consider the range of momenta from 0 to say $2\sqrt{2V_o}$, as negative momenta propagate away from the interaction region and higher positive momenta propagate forward so fast that they are also irrelevant in the problem of getting to x_2 in a time T. At first the initial distribution shears, as it would for free particle motion, and only extremely high energy trajectories have had time to reach the interaction region. As time evolves the density at the intersection $x = 0$ falls away rapidly. In the plots we have evolved a finite density of trajectories along the initial line; the density is so low for the classically allowed above barrier trajectories that after some time the manifold is not seen at $x = 0$. At the same time however, we see a below barrier loop structure developing with a healthy density. In time $T/2$, trajectories of a certain energy (below V_o) have just reached the barrier and are beginning to turn around. These loops are strongly suggestive of a tunneling contribution to the propagator, since they are very similar to the fixed energy manifolds (Fig. 1) which tunnel across the barrier in the energy domain. In fact we see that for long times and

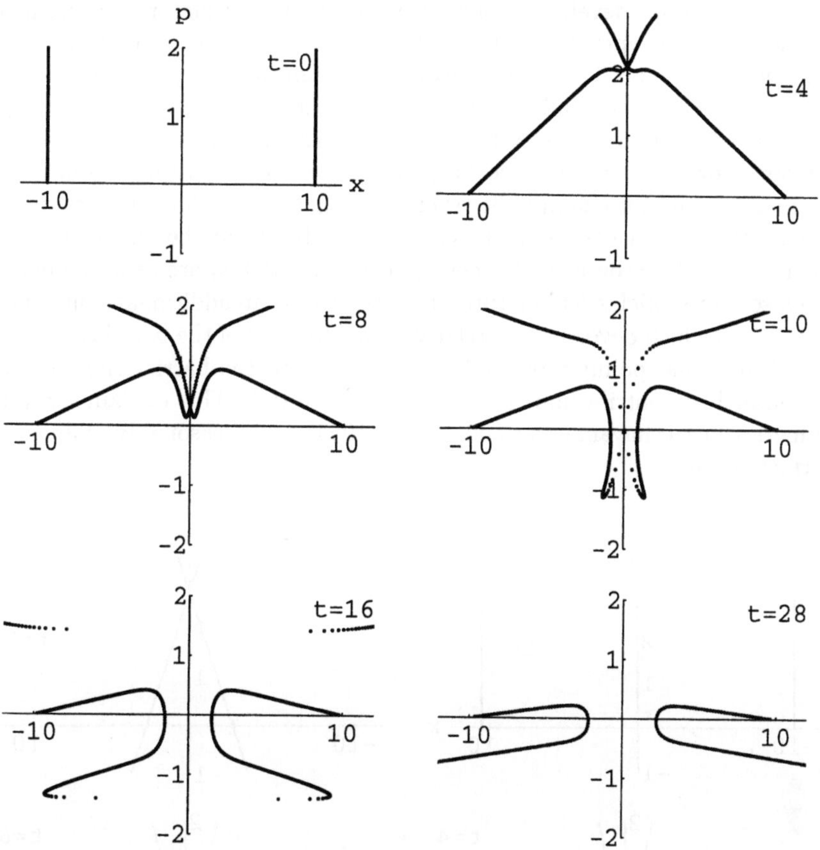

Fig. 2. A quasiclassical view of the propagator in phase space (see text). As time evolves the manifold at the intersection becomes very thin as the density dies exponentially; only at very dense seeding of initial conditions will they be seen. "Tunneling loops" develop however, which tend towards energy contours at long times.

endpoints far from the interaction region, the loops asymptotically approach the energy contours. The issue of getting across the barrier region becomes identical to the energy tunneling problem, in the long time, large $|x_1|, |x_2|$ limit. In this limit the energy of trajectories arriving at $x = 0$ changes only slowly with time. Moreover as time increases the trajectories arriving from the distant points are of lower and lower energy, so the loop "falls through" successively lower energies as time increases. It seems intuitively obvious that the computation of the tunneling part of the amplitude at a particular time should follow from the analogous problem at the current energy of the arriving trajectories, suitably weighted by the strength or number density of arriving trajectories. This "correction" to the conventional semiclassical propagator (4) is in fact dominant at long times.

No such loops develop for the inverse harmonic oscillator - the harmonic barrier has the property of all linear dynamical systems that an initial vertical line simply shears. Thus the conventional semiclassical propagator (which picks up the classically allowed intersection at $x = 0$) is in fact complete (see Fig. 3), which is as it must be since (4) is exact in this case. For generic barriers however, phase space loops will develop as described above. The harmonic oscillator example however does suggest that for a generic barrier the contribution to the semiclassical propagator from the classical overlap might *partially* account for tunneling effects if $|x_1|, |x_2|$ are near the barrier. Even so, our semiclassical picture suggests there is an additional contribution from loops which develop. The relative contributions of the two change as the initial and final points move out, so that close to the barrier the harmonic oscillator-like contributions give most of the result. This explains why the numerics in Keshavamurthy and Miller (1994) were reasonably close to the correct result.

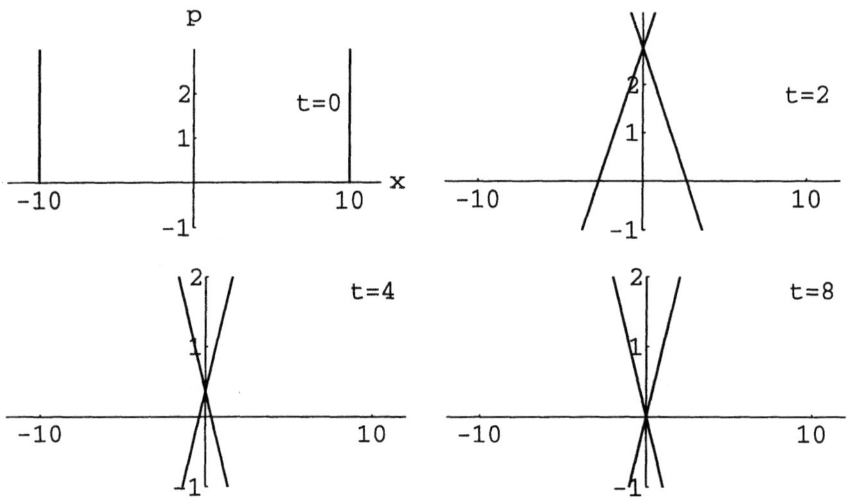

Fig. 3. The phase space evolution through the inverse harmonic oscillator $V(x) = 1 - x^2$, the harmonic approximation to the top of the sech^2 barrier in the previous picture. Here the initial vertical line simply shears. The density variation is not shown in this picture - the manifold becomes very thin very rapidly at the intersection just like the sech^2 case. No loops develop.

Classically forbidden processes normally contribute exponentially small amplitude compared with classically allowed processes. It may seem meaningless to include exponentially small terms in the presence of the dominant classical contribution: we expect the contribution of tunneling to be of order $\exp(-O(1/\hbar))$ and including this while neglecting terms of higher polynomial order in $\hbar$ arising from carrying out the stationary phase expansion near the classically allowed path to higher order is perhaps meaningless. However it also must be remembered that the semiclassical approximation of G above is the first term in an asymptotic expansion, and it is typical for such series that exponentially small terms become important in certain parameter domains (see also the book by Dingle (Dingle (1973)), and the paper of Balian et. al. (1978)). In our case there are two situations where the loop contribution has a significant, even dominant effect. The first is at long times; we shall discuss this now.

Classical analysis shows the phase space density at the intersection (the classical contribution) dies away as time evolves much faster than the density on the loops. The semiclassical expressions always rely on the Van Vleck prefactor determinants, which are the classical probabilities (or rather their square root). At long times, it then becomes a battle between two exponentially small quantities: the Van Vleck prefactor versus the tunneling integral. As $\hbar \to 0$ the Van Vleck prefactor is larger for any fixed time; however for fixed $\hbar$ there comes a time when the Van Vleck prefactor is smaller. Then the tunneling paths dominate over the above barrier paths. Turning to the semiclassical expression for the propagator, (4), we find the Van Vleck prefactor behavior at long times is exponential: There is just one classical path to include in the sum in the case of propagation across a barrier, and the solution of (5) for E_{cl} (and hence the quantities in (4)) given x_1, x_2 and T must be found. For the Eckart barrier, where

$$T = \frac{1}{\alpha\sqrt{2mE_{cl}}} \log\left(\frac{\sinh(\alpha x_2) + \sqrt{\cosh^2(\alpha x_2) - V_o/E_{cl}}}{\sinh(\alpha x_1) + \sqrt{\cosh^2(\alpha x_1) - V_o/E_{cl}}}\right), \tag{6}$$

a root search for E_{cl} must be done numerically in general. However, for long times and endpoints not in the interaction region, we find the magnitude of the semiclassical propagator is well approximated by ($V_o = \alpha = m = 1$)

$$|G^{sc}(x_2, T; x_1, 0)| \sim \left|\frac{\sqrt{2}e^{2x}}{4\pi i\hbar}\right|^{\frac{1}{2}} e^{-T/\sqrt{2}} \tag{7}$$

which shows the density of trajectories at the classical overlap dies away exponentially with time. On the other hand, the density on the loops decreases much slower; for large x_1, it goes as T^{-1} which may be seen by the following argument: The initial phase space distribution is composed of N initial conditions evenly distributed in momenta ranging from 0 to $2\sqrt{2}$ and each with x_1 as position coordinate. For x_1 far from the interaction region and long times, where the loops resemble below-barrier energy contours, the density on the loop is proportional to $\sim p_{turn}/x_1$, where p_{turn} is the initial momentum that the trajectory which has just hit the barrier and is beginning to turn around had. This may be seen by considering the top half of the loop which we approximate as a horizontal line from x_1 to the turning point at the barrier.

As time increases, the energy needed to reach the barrier decreases and p_{turn} falls roughly as x_1/t, with $t = T/2$ (free-particle motion). Hence the density on the loops falls as $1/T$, much slower than the exponential fall-off at the direct overlap. This suggests that for long times the loops have a significant effect and their contribution dominates the propagator. Indeed, from quantum mechanics we expect non-classical effects to become important at longer times and indeed the conventional semiclassical propagator becomes a worse approximation at longer times.

Ideally we would like to "add" the loop contribution to the conventional semiclassical propagator. One way of doing this is to write

$$\langle x_2 | x_1(T) \rangle = \langle x_2(-T/2) | x_1(T/2) \rangle \tag{8}$$

$$= \int dp \langle x_2(-T/2) | p \rangle \langle p | x_1(T/2) \rangle, \tag{9}$$

where semiclassically

$$\langle p | x_1(T/2) \rangle \propto \sum \left| \frac{\partial^2 S}{\partial p \partial x_1} \right|^{\frac{1}{2}} e^{\frac{i}{\hbar} S(x_1, p, T/2) - \frac{i\nu\pi}{2}}, \tag{10}$$

the sum being over all classical paths leaving x_1 and having momentum p after time $T/2$, $S(x_1, p, T/2) = \int^p x dp' - ET/2$, E being the classical energy for the path. From a phase space viewpoint, we are picking up each intersection of a horizontal line at p with the phase space distribution at time $T/2$. The integrand $\langle x_1(-T/2) | p \rangle \langle p | x_2(T/2) \rangle$ may be well behaved as a function of p in the region required. The integral would be performed numerically (not by stationary phase), or by saddle point methods, looking for stationary phase points in the complex p-plane. There is one real saddle point corresponding to the classical trajectory of energy above the barrier, and the tunneling loops correspond to stationary points at complex p and thus complex E. (We have not yet searched for them). However in the limit of large $|x|$ and $|x'|$, and at long enough times, there is a particularly simple way of evaluating the loop contribution, because the loop approximates an energy contour, the energy being that which at time $T/2$ just hits the barrier and begins to turn around. We are then able to analytically continue the expression (4) to this below-barrier energy.

In Fig. 4, we plot the result of such an extension and we find that in fact the loop contribution dominates the conventional semiclassical contribution at long times, and indeed provides a better approximation to the exact propagator (obtained by numerical Fourier transform of the exact energy Green function for the Eckart barrier; the latter was calculated in Grosche and Steiner (1995)). In the figures, we take $\hbar = 1$ unless otherwise specified.

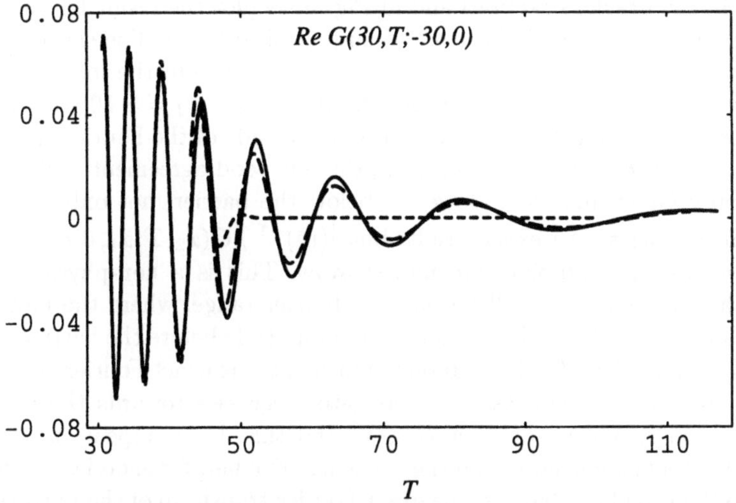

Fig. 4. Transmission through the Eckart barrier, $V(x) = \mathrm{sech}^2(\mathrm{x})$, $x_1 = -30 = -x_2$. Here the small-dashed line is the semiclassical propagator, (4), which is a good approximation to the exact (solid line) except at larger times where it collapses to zero. In this region however, our tunneling loop contribution (large-dashed line) is readily evaluated and gives a good result.

3 The Energy Domain

The second case where the loop contribution is important is in calculating the probability of going from x_1 to x_2 at energy E. This is given by the energy Green function, which may be written as the Fourier transform:

$$G(x_2, x_1; E) = (i\hbar)^{-1} \int_0^\infty e^{\frac{i}{\hbar}ET} G(x_2, T; x_1, 0) dT \qquad (11)$$

Let us first see how effective the *conventional* semiclassical propagator (4) is in predicting tunneling coefficients. We shall find that there are some discrepencies which we propose may be rectified by the loop correction of G in the time domain.

It seems we have two choices: either we perform the integral exactly numerically (using the semiclassical propagator in the integrand), or we use stationary phase. Below we shall investigate both choices for a barrier of form $V(x) = \mathrm{sech}^2 x$, where fortunately the exact answer is known (Grosche and Steiner (1995)) and will be used to compare our results with.

3.1 The Exact Fourier Transform of the Conventional Semiclassical Propagator

First, we perform the integral exactly (numerically, using an FFT routine, after finding $G^{sc}(x_2, T; x_1, 0)$ with the help of a numerical root search, as mentioned in the previous section) and in Fig. 5 we plot the result for the sech^2 barrier (the dotted, small-dashed and large-dashed lines). The endpoints are taken to be on either side of the hump, well away from the interaction region: we take $x_1 = -x_2$ and consider the three cases $x_2 = 5, 10$, and 30. We first note that well above the barrier the magnitude of the Fourier transform falls as $1/\sqrt{2mE_{cl}}$ and in this domain, there is good agreement with the exact energy Green function. However, below the barrier, not only does this method fall below the exact result, but $\left|(i\hbar)^{-1}\int G(x_2,T;x_1,0)e^{iET/\hbar}dT\right|^2$ *decreases* as the endpoints are moved away. This is a non-physical effect since the endpoints are well beyond the barrier range, where the motion is essentially free-particle: there is no loss of amplitude before the barrier region is reached, nor after. As the endpoints move out, the exact Fourier transform of the uncorrected semiclassical propagator regresses towards the classical transmission coefficient of 0 for $E < V_o$ and this decay appears to depend approximately exponentially on the endpoint. Further, for energies above but close to the top of the barrier, the exact Fourier transform of the uncorrected semiclassical propagator overshoots the exact result, and this effect increases as the endpoints are moved out. This effect was also observed in a recent paper (Grossman and Heller (1995)), where the S-matrix for the Eckart barrier was computed, involving the Fourier transform of the correlation function between Gaussian wavepackets. We also note here that as $\hbar$ is decreased, the endpoint dependence persists at energies below the barrier but dies away for energies above (Fig. 6).

3.2 Stationary Phase Fourier Transform of the Conventional Semiclassical Propagator

We now turn to the second option for evaluating (11). For a general potential $V(x)$, we attempt to perform the integral by stationary phase, the critical point T_{sp} given by the solution to

$$-\left.\frac{\partial S_{cl}}{\partial T}\right|_{T_{sp}} = E_{cl}(x_2, x_1, T_{sp}) = E \tag{12}$$

where E_{cl} is the classical energy function for the trajectory, determined by (5). For energies above any potential maximum between the endpoints, a real stationary phase point may be found and we obtain

$$G^{sc}(x_2, x_1; E) = \frac{1}{i\hbar\sqrt{p(x_2)p(x_1)}} e^{\frac{i}{\hbar}\int_{x_1}^{x_2} p(x)dx}, \tag{13}$$

where $p(x) = \sqrt{2m(E - V(x))}$. However if the transform variable E is less than any potential maximum between the endpoints, then there is no real stationary phase point satisfying (12) as there are no classical paths which can overcome the potential barrier. We may obtain a non-zero value for the

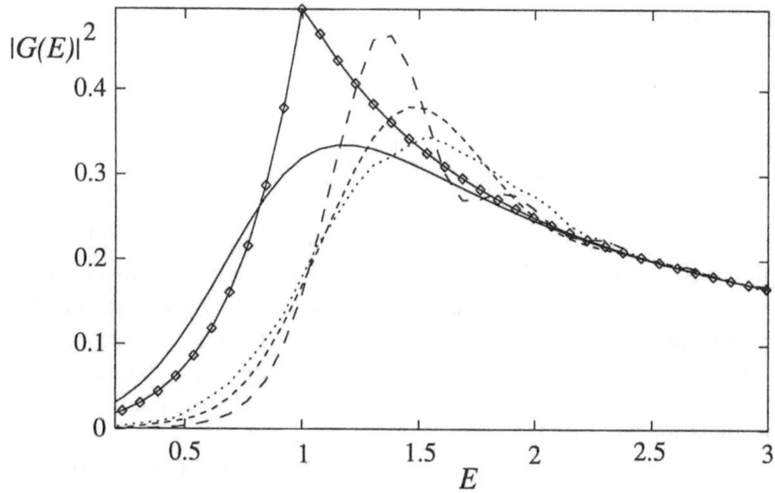

Fig. 5. Modulus-squared of the energy Green function $G(-x_2, x_2; E)$ for the potential $V(x) = \text{sech}^2 x$ (barrier top is at 1). The solid line is the exact, the -◇-line is the WKB result (stationary phase contribution of the Fourier transform of the semiclassical propagator), the dotted, small-dashed and larger-dashed lines are the numerical Fourier transforms of the semiclassical propagator with endpoints $x_2 = 5, x_2 = 10, x_2 = 30$ respectively. The exact and WKB results are independent of the endpoint, as required.

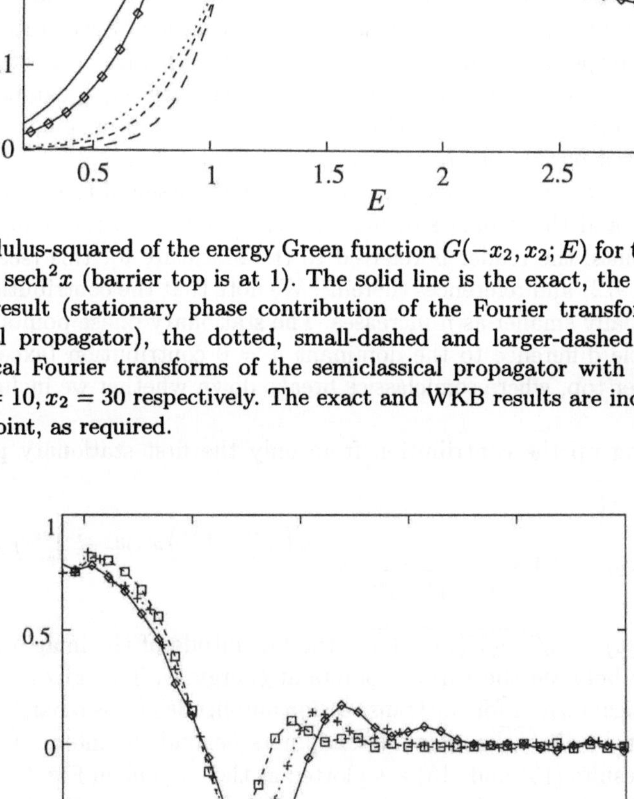

Fig. 6. $(|G(30,-30;E)|^2 - |G(5,-5;E)|^2)/|G(5,-5;E)|^2$ (which is zero for the exact Green function), for $\hbar = 1$ (-◇-), 0.75 (-+-),0.5(-□-). Above the barrier top (at 1), the endpoint dependence vanishes as $\hbar \longrightarrow 0$, in accordance with the stationary phase evaluation at energies above V_o being independent of $x_{1,2}$. However below the barrier top the endpoint dependence persists.

transmission by extension into the complex-time plane, where there are many stationary phase points:

$$T_{sp,n} = \left(\int_{x_1}^{a_1} + \int_{a_2}^{x_2}\right) \frac{dx}{\sqrt{2m(E - V(x))}} \\ -i(2n+1)\int_{a_1}^{a_2} \frac{dx}{\sqrt{2m(V(x) - E)}} \tag{14}$$

where $n = 0, 1, 2...$ and $a_{1,2} = V^{-1}(E)$ are the turning points. The resulting complex time stationary phase points are in accord with the idea that energy is fixed and therefore time is arbitrary, even complex. Each stationary phase point corresponds to analytically continuing a classical trajectory through the barrier where its momentum becomes imaginary. $T_{sp,n}$ can be thought of as representing a trajectory which bounces n times in imaginary time inside the inverted barrier before emerging out from the barrier.

It is not the aim of this paper to discuss the issue of the "multiple inside bounces" and the debate surrounding whether to sum the many stationary phase points; the reader is referred to the literature for this (see Berry and Mount (1972) and references therein). We note that the contributions become exponentially smaller as n increases. The stationary phase points with $n \neq 0$ make little difference to the dominant $n = 0$ contribution (except for near the barrier top, where semiclassics breaks down whether we include them or not).

Picking up the contribution from only the first stationary phase point gives

$$G^{sc}(x_2, x_1; E) = \frac{1}{i\hbar\sqrt{p(x_2)p(x_1)}} e^{\frac{i}{\hbar}\left(\int_{x_1}^{a_1} + \int_{a_2}^{x_2}\right)p(x)dx - \frac{1}{\hbar}\int_{a_1}^{a_2}|p(x)|dx} \tag{15}$$

where $|p(x)| = \sqrt{2m(V(x) - E)}$ is the magnitude of the imaginary momentum at x between the turning points at energy E. This gives the standard semiclassical answer for the transmission amplitude across a barrier after multiplication by $i\hbar\sqrt{2E}$ and taking endpoints beyond the interaction region.

The results (13) and (15) are plotted as the –◇– line in Fig. 5. We see that for energies above the barrier the stationary phase evaluation does very well, except close to the barrier top. For energies below, although it falls below the exact result, the stationary phase evaluation does do a better job than the numerical Fourier transform of the semiclassical propagator. It does not display the peculiar endpoint dependence of the latter.

We now come to a subtle, but fundamental, point of this paper: the integral through the complex stationary phase point is *not* an approximation to the Fourier integral (11) of the semiclassical propagator. We cannot interpret evaluation of the complex stationary phase point as arising from a deformation of the real time contour, because of an intervening (and non-rotatable) branch cut singularity. In deforming the time contour to pick up a saddle "hiding" behind a branch cut, one is not performing the same integral as in the original real time contour.

Let us consider again the Eckart barrier, $V(x) = V_o \text{sech}^2(\alpha x)$, and investigate the analytic structure of the time domain propagator in more detail. Branch point structure of $G^{sc}(x_2, T; x_1, 0)$ in the complex time plane arises from that of the action $S_{cl}(x_2, x_1, T)$ which in turn arises from that of the classical energy function $E_{cl}(x_1, x_2, T)$ defined implicitly by (5). In the case of the Eckart barrier, we have (6). As this expression does not lend itself easily to inversion we show there is a branch cut singularity as asserted above in a more indirect fashion. First we argue that the integral of $e^{iET/\hbar}G^{sc}$ along the contours C_1 (real-time) and C_2 (through the complex stationary phase point) shown in Fig. 7 are different, with the help of an observation made in Sect. 3.1. We assert that the non-physical endpoint-dependence of the exact integral along the real-time contour C_1 discussed earlier, which persists at $E < V_o$ as $\hbar$ becomes small (Fig. 6), is not present in the exact integral along contour C_2: it does not appear in the stationary phase evaluation along that contour which the exact integral along C_2 must tend towards as $\hbar$ becomes small. Thus the integrand cannot be analytic between C_1 and C_2. We note in contrast that the endpoint dependence for $E > V_o$ diminishes to zero as $\hbar \to 0$, in accordance with the fact that the evaluation by stationary phase (where, for such energies, the critical point lies on the real axis) is endpoint independent. Thus a singularity of the semiclassical propagator lies between the contours.

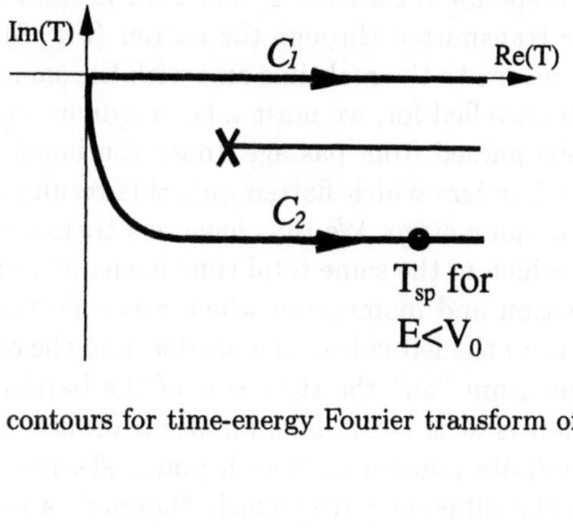

Fig. 7. Integration contours for time-energy Fourier transform of the semiclassical propagator.

Next we deduce this singularity is a branch cut, by pointing out that $E_{cl}(t)$ (and therefore $G^{sc}(x_2, T; x_1, 0)$) is a multi-valued function of time. It is clear that time is a multi-valued function of energy in the classical time-energy relation for the sech^2 barrier (6), with branches differing by $2\pi i/(\alpha\sqrt{2mE_{cl}})$, reflecting the behavior of the trajectory through (possibly complex) turning points. What is perhaps not so clear, is that the inverse function, energy as a function of time, is also multi-valued. To see this, due to the lack of a closed form for the inverse function, we must resort to a numerical search for the E_{cl} which solve (6) for a given T. This reveals in fact a countably infinite number of possible complex roots: for example, for real time, there is a unique real $E_{cl} > V_o$ representing the classically allowed path and the complex solutions represent trajectories with complex initial momentum leaving x_1, then bouncing between complex classical turning points before reaching x_2.

We emphasize that there are no such branch points in the case of the inverse harmonic oscillator. For the potential $V(x) = -\frac{1}{2}\omega^2 x^2$ we may solve exactly for the action as a function of time:

$$S(x_2, x_1, T) \tag{16}$$
$$= \sqrt{\frac{m\omega^2}{4}} \left((x_1^2 + x_2^2)\coth(\omega T/\sqrt{m}) - 2x_1 x_2 \text{cosech}(\omega T/\sqrt{m})\right) .$$

As semiclassics in the time domain is exact for this potential, an exact Fourier transform of the semiclassical propagator gives the exact result. Also the stationary phase evaluation of (11) may be done and yields a good and legal approximation as there are no intervening branch points to prevent the deformation of contour into the complex plane.

The analytic structure of the trajectory of a particle of energy E gives us some insight into the different, but related, analytic structure of the action which we have been discussing above. Approaching the barrier from the left at energy $E < V_o$, there are two choices at the barrier: either to continue taking real time steps for total time T and turn around or take imaginary time steps and be transmitted through the barrier (Fig. 8). Once through to the other side, to return to the real time axis with the same total time T that the reflected path travelled for, we must take imaginary time steps to cancel the imaginary time gained from passage under the barrier (the dashed line in the figure). For barriers which flatten out, this results in a complex final endpoint and final momentum. We thus have two trajectories with the same initial conditions which in the same total time have different final conditions: one with real position and momentum which was reflected from the barrier and after time T is on the left side of the barrier, and the other with complex position and momentum "on" the right side of the barrier (that, is the real part of the position is at $x > 0$). This implies a branch point as shown. In fact, there is an infinite number of branch points stacked above each other, corresponding to the different paths which "bounce" a different number of times inside the inverted barrier, as mentioned earlier (see also Miller (1974)).

In the case of a harmonic barrier $-\frac{1}{2}\omega^2 x^2$, there is no branch point; the trajectory that has penetrated the barrier will in fact return to the left side of the barrier in imaginary time due to the non-zero slope of the potential: explicitly, the equation for the trajectory is $x(t) = x(0)\cosh(\omega t/\sqrt{m}) + (p(0)/m\omega)\sinh(\omega t/\sqrt{m})$, free of any singularities. This is quite in contrast to

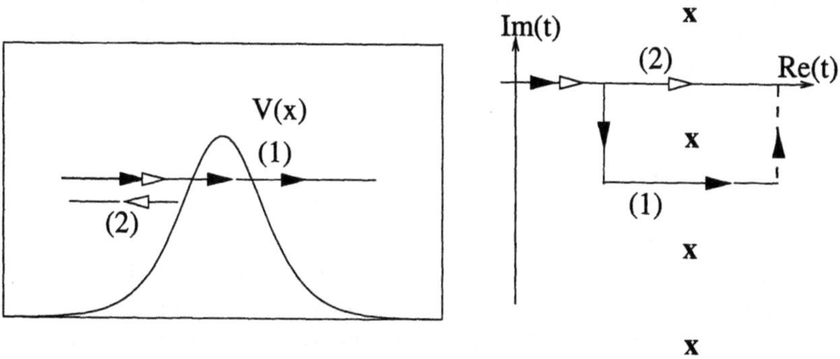

Fig. 8. Trajectories of $E < V_o$ and branch structure of $x^E(t)$. $x^E(t)$ is multi-valued: shown explicitly are a tunneling path (1), complex-time, and a reflected path (2), real-time. (1) returns to the real time axis through the dashed line (see text).

the solution for the Eckart barrier $V_o\text{sech}^2(\alpha x)$,

$$x(t) = -\sinh^{-1}\left(\sqrt{Vo/E-1}\cosh(\sqrt{2mE}\alpha t)\right)/\alpha + \sinh^{-1}\left(\sqrt{Vo/E-1}\right)/\alpha + x(0), \tag{17}$$

which has branch points in the complex-time plane at times ($n = 0, 1, 2...$),

$$t = \pm\frac{1}{\alpha\sqrt{2mE}}\left(\tanh^{-1}\sqrt{\frac{E}{V_o}} + i\pi(2n+\frac{1}{2})\right). \tag{18}$$

Branch points of the trajectory in the complex-time plane arise from singularities of the potential in the complex x-plane, causing singular behavior in the trajectory of a particle which has approached this divergence after a particular series of complex time steps. Whether a certain potential gives branch structure may be deduced from the relation (5): if, at energy E, we can reach a final x_2 where $V(x_2)$ blows up, in a finite (complex) time, by taking a certain sequence of complex dt and dx, then the trajectory has branch structure. We see then for the sech^2 barrier, equation (6), we can have $x_2 = i\pi/2$ and a finite complex time, but for the inverse harmonic oscillator a particle cannot reach the blow up at $x = \infty$ except at infinite times. An interesting case is a Gaussian potential $V(x) = V_o e^{-x^2}$, which blows up at $x = \pm i\infty$: this point may in fact be reached in a finite time, which can be roughly seen by the relation (5): as $x \to \pm i\infty$, the integrand vanishes and the resulting time is finite.

In summary, performing the Fourier transform of the uncorrected semiclassical propagator exactly at energies below the barrier is unsatisfactory in that it not only underestimates the transmission but also demonstrates a non-physical dependence on the endpoints such that as the endpoints move away, the tunneling probability vanishes. A stationary phase evaluation at these energies to obtain the WKB result is really not a deformation of the contour so much as a discovery of an independent classical path.

We also note that the WKB tunneling factor emerges from other semiclassical approaches, e.g. taking the classical limit to Feynman's path integral directly in the energy domain, or considering WKB wavefunctions and connection formulae. We observe that (15) is a stationary phase approximation to a Fourier integral along the complex contour C_2. This motivates us to suggest that in working from the time domain, we may use the *uncorrected* semiclassical propagator effectively provided the "Fourier integral" to the energy domain is performed along the real-time contour C_1 for $E > V_{max}$ and the complex-time contour C_2 for $E < V_{max}$. That is, the contour taken is dictated by the position of the saddle point.

4 Conclusion

We have shown that the semiclassical path integral provides an incomplete description of tunneling through a potential barrier in both the time and the energy domains. At long times conventional semiclassics, which uses above-barrier trajectories, begins to fail. We are able to calculate a correction, involving below-barrier trajectories, which gives a good improvement to the propagator, for endpoints which are not close to the barrier.

Our observations in the time domain help us to understand the energy domainGreen's function!energy-domain, where the conventional treatments are not satisfactory as discussed above. Our time domain picture shows that the integrand is at fault: the above-barrier trajectories of the time domain semiclassical propagator (direct overlap of the phase space manifolds) describe tunneling poorly and the below-barrier trajectories (the loops) are needed. Both contributions are be needed in a numerical Fourier transform to accurately predict tunneling coefficients.

5 Acknowledgments

EJH thanks Bill Miller for energetic discussions over the course of several years, which inspired the present study. NTM thanks Adam Lupu-Sax for help with numerical work and our group for useful discussions. This research was supported by the National Science Foundation under grant number CHE-9321260, and through ITAMP at the Harvard-Smithsonian Center for Astrophysics.

References

Balian R., Parisi G. and Voros A. (1978): Phys. Rev. Lett. **41**, 1141.

Berry M.V. and Mount K.E. (1972): Rep. Prog. Phys. **35**, (1972).

Dingle R.B. (1973): *Asymptotic Expansions: Their Derivation and Interpretation*, (Academic Press Inc., London).

Feynman R. and Hibbs A., (1965): *Quantum Mechanics and Path Integrals*, (McGraw-Hill, New York).

Grosche C. and Steiner F. (1995): J. Math. Phys. **36**, 2354.

Grossman F. and Heller E.J. (1995): Chem. Phys. Lett. **241**, 45.

Holstein B. R. and Swift A.R. (1982): Am.J.Phys. **50**, 829.

Keshavamurthy S. and Miller W. H. (1994): Chem. Phys. Lett. **218**, 189.

McLaughlin D.W. (1972): J. Math. Phys. **13**, 1099.

Miller W.H. (1974): Adv. Chem. Phys. **25**, 69.

Rajaraman R. (1975): Phys. Rep. **21c**, 227.

Schulman L.S. (1981): *Techniques and Applications of Path Integration*, (Wiley, New York).

Tanner G. and Wintgen D. (1995): "Semiclassical quantization of intermittency in helium." Phys. Rev. Lett. **75**, 2928-31.

Semiclassical Floquet Spectra Without Periodic Orbits from $\hbar$-Expansion of Floquet Matrix

Peter Gerwinski and Fritz Haake

Fachbereich Physik, Universität-Gesamthochschule Essen, 45117 Essen, Germany

Abstract. A full semiclassical quasienergy spectrum of the kicked top was recently calculated by diagonalizing the semiclassically approximated Floquet matrix. Classical orbits between prescribed initial and final points, real or of the ghost type, were incurred for the matrix elements rather than periodic orbits which arise in the Gutzwiller approach. In lowest order in $\hbar$ an accuracy of roughly 3 % of the mean spacing was obtained for the quasienergies. We here extend the analysis to the next-to-leading order in $\hbar$ for the Floquet matrix. The accuracy for the quasienergy spectrum improves accordingly. Moreover, we thus explain why the previous leading-order treatment gave the mentioned 3 % accuracy.

1 Introduction

Semiclassical efforts at calculating quantum mechanical energy or quasienergy spectra may be based on Gutzwiller's periodic-orbit theory (Gutzwiller 1990). There is evidence that such efforts must, in order to achieve reasonable accuracy, include ghost orbits in addition to the usual real ones (Kuś et al. 1993a). Other semiclassical methods like the one based on the so-called generalized Gaussian wave packets (Sepúlveda et al. 1992) have proven to be promising alternatives.

An interesting semiclassical method with some history in physical chemistry (Mandelshtam and Taylor 1995), (Mandelshtam et al. 1995) approximates the matrix elements of the propagator or Floquet operator in a suitable basis (Gerwinski et al. 1995), (Braun et al. 1996). Each matrix element is written as an integral (for instance as a path integral according to Feynman) which is approximated using the stationary-phase or saddle-point method. Not too surprisingly, the saddle-point condition turns out to be connected with the classical time evolution. In the resulting semiclassical matrix element the classical action plays the role of a quantum mechanical phase.

As in Ref. (Gerwinski et al. 1995), (Braun et al. 1996), we here propose to apply the method just described to a periodically kicked top (Haake 1991). We shall employ a basis of coherent states and improve the result of Ref. (Gerwinski et al. 1995) by implementing the second order of an asymptotic $\hbar$-expansion (Gaspard 1994). We shall thus be able to explain the surprisingly high accuracy of about 3 % obtained in Ref. (Gerwinski et al. 1995) and to further improve the semiclassical quasienergies such that the typical error, in units of the mean quasienergy spacing, becomes proportional to $\hbar$.

2 The Kicked Top

The kicked top is an angular momentum $\mathbf{J}$ periodically exposed to linear rotations about the y-axis by the angle $\pi/2$ and subsequent nonlinear torsions about the z-axis proportional to the z-component of the angular momentum vector. The squared angular momentum is preserved, $\mathbf{J}^2 = \text{const}$. The classical phase space is the sphere $\mathbf{J}^2 = \text{const}$. Depending on the values of a suitable control parameter, the torsion strength k, the motion can be regular ($k \lesssim 2$) dominated by chaos ($k \gtrsim 6$) or mixed (in between). For this article, we choose $k = 8$, thus focussing on the fully chaotic regime.

To denote points in the spherical phase space we employ a stereographic projection variable which combines the usual azimuthal and polar angles ϕ, θ into a single complex number $\gamma \equiv \tan \frac{\theta}{2} e^{i\phi}$. Resorting for convenience to a stroboscopic period-to-period description of the classical dynamics we are led to a map $\gamma \mapsto \Gamma$ for our stereographic-projection variable which reads

$$\Gamma = \frac{1+\gamma}{1-\gamma} \exp\left(-ik\frac{\gamma+\gamma^*}{1+\gamma\gamma^*}\right) . \tag{1}$$

This map actually is a canonical transformation whose generating function (Kús et al. 1993b)

$$W(\Gamma^*, \gamma) = \ln(1-\gamma) + \ln(1+\Gamma\Gamma^*) + ik\left(\frac{\Gamma\Gamma^*}{1+\Gamma\Gamma^*}\right)^2 , \tag{2}$$

has the properties

$$\frac{\partial W}{\partial \Gamma^*} = \frac{\Gamma}{1+\Gamma\Gamma^*} , \quad \frac{\partial W}{\partial \gamma} = \frac{\gamma^*}{1+\gamma\gamma^*} . \tag{3}$$

If the two phase space points γ and Γ are related by the classical map $\gamma \mapsto \Gamma$, the generating function $W(\Gamma^*, \gamma)$ has a vanishing real part while the imaginary part is the action of the trajectory.

Quantum mechanically, the stroboscopic dynamics of our kicked top is characterized by its time evolution operator taken over one period, the so-called *Floquet operator* [1]

$$F = e^{-i\frac{k}{2j}J_z^2} e^{-i\frac{\pi}{2}J_y} . \tag{4}$$

[1] When taking $2j+1$ instead of $2j$ in the exponent of $e^{-i\frac{k}{2j}J_z^2}$, semiclassical calculations become easier and the phase factor $\exp\left[-i\frac{k}{4}\left(\frac{1-v}{1+v}\right)^2\right]$ in (16) vanishes. We here take $2j$ to be consistent with Ref. (Gerwinski et al. 1995).

This actually reveals more intuitively the period-to-period motion as the sequence of a rotation by $\pi/2$ about the y-axis and a torsion of stregth k about the z-axis than does the above classical map for the somewhat unfamiliar stereographic projection variable γ. The angular momentum operators now obey the well-known commutation relations $[J_x, J_y] = \mathrm{i}J_z$ etc. Note that we employ a scaling such that the angular momentum $\mathbf{J}$ as well as Planck's constant $\hbar = 1/j$ are dimensionless. The quantity j is the familiar angular-momentum quantum number which specifies the conserved squared angular momentum as $\mathbf{J}^2 = j(j+1)$. The classical limit is obtained as j grows large, $j \to \infty$. Finally, j assigns to the dimension of the Hilbert space the finite value $2j+1$.

As suitable basis states we choose angular momentum coherent states $|\gamma\rangle$ (Arecchi et al. 1972), (Glauber and Haake 1976); the state $|\gamma\rangle$ assigns the "direction" $\gamma = \tan\frac{\theta}{2}\mathrm{e}^{\mathrm{i}\phi}$ to the mean $\langle\mathbf{J}\rangle$ with the minimum uncertainty allowed by the commutation relations. The scalar product of two coherent states reads

$$\langle\Gamma|\gamma\rangle = \left(\frac{(1+\Gamma^*\gamma)^2}{(1+\Gamma\Gamma^*)\,(1+\gamma\gamma^*)}\right)^j \quad . \tag{5}$$

Evidently, the coherent states are normalized, but not mutually orthogonal. Nevertheless, $2j+1$ linear independent coherent states $|\gamma_n\rangle$ with $n = 1, \ldots,$ $2j+1$ can serve as a basis.

Let λ be an eigenvalue of F with eigenstate $|\psi\rangle$. We expand $|\psi\rangle$ in the basis chosen

$$|\psi\rangle = \sum_{n=1}^{2j+1} \psi_n \, |\gamma_n\rangle \quad . \tag{6}$$

(Note that $\psi_n \neq \langle\gamma_n|\psi\rangle$ due to the nonorthogonality of coherent states.) Then we can extract the exact spectrum of the Floquet operator out of its "matrix elements" between coherent states [2] by solving the *generalized eigenvalue problem*,

$$\sum_{n=1}^{2j+1} \langle\gamma_m|F|\gamma_n\rangle\,\psi_n = \sum_{n=1}^{2j+1} \lambda\,\langle\gamma_m|\gamma_n\rangle\,\psi_n \quad . \tag{7}$$

The matrix element $\langle\Gamma|F|\gamma\rangle$ between two coherent states can be calculated rigorously,

$$\langle\Gamma|F|\gamma\rangle = \left[\frac{(1-\gamma)^2\mathrm{e}^{-\mathrm{i}k/2}}{2(1+\Gamma\Gamma^*)(1+\gamma\gamma^*)}\right]^j \sqrt{\frac{\mathrm{i}j}{2\pi k}} \int\limits_{-\infty}^{+\infty} \mathrm{d}t\, \mathrm{e}^{jV(t)} \tag{8}$$

with the exponent

[2] Due to the non-orthogonality of coherent states, $\langle\Gamma|F|\gamma\rangle$ is *not* the *matrix element* of F between two states in the sense of linear algebra.

$$V(t) = -\frac{\mathrm{i}}{2k}t^2 + 2\ln(1+v(t)) \ , \tag{9}$$

$$v(t) = \Gamma^* \frac{1+\gamma}{1-\gamma}\mathrm{e}^{\mathrm{i}k+t} \ . \tag{10}$$

3 First-Order Approximation

We proceed by applying a saddle-point approximation to the integral in (8). In anticipation of a later generalization we here give a short sketch of a derivation of the saddle-point expansion. To approximate $\int \mathrm{d}t\, \mathrm{e}^{jV(t)}$, we first expand the exponent $jV(t)$ around a saddle-point t_0 with $V'(t_0) = 0$ and write $\mathrm{e}^{jV(t)}$ as a product of two exponentials,

$$\begin{aligned}\mathrm{e}^{jV(t)} = &\exp\left(jV(t_0) + \tfrac{1}{2!}jV''(t_0)(t-t_0)^2\right) \\ &\exp\left(\tfrac{1}{3!}jV'''(t_0)(t-t_0)^3 + \tfrac{1}{4!}jV''''(t_0)(t-t_0)^4 + \ldots\right) \ .\end{aligned} \tag{11}$$

Next, we expand the second of these exponentials in powers of its argument and then integrate along the real axis. Only even powers of $(t-t_0)$ survive and we get

$$\begin{aligned}\int_{-\infty}^{+\infty} \mathrm{d}t\, \mathrm{e}^{jV(t)} = &\mathrm{e}^{jV(t_0)} \int_{-\infty}^{+\infty} \mathrm{d}t\, \mathrm{e}^{\frac{1}{2}jV''(t_0)(t-t_0)^2} \\ &\cdot \left[1 + \tfrac{1}{4!}jV''''(t_0)(t-t_0)^4 + \tfrac{1}{2}\left(\tfrac{1}{3!}jV'''(t_0)(t-t_0)^3\right)^2 + \ldots\right] \ .\end{aligned} \tag{12}$$

The integration over $\mathrm{e}^{\frac{1}{2}jV''(t_0)(t-t_0)^2}(t-t_0)^{2n}$ yields a factor of $j^{-(2n+1)/2}$ which competes with the powers of j in the expansion. For a first-order approximation in $1/j$ we drop everything but the 1 in the square brackets in (12) and thus arrive at

$$\int_{-\infty}^{+\infty} \mathrm{d}t\, \mathrm{e}^{jV(t)} = \sum_{\text{saddles}} \mathrm{e}^{jV(t_0)} \sqrt{\frac{\pi}{-\frac{1}{2}jV''(t_0)}} \left[1 + \mathcal{O}(\tfrac{1}{j})\right] \ , \tag{13}$$

i.e., the familiar formula for the first-order term of a saddle-point expansion of the integral $\int \mathrm{d}t\, \mathrm{e}^{jV(t)}$. We have allowed for several contributing saddles by indicating a sum.

For the given problem, the saddle-point condition reads

$$V'(t_0) = -\frac{\mathrm{i}}{k}t_0 + \frac{2v(t_0)}{1+v(t_0)} = 0 \ . \tag{14}$$

Using the definition (10) of $v(t)$ we can write this as

$$v = \Gamma^* \frac{1+\gamma}{1-\gamma}\exp\left(\mathrm{i}k\frac{1-v}{1+v}\right) \ . \tag{15}$$

When Γ is the image of γ under the classical time-evolution map (1), we have $v = \Gamma\Gamma^*$ as a positive solution. In this sense, (15) is equivalent to the classical map, and real positive solutions correspond to classical trajectories. Complex solutions correspond to *ghost trajectories* (Kuś et al. 1993a), (Gerwinski et al. 1995) with complex action, i. e. values of the generating function $W(\Gamma^*, \gamma)$, (2), with nonzero real part.

For given Γ and γ, each solution v of (15) gives a contribution

$$\langle\Gamma|\,F\,|\gamma\rangle_{\mathrm{SC}} = \sqrt{\frac{\partial^2 W(\Gamma^*,\gamma)}{\partial\Gamma^*\partial\gamma}}\,\mathrm{e}^{(2j+1)W(\Gamma^*,\gamma)} \cdot [(1+\Gamma\Gamma^*)(1+\gamma\gamma^*)]^{-j}\exp\left[-\mathrm{i}\frac{k}{4}\left(\frac{1-v}{1+v}\right)^2\right] \tag{16}$$

to the semiclassical matrix element where $W(\Gamma^*, \gamma)$ is the generating function of the classical map introduced in (2).

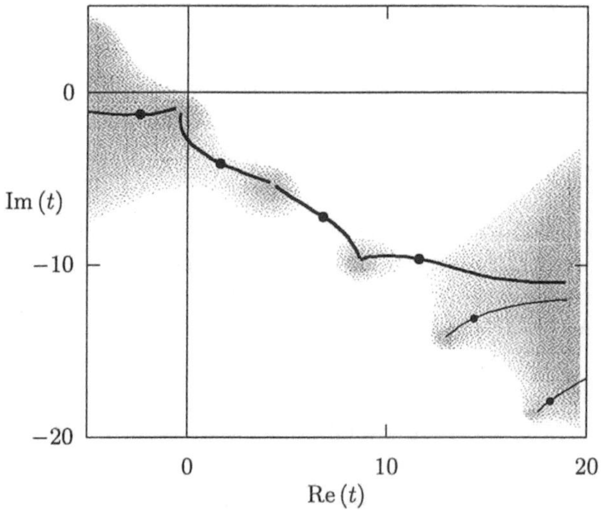

Fig. 1. Gray-scale coded contour plot of $|\mathrm{e}^{jV(t)}|$ in the complex t-plane. White regions indicate $|\mathrm{e}^{jV(t)}| \gg 1$ and dark ones $|\mathrm{e}^{jV(t)}| \ll 1$. The thick black line shows the path of constant phase from $-\infty$ to $+\infty$ through saddles (thick dots) and interlacing zeros of $\mathrm{e}^{jV(t)}$. Small dots refer to saddles not on allowed paths of integration.

It is important that not all solutions of (15) must be taken into account. To find the actually contributing solutions we must, for each pair (Γ, γ), explicitly construct paths of constant phase through saddles and zeros of $e^{jV(t)}$ in the complex t-plane [see (8) and Fig. 1]. For large j it suffices to identify the dominant ones among these saddles, i. e. the ones with largest $|e^{jV(t)}|$. The actual numerical search for the contributing saddles for each of the $(2j+1)^2$ matrix elements is the hard part of the work presented here. We should note that the $2j+1$ phase space points γ_n "hosting" the coherent states are chosen at random so as to cover the spherical phase space with roughly uniform density. No single pair (γ_n, γ_m) of these points can be expected to be connected by the classical map (1), simply because by specifying both an initial and a final phase space point one overdetermines a classical trajectory. Consequently, a saddle point t_0 contributing to a generic matrix element between coherent states corresponds to a ghost trajectory rather than to a real classical trajectory.

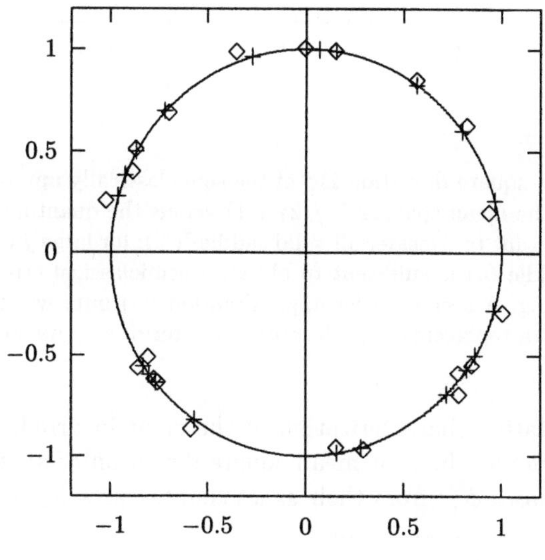

Fig. 2. Exact + and semiclassically approximated ◇ eigenvalues for $j = 10$.

Once all of the matrix elements $\langle\gamma_n|F|\gamma_m\rangle$ are evaluated through the procedure outlined one proceeds to the quasienergy spectrum from (7). Fig. 2 shows a semiclassical spectrum obtained with this method for $j = 10$. Obviously, the error is a small fraction of the mean spacing $2\pi/(2j+1)$ of neighboring eigenphases.

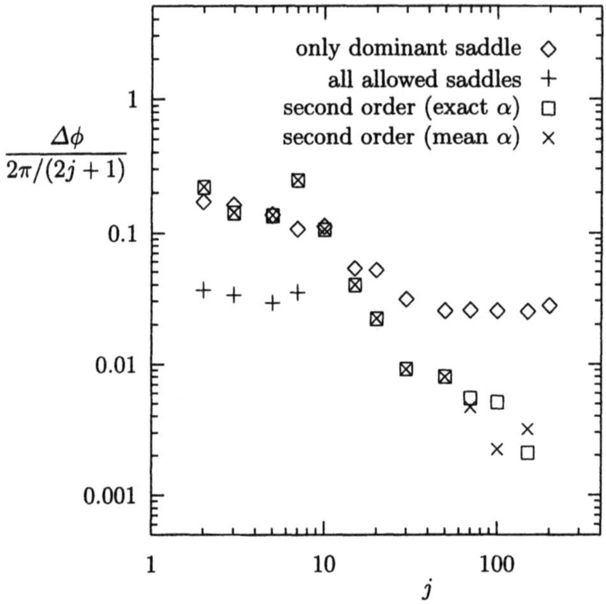

Fig. 3. Root mean square deviation $\Delta\phi$ of the semiclassically approximated eigenphases scaled by the mean spacing $2\pi/(2j+1)$ versus the quantum number j. For small j, it is important to consider all valid saddles (+); for large j the contribution of the leading saddle (⋄) is sufficient to obtain a semiclassical error of about 3 % of the mean spacing. A second-order approximation (□) improves the result. It is insensitive against a replacement of the correction term α by its mean value (×).

For a quantitative characterization of the error incurred, we define the *overall phase error* as the root mean square deviation of all semiclassically approximated phases ϕ_n^{SC} from their exact counterparts ϕ_n as

$$\Delta\phi = \sqrt{\frac{1}{2j+1}\sum_{n=1}^{2j+1}(\phi_n^{\mathrm{SC}} - \phi_n)^2} \ . \tag{17}$$

Upon referring this overall phase error $\Delta\phi$ to the mean spacing $2\pi/(2j+1)$ of neighboring eigenphases we get the *overall relative error* which is plotted in Fig. 3 in dependence of j. We show the result for two strategies, one accounting for all allowed saddles and the other only for the dominating saddle. We infer that for small values of j it is necessary to include all (or at least the few most important) saddles while for j upwards of about 10 the single dominant saddle suffices to get the fine accuracy of about 3 % of the mean spacing.

4 Second-Order Approximation

Inasmuch as the overall phase error turns out to be a constant fraction of the mean spacing, i. e. independent of $j \propto 1/\hbar$, we should think about the prefactor and aim at explaining its gratifyingly small value of about 0.03. To that end we now proceed to calculating the next order of the saddle-point expansion (13) which underlies the semiclassical matrix element (16).

Once more considering the saddle-point expansion (12), we now include the terms of first order in $1/j$ in the square brackets and drop only corrections of second and higher orders. This immediately yields an improved version of (13),

$$\int_{-\infty}^{+\infty} \mathrm{d}t\, \mathrm{e}^{jV(t)} = \sum_{\text{saddles}} \mathrm{e}^{jV(t_0)} \sqrt{\frac{\pi}{-\frac{1}{2}jV''(t_0)}} \left[1 + \frac{\alpha(\Gamma,\gamma)}{j} + \mathcal{O}\left(\tfrac{1}{j^2}\right)\right] \quad ; \tag{18}$$

the additionally included term involves the quantity

$$\alpha(\Gamma,\gamma) = \frac{1}{8}\frac{V''''(t_0)}{V''(t_0)^2} + \frac{5}{24}\frac{V'''(t_0)^2}{V''(t_0)^3} \tag{19}$$

with t_0 the value of t in (8) at the saddle point of $\mathrm{e}^{jV(t)}$.

By including the correction term just found in all semiclassical matrix elements we improve the result for our semiclassical spectrum by one order of $1/j$ (see Fig. 3). Note that this improvement only becomes visible for large values of j where it is sufficient to take only the dominant saddle.

In order to gain a better appreciation of the role of the $1/j$-correction to the semiclassical matrix elements we propose to inspect the distribution of the quantity $\alpha(\Gamma,\gamma)$ as numerically obtained for the various matrix elements. It turns out that the value of $\alpha(\Gamma,\gamma)$ varies only rather little over the $(2j+1)^2$ matrix elements, i. e. is practically independent of Γ and γ. The histogram in Fig. 4 illustrates the sharpness of the distribution of the imaginary part of α; the underlying data imply the mean value -0.065 for $\mathrm{Im}\,\alpha$.

Momentarily taking α as a sharp constant (independent of Γ and γ), we can easily calculate the $1/j$-correction of the eigenphases. That correction arises from multiplication of the whole matrix $\langle\Gamma|F|\gamma\rangle$ by the constant factor $1+\alpha/j$. The eigenvalues of F then acquire this prefactor as well. This means that the phase of $1+\alpha/j$ is added to the eigenphases calculated without the $1/j$-correction. Since $j \gg 1$, we get the common shift

$$\arg\left(1 + \frac{\alpha}{j}\right) \approx \mathrm{Im}\,\frac{\alpha}{j} \tag{20}$$

for *all* eigenphases. Disregarding second and higher orders of the $1/j$-expansion we may identify the shifted eigenphases with the exact ones and thus estimate the overall phase error incurred in the leading order as $\Delta\phi \approx |\mathrm{Im}\,\alpha/j|$. Then the overall *relative* error is approximated by

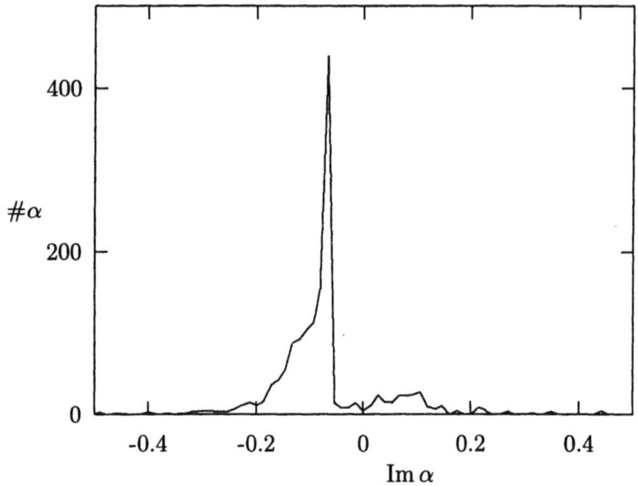

Fig. 4. Histogram of the imaginary part of $\alpha(\Gamma,\gamma)$ for 1681 pairs of Γ and γ when taking only allowed solutions of the saddle-point equation (15) into account. The distribution has a sharp maximum. The mean value is -0.065 which allows for an estimation of the semiclassical error to be $\langle \mathrm{Im}\,\alpha\rangle/\pi = 0.021$. For the real part of $\alpha(\Gamma,\gamma)$ (not shown) we get a similar picture.

$$\frac{\Delta\phi}{2\pi/(2j+1)} \approx \frac{|\mathrm{Im}\,\alpha|}{\pi} \; . \tag{21}$$

Upon taking the distinguished value of $\mathrm{Im}\,\alpha$ from the distribution depicted in Fig. 4, $\mathrm{Im}\,\alpha = -0.065$, we can expect the leading order of the saddle-point approximation to yield the quasienergy spectrum with an overall relative error of $0.065/\pi \approx 0.02$. The actually incurred overall relative error of about 0.03 is thus amazingly well reproduced and "explained".

As an independent control of the effective sharpness of the quantity $\alpha(\Gamma,\gamma)$ we have replaced $\alpha(\Gamma,\gamma)$ in the $1/j$-correction of each matrix element with its mean $\langle\alpha\rangle$ over all matrix elements. The spectrum so obtained through (7) hardly differs from the one calculated with the true values of $\alpha(\Gamma,\gamma)$, as shown in Fig. 3.

We would like to thank Petr A. Braun, Anja Drewitz, and Carsten Seeger for fruitful discussions about the second-order saddle-point approximation. Financial support by the Sonderforschungsbereich 237 *Unordnung und große Fluktuationen* of the Deutsche Forschungsgemeinschaft is gratefully acknowledged.

References

F. T. Arecchi, E. Courtens, R. Gilmore, and H. Thomas, Phys. Rev. A **6**, 2211 (1972)
P. A. Braun, P. Gerwinski, F. Haake, and H. Schomerus, Z. Phys. B **100**, 115 (1996)
P. Gaspard, Prog. Th. Phys. Supp. **116**, 59 (1994).
R. Glauber and F. Haake, Phys. Rev. A **13**, 357 (1976)
P. Gerwinski, F. Haake, M. Kuś, H. Wiedemann, and K. Życzkowski, Phys. Rev. Lett. **74**, 1562 (1995)
M. C. Gutzwiller, Chaos in Classical and Quantum Mechanics, Springer, New York (1990)
F. Haake, Quantum Signatures of Chaos, Springer, Berlin (1991)
M. Kuś, F. Haake, D. Delande, Phys. Rev. Lett. **71**, 2167 (1993)
M. Kuś, F. Haake, and B. Eckhardt, Z. Phys. B **92**, 221 (1993)
V. A. Mandelshtam and H. S. Taylor, J. Chem. Phys. **103**, 2903 (1995)
V. A. Mandelshtam, T. P. Grozdanov, and H. S. Taylor, J. Chem. Phys. **103**, 10074 (1995)
M. A. Sepúlveda, S. Tomsovic, E. J. Heller, Phys. Rev. Lett. **69**, 402 (1992)

Spectral Statistics of Chaotic and Disordered Systems

Oded Agam[1,3] and Shmuel Fishman[2,3]

[1] NEC Research Institute Inc. 4 Independence Way, Princeton NJ 08540, USA
[2] Department of Physics, Technion, Haifa 32000, Israel
[3] Institute for Theoretical Physics, University of California, Santa-Barbara, CA 93106, USA

Abstract. In this review the two point correlation function, $K(s)$, and the form factor $\hat{K}(t)$ are calculated for a finite volume disordered Lorentz gas. This is a hypercubic billiard with small hyperspheres placed at random positions in its interior. The deviations from the random matrix theory universal behavior are described in detail. These are characterized by the spectrum of the Perron–Frobenius operator associated with relaxational dynamics in the corresponding classical system. Various regimes of the two point correlation function, from the ballistic chaotic regime to the diffusive one, are explored and characterized. The derivation of the relation between $K(s)$ and the Perron–Frobenius operator, which was obtained in the framework of the supersymmetric σ-model, is outlined.

1 Introduction

Chaotic systems are characterized by motion that looks random although the systems are described by simple models. It resembles, to some extent, a random walk in phase space. On the other hand motion in a random potential is typically chaotic. In some aspects the quantum mechanical behavior of chaotic systems is similar to the one of random ones. In particular the level statistics over some energy scales are described by random matrix theory (RMT)(Mehta 1991). Recently the nonlinear σ-model, that was applied in the past to disordered systems (Efetov 1983), was extended to chaotic ones (Andreev et al. 1996). The RMT behavior as well as the corrections to it were obtained in the semiclassical limit. In particular the level-level correlations are given in terms of the spectral determinant of the Perron–Frobenius (PF) operator (Agam, Altshuler, and Andreev 1995), while for disordered systems they are given in terms of the diffusion operator. In this review these correlations are presented explicitly for a class of chaotic systems, where as the disorder is varied the motion changes from diffusive to ballistic. These are billiards with spherical impurities located at random positions.

Among the simplest chaotic systems are two dimensional billiards, such as the stadium billiard or the Sinai billiard. For such systems scattering by the boundary results in chaotic motion for nearly all trajectories. There are however trajectories, between flat sides of the billiard that look like trajectories of an integrable system. Let us focus now on the Sinai billiard (where the boundary consists of a square and a circle in its interior). The chaotic trajectories are scattered by the circle, while the trajectories that are not scattered by the circle are identical to those of a square billiard that is integrable. If one starts with some distribution or probability density of initial points in phase space it will deform and spread as a result of this scattering. Because of the finiteness of the system it will approach the invariant density. The operator that describes this evolution of the initial density, is the PF operator (Ruelle 1986). Its eigenvalues determine the decay rate of general smooth densities to the invariant density. Let us assume now that the circle of the Sinai billiard is replaced by a random configuration of many small circles. This is the finite volume disordered Lorentz gas. An initial distribution in phase space will spread also in this case and will approach the invariant density. In the limit of high density of the circles the distribution of initial conditions will spread diffusively. In this case the PF operator reduces to the diffusion operator. As the density of impurities decreases the PF operator deviates from the diffusion operator. For these systems the classical behavior is determined by the elastic mean free path (Agam and Fishman 1996a):

$$l = \frac{1}{\sigma n}, \tag{1}$$

where σ is the total cross-section of an impurity and n is the density of impurities. When l is much smaller than the system size the motion is diffusive. This qualitative pattern is common for billiards in all dimensions, as well as to other systems obtained by placing scatterers at random in a system that is otherwise integrable.

The specific quantity that will be explored in detail is the dimensionless density–density correlator. It is defined to be

$$K(s) = \Delta^2 \langle \rho(\epsilon)\rho(\epsilon + s\Delta)\rangle_\epsilon - 1, \tag{2}$$

where $\rho(\epsilon)$ is the density of states at energy ϵ, Δ is the mean level spacing, and $\langle\cdots\rangle_\epsilon$ represents an averaging over some interval of the energy ϵ for a specific realization of disorder. An ensemble average over realizations of disorder is denoted hereafter by $\langle\cdots\rangle$. In what follows the ensemble averaged correlator, $\langle K(s)\rangle$, will be calculated. A related function which is usually used in this context is the dimensionless form factor which is the Fourier transform of $K(s)$. Denoting it by $\hat{K}(t)$ it is given by

$$\hat{K}(t) = \int ds K(s) e^{-ist}. \tag{3}$$

For short times (compared to the Heisenberg time) the form factor can be calculated either in perturbation theory, expanding in $1/k_F l$ (the momentum is $\hbar k_F$), or in the framework of periodic orbit theory, starting with the Gutzwiller trace formula (Gutzwiller 1990) and applying the diagonal approximation. For longer times this approximation fails and additional quantum properties (for which the discreteness of the spectrum is important) should be taken into account (Berry 1985). The nonlinear σ-model (Andreev et al. 1996), as well as a novel semiclassical method that imposes the discreteness of the spectrum (Bogomolny and Keating 1996), enable to study the form factor in this regime. In particular a relation between the perturbative and the nonperturbative contributions was found as the leading order of an expansion in a parameter that reduces to the Thouless conductance for disordered systems. Both the perturbative and the nonperturbative contributions are expressed in terms of the spectral determinant of the PF operator, that is a purely classical quantity.

The various regimes are characterized by the relevant time scales. The scattering by impurities is characterized by the elastic time $\tau = l/v$, where l is the elastic mean free path and v is the velocity of the particles. Another time scale is the time of flight $\tau_f = L/v$ where L is the size of the billiard. If $\tau \ll \tau_f$, the billiard is diffusive while in the opposite limit it is ballistic and chaotic if τ is finite. In the diffusive case the relevant time scale is the Thouless time $\tau_c = L^2/D$, where D is the diffusion coefficient. The time scales that were mentioned so far are classical time scales. The quantum time scale is the Heisenberg time, τ_H. The form factor $\hat{K}(t)$ will be presented in the semiclassical approximation in various regimes of τ/τ_f. The various regimes of its behavior are determined by the value of t compared to τ, τ_f and τ_H. Universal RMT behavior is found for $\tau_f, \tau \ll t$ (Berry 1985), but some deviations from universality are found in the vicinity of τ_H (Agam, Altshuler, and Andreev 1995), (Agam and Fishman 1996a).

In the ballistic regime $\tau_f \ll \tau$ for $\tau_f < t \ll \tau$ there is a large fraction of periodic orbits of period t that are not scattered by impurities and therefore are identical to those of the corresponding integrable system (provided that in absence of impurities the system is indeed integrable). Consequently for $t \ll \tau$ a value of $\hat{K}(t)$ corresponding to Poisson statistics, characteristic of integrable systems, is approached.

The advance in fabrication techniques enables to manufacture mesoscopic systems that are relatively clean, namely their elastic mean free path can be larger than the size of the system. Such systems were prepared with boundaries of different shapes and used to explore the difference between transport in chaotic and in integrable systems (Marcus et al. 1992), (Chang et al. 1994), (Fromhold et al. 1994). Theoretical explorations were performed in parallel with the experimental investigations (Doron and Smilansky 1990), (Doron and Smilansky 1991), (Jalabert, Baranger, and Stone 1990), (Baranger, Jalabert, and Stone 1993). Several experimental results (Weiss et al. 1991), (Levy et al. 1993) on relatively clean systems were explained in the framework of the semiclassical methods that were developed in the field of "quantum chaos" (Fleischmann, Geisel, and Ketzmerick 1992), (Gefen, Braun, and Montambaux 1994), (Ullmo, Richter, and Jalabert 1995). One can prepare nanostructure devices so that their boundaries induce integrable classical dynamics if there is no disorder, e.g. when the boundary is a circle or a rectangle.

Since the level statistics of generic integrable systems is Poissonian (Berry and Tabor 1977a), one expects that the results of RMT cannot be applied directly for such systems. Yet, mesoscopic systems, even if prepared with a very high degree of cleanliness, inevitably contain some impurities. The model presented in this review is natural for the description of such systems. A system described by our model was realized experimentally using microwave cavities (Stein and Stöckman 1992), (Sridhar and Heller 1992), (Kudrolli, Kidambi, and Sridhar 1995), and the form factor $\hat{K}(t)$ can be obtained. The advantage of microwave experiments (compared to experiments in mesoscopic physics) is the high control and the fact that the system is described by a wave equation corresponding to a single particle Schrödinger equation.

In this review the specific results will be presented for a d dimensional disordered Lorentz gas (Hauge 1974) in a finite volume. It was studied in the past by the present authors (Agam and Fishman 1996a), (Agam and Fishman 1996b). The boundary conditions are assumed to be periodic, so the geometry is that of a d dimensional torus. Such a system is illustrated in Fig. 1 for $d = 2$. This model is a prototype system for integrable systems (where the motion in phase space is on tori) doped with impurities that scatter particles only within some finite range of their center. The trajectories that are not scattered by impurities are identical to those of the corresponding integrable system. The trajectories that are scattered by impurities are chaotic. The calculations were done in the framework of the semiclassical approximation. For the validity of this approximation it was assumed that the energy is sufficiently high so that the wavelength of the particles is much smaller than the radius of the spheres. It was assumed also that all the spheres have the same radius and that it is much smaller than the size of the system. The classical motion

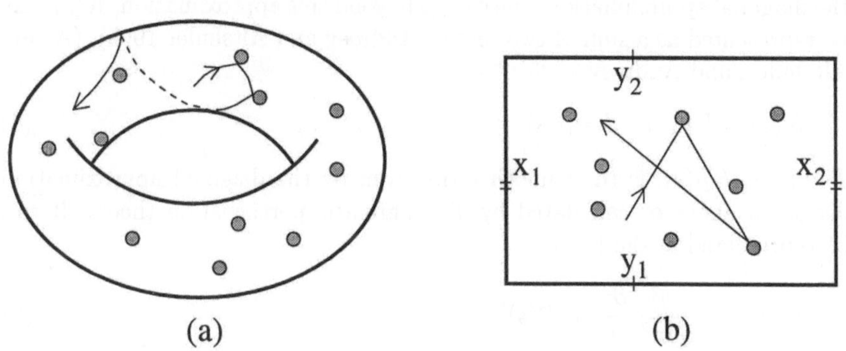

Fig. 1. (a) Torus with impurities and (b) the corresponding billiard with points on opposite sides (x_1 and x_2, y_1 and y_2) identical.

for this prototype system resembles the motion of a large variety of classical systems in angle-action variables. This similarity holds also in the framework of the semiclassical approximation (Berry and Tabor 1977b). However, it is proper to mention here that even with only one impurity our model system is completely chaotic. In fact it is the d dimensional Sinai billiard which was proved to be a K system. This model is somewhat similar to the model studied by Altland and Gefen (1993,1995). In their model pointlike impurities were introduced into an integrable system. Both the diffusive regime, where the elastic mean free path is much smaller than the size of the system, and the clean regime, where it is much larger than the system were explored. In their calculations diagrammatic techniques were used and very similar results were found. The calculations presented in this review were performed in the framework of the semiclassical approximation, hence, effects of diffraction, tunneling, and ghost orbits that were found important in various studies (Haake, Kuś, and Delande 1993), (Kuś, Haake, and Eckhardt 1993), (Vattay, Wirzba, and Rosenquist 1994), are not taken into account here.

The results presented in this review rely on a relation between the form factor $\hat{K}(t)$ and the spectral determinant of the PF operator (Agam, Altshuler, and Andreev 1995). These were obtained (Andreev and Altshuler 1995) to the leading order in $1/g$, where $g = \tau_H/\tau_c$, while τ_c is the time scale related to the first nonvanishing eigenvalue of the PF operator, corresponding to the slowest decay. For a disordered system τ_c is the Thouless time while g is the Thouless dimensionless conductance.

Throughout this review dimensionless quantities will be used. In particular, the length L of the billiard and the mass of the particle are chosen to be equal to unity; energy is measured in units of the mean level spacing ($s = \epsilon/\Delta$) and time is measured in units of $\hbar/\Delta$. In these units the Heisenberg time is $\tau_H = h\rho_0 = 2\pi$, where $\rho_0 = 1/\Delta$.

Having the semiclassical density–density correlator, that was obtained in the diagonal approximation, one can go beyond this approximation. $K(s)$ may be represented as a sum of two terms (Andreev and Altshuler 1995), (Agam, Altshuler, and Andreev 1995):

$$K(s) = K_P(s) + K_{NP}(s). \tag{4}$$

The first, $K_P(s)$, is the smooth term given by the diagonal approximation discussed above or calculated by diagrammatic perturbation theory. It can be represented in the form

$$K_P(s) = -\frac{\beta}{4\pi^2}\frac{\partial^2}{\partial s^2}\ln[\mathcal{D}(s)], \tag{5}$$

where β equals two for systems which are invariant with respect to time reversal and unity for the unitary ensemble. $\mathcal{D}(s)$ is the spectral determinant of the classical PF operator, associated with the decaying modes of a disturbance in the density of particles in the system. It is given by

$$\mathcal{D}(s) = \prod_{\mu} \mathcal{N}_\mu (s^2 + \gamma_\mu^2)^{-1}, \tag{6}$$

where γ_μ are the eigenvalues (in units of Δ) of the PF operator, and $\mathcal{N}_\mu$ are regularization factors introduced to make the product converge. In the diffusive limit the γ_μ are the eigenvalues of the diffusion operator. The spectral determinant $\mathcal{D}(s)$ is normalized such that

$$\lim_{s\to 0} s^2\mathcal{D}(s) = 1 \tag{7}$$

The second term, K_{NP}, cannot be calculated by usual perturbation theory. It was obtained for hyperbolic systems in the limit of large s retaining the non-perturbative terms of the correlator (Andreev and Altshuler 1995). It turns out that K_{NP} is also governed by the same classical spectral determinant. For the unitary (u) and orthogonal (o) cases it has the forms:

$$K^u_{NP}(s) = \frac{\cos(2\pi s)}{2\pi^2}\mathcal{D}(s), \qquad \text{and} \qquad K^o_{NP}(s) = \frac{\cos(2\pi s)}{2\pi^4}\mathcal{D}^2(s). \tag{8}$$

These formulas are crucial for the extension beyond the diagonal approximation or perturbation theory. Their derivation in the framework of the σ-model will be outlined in Sec. 2 of this review. An alternative derivation, within the semiclassical theory, was developed in (Bogomolny and Keating 1996). Recently, the formalism was also extended to maps (Zirnbauer 1996).

In this review, periodic orbit theory will be used (see Sec. 3) in order to calculate the perturbative part of the density correlator $K_P(s)$. Then in Sec. 4, by identifying the spectral determinant $\mathcal{D}(s)$ we shall be able to extend the results to include also the non–perturbative part $K_{NP}(s)$. In this way we will be able to go beyond the diagonal approximation. The resulting form factor $\hat{K}(t)$ is presented in detail in Sec. 5 for the two dimensional case. Finally the results are summarized and discussed in Sec. 6.

2 Semiclassical Field Theoretic Approach to Quantum Chaos

The purpose of this section is to summarize and outline the derivation of some results obtained from a field theoretic approach to the characterization of spectral statistics of systems which are chaotic in their classical limit. In particular, formulae (4) (5) and (8) for the two point correlation function are derived. This section will be divided into two parts. In the first we review the derivation of the super-symmetric non-linear σ-model for chaotic systems, and identify its low lying degrees of freedom as the Perron–Frobenius eigenmodes of the corresponding classical system. In the second part, we discuss the perturbation theory used in order to calculate $K(s)$.

To keep the discussion as simple as possible, we restrict our attention to hard chaos systems which belong to the unitary ensemble and having no discrete symmetries. Energy will be measured in units of the mean level spacing Δ, and time in units of $\hbar/\Delta$.

2.1 The Non-linear σ-Model for Chaotic Systems

The basic object needed for calculating spectral statistical characteristics is the generating function $\langle \mathcal{Z}(J)\rangle$, where $\langle\cdots\rangle$ denotes an averaging with respect to energy, disorder or both. Any n-point correlation function can be obtained by taking its derivatives with respect to the various components of the source J. Focusing on the two point correlation function, $\mathcal{Z}(J)$ can be expressed as a functional integral (Efetov 1983). Introducing the four component supervector field $\Psi^T(\mathbf{q}) = (\chi^A, \psi^A, \chi^R, \psi^R)$, where ψ (χ) denote commuting (anti-commuting) components, and the superscript A (R) denotes the advanced (retarded) components (Efetov 1983),

$$\mathcal{Z}(J) = \int \mathcal{D}\Psi \, \exp\{-\mathcal{S}[\Psi,\epsilon]\}, \tag{9}$$

with action given by

$$\mathcal{S}[\Psi,\epsilon] = i\int d^d q \bar{\Psi}(\mathbf{q}) \left[\epsilon - \frac{s^+}{2}\Lambda - \hat{H} - Jk\Lambda\right] \Psi(\mathbf{q}). \tag{10}$$

Here $s^+ = s + i0$, $\bar{\Psi} = \Psi^\dagger \Lambda$, while $\Lambda = \mathrm{diag}(1,1,-1,-1)$ and $k = \mathrm{diag}(1,-1,1,-1)$ break the symmetry between advanced/retarded field components and super-symmetry respectively. The use of fermionic as well as bosonic components ensures the normalization $\mathcal{Z}(0) = 1$. For calculating the two point density correlator one can choose J to be constant, then

$$K(s) = -\frac{1}{16\pi^2}\mathrm{Re}\left.\frac{\partial^2\langle\mathcal{Z}(J)\rangle}{\partial J^2}\right|_{J=0}. \tag{11}$$

In the case of disordered metallic grains, the averaging of the $\mathcal{Z}(J)$ is performed over realizations of the disordered potential. However, for general chaotic systems there is no ensemble to average over and one can only average over an energy interval. Such an averaging is also required for disordered systems when the elastic mean free path is larger than the system size. Here, we therefore employ an energy averaging over a wide energy band centered at ϵ_0 and containing a large number of levels N such that $1 \ll N \ll \epsilon_0$:

$$\langle\cdots\rangle_{\epsilon_0} = \int \frac{d\epsilon}{\sqrt{2\pi}N} \exp\left[-\frac{(\epsilon-\epsilon_0)^2}{2N^2}\right](\cdots). \tag{12}$$

The results are independent of this particular Gaussian form of energy averaging as long as s is sufficiently small compared to the bandwidth N.

Performing the energy averaging of (9) generates a non–local quartic interaction term among the Ψ fields,

$$\langle\mathcal{Z}(J)\rangle_{\epsilon_0} = \int \mathcal{D}\Psi \, \exp\{-\mathcal{S}[\Psi,\epsilon_0] - \mathcal{S}_{\mathrm{int}}\}, \tag{13}$$

where

$$\mathcal{S}_{\rm int} = \frac{N^2}{2}\left(\int d^d q \bar{\Psi}(\mathbf{q})\Psi(\mathbf{q})\right)^2 . \tag{14}$$

This interaction term can be decoupled by means of the Hubbard–Stratonovich transformation. Introducing the 4×4 supermatrix fields $Q(\mathbf{q}, \mathbf{q}')$ one has,

$$\exp\left(-\mathcal{S}_{\rm int}\right) = \int \mathcal{D}Q \exp\left[-\mathrm{STr}_{\mathbf{q}}\left(\frac{Q^2}{2} - iNQ\Psi\bar{\Psi}\right)\right], \tag{15}$$

where $\mathrm{STr}_{\mathbf{q}}$ denotes the trace operation for supermatrices[1], while the subscript $\mathbf{q}$ implies a further extension of the trace to include integration over all spatial variables, e.g. $\mathrm{STr}_{\mathbf{q}} Q^2 = \int d^d q d^d q' \mathrm{STr}\, Q(\mathbf{q}, \mathbf{q}')Q(\mathbf{q}', \mathbf{q})$. Integrating over Ψ we obtain:

$$\langle \mathcal{Z}(J) \rangle_{\epsilon_0} = \int \mathcal{D}Q \ \exp\left[-\frac{1}{2}\mathrm{STr}_{\mathbf{q}} Q^2 + \mathrm{STr}_{\mathbf{q}} \ln \mathcal{G}\right], \tag{16}$$

with

$$\mathcal{G}^{-1}(Q) = \epsilon_0 - \frac{s^+}{2}\Lambda - \hat{H} - Jk\Lambda - NQ. \tag{17}$$

This integral can be evaluated now by the saddle point method where the small parameter is the inverse band width $1/N$. However, the fluctuations around the saddle point manifold are not isotropic. These can be divided into two categories according to their influence on the exponent in (16). Those fluctuations on which this exponent term has strong dependence define the "steepest descent" directions and are called massive modes. They can be integrated out by the standard saddle point approximation. The integration over the remaining degrees of freedom called massless or Goldstone modes should be performed exactly. The latter fluctuations can be viewed as rotations, $Q = T^{-1} Q_0 T$ where

$$Q_0(\mathbf{x}) = \frac{\epsilon_0 - H}{2N} + i\left[1 - \left(\frac{\epsilon_0 - H}{2N}\right)^2\right]^{\frac{1}{2}} \Lambda, \tag{18}$$

is a solution of the saddle point equation $Q_0 \mathcal{G}^{-1}(Q_0) = N$, while T are matrix operators which leave the interaction term (14) invariant under the transformation $\Psi \to T\Psi$. Substituting this form into the logarithm term of the generating function (16), and using the invariance of the STr operation with respect to cyclic permutation yields

$$\begin{aligned} \mathrm{STr}_{\mathbf{q}} \ln\left[\mathcal{G}(Q)\right] &= \mathrm{STr}_{\mathbf{q}} \ln\left[\mathcal{G}(T^{-1}Q_0 T)\right] = \\ &= -\mathrm{STr}_{\mathbf{q}} \ln\left(\epsilon_0 - \hat{H} - NQ_0 - \frac{s^+}{2} T\Lambda T^{-1} - T[H, T^{-1}] - TJk\Lambda T^{-1}\right). \end{aligned} \tag{19}$$

---

[1] The super-trace operation over a super–matrix M is defined as $\mathrm{STr} M = \mathrm{Tr} M_{\rm ff} - \mathrm{Tr} M_{\rm bb}$ where $M_{\rm ff}$ ($M_{\rm bb}$) denote the fermion-fermion (boson–boson) block of M.

The last three terms in the logarithm are small. It can, therefore, be expanded as a power series in these terms. This is actually an expansion in $1/N$. Thus keeping only the leading term we have,

$$\mathrm{STr}_{\mathbf{q}} \ln\left[\mathcal{G}(Q)\right] = \frac{1}{N}\mathrm{STr}_{\mathbf{q}}\left\{Q\left(\frac{s^+}{2}\Lambda + Jk\Lambda + T^{-1}[H,T]\right)\right\} + O\left(\frac{1}{N^2}\right) \quad (20)$$

where we have used the property that the super-trace of a matrix proportional to a unit matrix vanishes, and the saddle point equation $\mathcal{G}(Q_0) = Q_0/N$.

Our next step involves the semiclassical approximation. For this purpose we express the various quantities of the action in terms of their Wigner representations. Given an operator $\hat{\mathcal{O}}$ as a matrix with elements $\mathcal{O}(\mathbf{q}_1, \mathbf{q}_2)$ between two position states at $\mathbf{q}_1$ and $\mathbf{q}_2$, its Wigner representation is a function of the phase space variables $\mathbf{x} = (\mathbf{q}, \mathbf{p})$ defined by

$$\mathcal{O}(\mathbf{x}) = \int d^d q' \; \exp(i\mathbf{p}\mathbf{q}'/\hbar)\; \mathcal{O}(\mathbf{q} + \mathbf{q}'/2, \mathbf{q} - \mathbf{q}'/2). \quad (21)$$

We will use the fact that, in the semiclassical limit, the Wigner transform of a product of operators is equal to the product of the Wigner transformed operators, $(\mathcal{O}_1\mathcal{O}_2)(\mathbf{x}) \to \mathcal{O}_1(\mathbf{x})\mathcal{O}_2(\mathbf{x})$, where $\mathcal{O}_{1,2}(\mathbf{x})$ are smooth slowly varying functions on the quantum scale (Moyal 1949). Similarly, the Wigner transform of a comutator, in the semiclassical limit, can be expressed in terms of Poisson brackets, $[\mathcal{O}_1, \mathcal{O}_2](\mathbf{x}) \to i\hbar\{\mathcal{O}_1(\mathbf{x}), \mathcal{O}_2(\mathbf{x})\}$ where

$$\{\mathcal{O}_1, \mathcal{O}_2\} = \sum_i \left[\frac{\partial \mathcal{O}_1}{\partial q_i}\frac{\partial \mathcal{O}_2}{\partial p_i} - \frac{\partial \mathcal{O}_2}{\partial q_i}\frac{\partial \mathcal{O}_1}{\partial p_i}\right] \quad (22)$$

Finally, we express traces of operators over the Hilbert space as phase space integrals over the corresponding Wigner representations, $\mathrm{Tr}_{\mathbf{q}}(\mathcal{O}) = h^{-d}\int d\mathbf{x}\; \mathcal{O}(\mathbf{x})$. The resulting generating function then takes the form,

$$\langle Z(J)\rangle_{\epsilon_0} = \int DQ \; \exp\left(-S_{\mathit{eff}}[Q]\right), \quad (23)$$

where

$$S_{\mathit{eff}}[Q] = \frac{1}{N}\int \frac{d\mathbf{x}}{h^d}\; \mathrm{STr}\left[Q\left(\frac{s^+}{2}\Lambda + Jk\Lambda + i\hbar T^{-1}\hat{\mathcal{L}}T\right)\right], \quad (24)$$

and $\hat{\mathcal{L}}$ is the Liouville operator defined via the Poisson brackets with the Hamiltonian $\hat{\mathcal{L}}\,\cdot = \{\,\cdot\,, H\}$. Here, arguments of Q and T are omitted. These should be understood as functions of $\mathbf{x}$. In the semiclassical limit only matrices T which are independent of the energy contribute to the functional integral. Thus the massless modes are given by

$$Q(\mathbf{x}) = T^{-1}(\mathbf{x}_\parallel) Q_0(H) T(\mathbf{x}_\parallel), \quad (25)$$

where $\mathbf{x}_\parallel$ denotes a phase space coordinate on the energy shell $\epsilon_0 = H(\mathbf{x})$, and $Q_0(H)$ given by (18) is independent of $\mathbf{x}_\parallel$.

The only dependence on the coordinate normal to the energy shell enters through $Q_0(H)$, therefore, the integral over this variable can be performed. Introducing the notation

$$\mathcal{Q}(\mathbf{x}_\parallel) = \frac{1}{\pi N}\int dH\, T^{-1}(\mathbf{x}_\parallel) Q_0(H) T(\mathbf{x}_\parallel) = T^{-1}(\mathbf{x}_\parallel)\Lambda T(\mathbf{x}_\parallel) \tag{26}$$

we obtain the final expression for the effective action,

$$S_{\mathit{eff}}[\mathcal{Q}] = \frac{i\pi}{h^d}\int d\mathbf{x}_\parallel\, \mathrm{STr}\left[\mathcal{Q}\left(\frac{s^+}{2}\Lambda + Jk\Lambda + i\hbar T^{-1}\hat{\mathcal{L}}T\right)\right]. \tag{27}$$

Here $\mathcal{Q}$ and T should be understood as functions of $\mathbf{x}_\parallel$.[2] The functional integral (23), thus, extends over all matrices $T(\mathbf{x}_\parallel)$ which parameterize the coset superspace manifold $U(1,1/2)/[U(1/1)\times U(1/1)]$ where $U(1,1/2)$ and $U(1/1)$ are the pseudo-unitary and unitary super-groups respectively (Zirnbauer 1986).

An action equivalent to (27) has been proposed for stochastic Hamiltonians using an argument which relies on disorder averaging in the limit of vanishing disorder (Muzykantskii and Khmelnitskii 1995). The results of the next section, however, imply that a statistical description of ballistic systems using this action applies only upon energy averaging.

An interpretation of the functional integral (23) is intimately related to the identification of the low lying modes of the action (27). In the case of disordered metallic grains, these are the diffusion modes in the grain which have a discrete spectrum. In the present case one may try to associate the low energy degrees of freedom with eigenmodes of the *reversible* evolution operator $e^{-\mathcal{L}t}$. Yet, according to Koopman's theorem this operator is unitary and therefore the spectrum of $\mathcal{L}$ is purely imaginary. This implies that there is no natural ordering among the different modes, all of them have an equally important contribution. Moreover, since for chaotic systems this spectrum is infinitely degenerate, evaluation of the functional integral even on a perturbative level seems to be impossible.

However, as we shall argue the spectrum associated with reversible classical dynamics is not the relevant one. Indeed, the physically relevant spectrum of a chaotic system is that describing relaxational dynamics. It has been shown in (Ruelle 1986) that for Axiom-A systems the position of poles of the correlation function of two differentiable observables $B(\mathbf{x})$ and $C(\mathbf{x})$,

$$\rho_{BC}(s) = \int_0^\infty dt\, e^{is^+t}\int d\mathbf{x}\, B(\mathbf{x})e^{-\hat{\mathcal{L}}t}C(\mathbf{x}) \tag{28}$$

are independent of BC, and except for one pole at the origin all the others have a finite real part. The spectrum determined by these poles therefore accounts for relaxational *irreversible* dynamics. It appears as soon as one introduces coarse graining on phase space (recall that both B and C are

[2] The phase space coordinates in our notation are normalized as $h^{-d}\int d\mathbf{x}_\parallel = 1$.

differentiable). It can also be obtained by adding an infinitesimal noise to the system: consider first the infinite time limit and only then take the noise to zero.

Now, since the functional integral (23), as any functional integral, is defined through a coarse graining procedure, one is forced to identify the low lying modes of the σ-model action (27) as the relaxational modes of the underlying classical dynamics. The eigenvalues associated with them, denoted as γ_μ, are known as Ruelle resonances or the Perron–Frobenius spectrum (Ruelle 1986), (Pollicott 1990). An exact formal expression for the corresponding spectral determinant, $1/Z(z) = \det(z - \hat{\mathcal{L}})$, is given in terms of the classical periodic orbits of the system. For two dimensional systems it is of the form (Cvitanovič and Eckhardt 1991)

$$1/Z(z) = \prod_p \prod_{k=0}^{\infty} \left(1 - \frac{e^{zt_p}}{|\Lambda_p|\Lambda_p^k}\right)^{k+1} \tag{29}$$

where t_p is the period of the p-th primitive orbit and Λ is the eigenvalue of the monodromy matrix with absolute value larger than one.

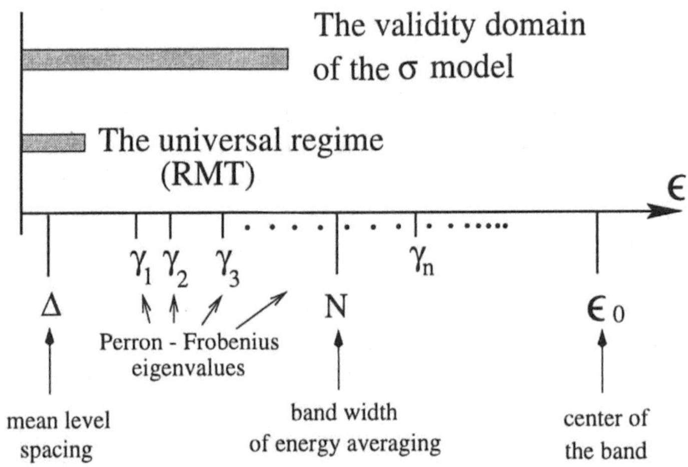

Fig. 2. A schematic picture of the various energy scales, the domain of validity of random matrix theory, and that of the σ-model.

The various energy scales of our theory are depicted in Fig. 2. The energy averaging is performed over a wide energy band of width N centered at ϵ_0, where $\epsilon_0 \gg N$. It is also required that N is much larger than the energy scale set by the real part of the first nonzero Perron–Frobenius eigenvalue γ_1. The mean level spacing Δ is assumed to be the smallest energy scale. The universal regime described by Random Matrix Theory corresponds to energies smaller than γ_1. The field theory described here accounts also for energies which are much larger than γ_1.

2.2 Perturbation Theory

In what follows we describe the perturbation theory introduced in (Andreev and Altshuler 1995) to calculate the two-point correlation function. We consider the high frequency asymptotics of $K(s)$ where the functional integral (23) can be evaluated by stationary point approximation.

Substituting (23) and (27) in (11) we have

$$K(s) = \frac{1}{16}\mathrm{Re}\int D\mathcal{Q}\,\left(\int \frac{d\mathbf{x}_\parallel}{h^d}\mathrm{STr}[k\Lambda\mathcal{Q}(\mathbf{x}_\parallel)]\right)^2 \times \exp\left\{\frac{-i\pi}{h^d}\int d\mathbf{x}_\parallel\;\mathrm{STr}\left[\mathcal{Q}\left(\frac{s^+}{2}\Lambda + i\hbar T^{-1}\hat{\mathcal{L}}T\right)\right]\right\}. \tag{30}$$

This integral, in the limit $s \gg 1$, has two stationary points: $\mathcal{Q} = \Lambda$ and[3] $\mathcal{Q} = -k\Lambda$. Consider first $\mathcal{Q} = \Lambda$ which corresponds to the ordinary perturbative diagrammatic expansion in the theory of disordered conductors (Efetov 1983), (Altshuler and Shklovskii 1986). A useful parameterization near this stationary point is

$$T = I + i\begin{pmatrix} 0 & B \\ \bar{B} & 0 \end{pmatrix}. \tag{31}$$

Expanding the various terms in (30) up to second order in B and $\bar{B}$ we have,

$$\mathrm{STr}(\mathcal{Q}T^{-1}\hat{\mathcal{L}}T) \simeq \mathrm{STr}(\bar{B}\hat{\mathcal{L}}B - B\hat{\mathcal{L}}\bar{B}) \qquad \mathrm{STr}(\Lambda\mathcal{Q}) \simeq -4\mathrm{STr}(\bar{B}B)$$
$$\mathrm{STr}(k\Lambda\mathcal{Q}) \simeq 4 - 2\mathrm{STr}(k'B\bar{B} + k'\bar{B}B)$$

where $k' = \mathrm{diag}(1,-1)$. Substituting them in (30) we obtain the perturbative part of the two point correlation function

$$K_P(s) = \mathrm{Re}\int D[B,\bar{B}]\;\left(1 - \frac{1}{2}\int\frac{d\mathbf{x}_\parallel}{h^d}\mathrm{STr}[k'B\bar{B} + k'\bar{B}B]\right)^2 \times \exp\left\{\frac{i2\pi}{h^d}\int d\mathbf{x}_\parallel\;\mathrm{STr}\left[\bar{B}(s^+ + i\hat{\mathcal{L}})B\right]\right\}. \tag{32}$$

where we have used the properties of $\hat{\mathcal{L}}$ in order to integrate by parts the term $\int d\mathbf{x}_\parallel \mathrm{STr}\bar{B}\hat{\mathcal{L}}B = -\int d\mathbf{x}_\parallel \mathrm{STr}B\hat{\mathcal{L}}\bar{B}$. Next step involves an expansion of $B(\mathbf{x}_\parallel)$ and $\bar{B}(\mathbf{x}_\parallel)$ in terms of the right $\phi_\mu(\mathbf{x}_\parallel)$ and left $\tilde{\phi}_\nu(\mathbf{x}_\parallel)$ eigenfunctions

[3] Due to the presence of the infinitesimal imaginary part in s^+, the second stationary point is unstable unless a regulator, $-u^2\mathrm{STr}(\Lambda\mathcal{Q})^2$, is introduced into the action. The parameter u should be kept finite throughout the calculation and set to zero only at the end. Here we will ignore this subtlety.

of the Perron–Frobenius operator. Notice that in our case the left and right eigenfunctions are not simply complex conjugate one of the other. Moreover, in general they lie outside the Hilbert space where B and $\bar{B}$ are defined. Nevertheless they form a biorthonormal set, $h^{-d}\int d\mathbf{x}_\parallel\, \tilde{\phi}_\nu(\mathbf{x}_\parallel)\phi_\mu(\mathbf{x}_\parallel) = \delta_{\nu,\mu}$ which spans this space. Thus

$$B(\mathbf{x}_\parallel) = \sum_\mu B_\nu \phi_\mu(\mathbf{x}_\parallel) \quad \text{and} \quad \bar{B}(\mathbf{x}_\parallel) = \sum_\nu \bar{B}_\nu \tilde{\phi}_\nu(\mathbf{x}_\parallel)\,. \tag{33}$$

Each matrix B_μ and $\bar{B}_\nu$ can be parameterized as

$$B_\mu = \begin{pmatrix} a_\mu & i\sigma_\mu \\ \eta^*_\mu & ib_\mu \end{pmatrix} \qquad \bar{B}_\nu = \begin{pmatrix} a^*_\nu & \eta_\nu \\ i\sigma^*_\nu & ib^*_\nu \end{pmatrix}, \tag{34}$$

were Roman letters denote ordinary commuting variables, while Greek letters denote anti-commuting variables. Substituting (33) and (34) in (32) we obtain

$$K_P(s) = \text{Re} \int \prod_\mu \left[d^2a_\mu d^2b_\mu d^2\sigma_\mu d^2\eta_\mu\right] \left(1 - \sum_\mu \left[a_\mu a^*_\mu - b_\mu b^*_\mu\right]\right)^2$$
$$\times \exp\left\{2\pi \sum_\mu (is^+ - \gamma_\mu)\left[a_\mu a^*_\mu + b_\mu b^*_\mu + \sigma^*_\mu \sigma_\mu + \eta_\mu \eta^*_\mu\right]\right\}.$$

The evaluation of this Gaussian integral is straightforward. Ignoring the constant term associated with the disconnected part of the correlator we obtain,

$$K_P(s) = \frac{1}{2\pi^2}\text{Re} \sum_\mu \frac{1}{(-is + \gamma_\mu)^2}.$$

Consider now the second stationary point $\mathcal{Q} = -k\Lambda$. The simplest way of evaluating its contribution is to perform a global rotation on the coset superspace $U(1,1/2)/[U(1/1) \times U(1/1)]$ which maps this stationary point back onto the first one $\mathcal{Q} = \Lambda$, and then to apply the same perturbation theory described above. Such a rotation does not change the invariant measure, however the integrand is transformed, so that

$$K(s) = \frac{1}{16}\text{Re} \int D\mathcal{Q} \left(\int \frac{d\mathbf{x}_\parallel}{h^d} \text{STr}[\Lambda \mathcal{Q}(\mathbf{x}_\parallel)]\right)^2$$
$$\times \exp\left\{\frac{-i\pi}{h^d} \int d\mathbf{x}_\parallel\ \text{STr}\left[\mathcal{Q}\left(-\frac{s^+}{2}k\Lambda + i\hbar T^{-1}\hat{\mathcal{L}}T\right)\right]\right\}.$$

Using now the procedure described above, namely, substituting the parameterization (31) with (34), keeping only the quadratic order in $B, \bar{B}$ we obtain

$$K_{NP}(s) = \mathrm{Re} \int \prod_{\mu} \left[d^2 a_\mu d^2 b_\mu d^2 \sigma_\mu d^2 \eta_\mu\right]$$

$$\left(\sum_{\mu} \left[a_\mu a_\mu^* + b_\mu b_\mu^* + \sigma_\mu^* \sigma_\mu + \eta_\mu \eta_\mu^*\right]\right)^2 \times$$

$$\exp\left\{2\pi i s^+ + \sum_{\mu} \left[i s^+ (a_\mu a_\mu^* - b_\mu b_\mu^*) - \gamma_\mu (a_\mu a_\mu^* + b_\mu b_\mu^* + \sigma_\mu^* \sigma_\mu + \eta_\mu \eta_\mu^*)\right]\right\}.$$

We have denoted this second saddle point contribution by $K_{NP}(s)$ and refer to it as "non-perturbative" since it cannot be obtained by the simple diagonal approximation.[4] To calculate the above integral one should notice that the Grassmanian variables σ_0, σ_0^* η_0, and η_0^* do not appear in the exponent since $\gamma_0 = 0$. The only non-vanishing contribution of the pre-exponent comes from the term $\sigma_0^* \sigma_0 \eta_0 \eta_0^*$, and the result of the integration is

$$K_{NP}(s) = \mathrm{Re} \frac{e^{i2\pi s^+}}{(s^+)^2} \prod_{\mu \neq 0} \frac{\gamma_\mu^2}{s^2 + \gamma_\mu^2}.$$

Introducing the spectral determinant

$$\mathcal{D}(s) = Z(is)Z(-is), \tag{35}$$

where $1/Z(z)$ is the Ruelle zeta function normalized such that in the limit $z \to 0$, $zZ(z) \to 1$, we obtain (5) and (8) for the unitary case,

$$K_P(s) = -\frac{1}{4\pi^2} \frac{\partial^2}{\partial s^2} \ln \mathcal{D}(s), \qquad K_{NP}(s) = \frac{\cos(2\pi s)}{2\pi^2} \mathcal{D}(s).$$

3 Periodic Orbit Analysis of $K_P(s)$

In this section we employ the periodic orbit theory in order to calculate the perturbative part of the two point correlation function $K_P(s)$. The semiclassical formula for the density of states $\rho(\epsilon)$, in our system, may be represented as a sum of three terms:

$$\rho(\epsilon) = \rho_0(\epsilon) + \rho_{po}^{(i)}(\epsilon) + \rho_{po}^{(c)}(\epsilon). \tag{36}$$

Here $\rho_0(\epsilon) = 1/\Delta$ is the Weyl term. To the leading order in $1/h$, it takes the form

$$\rho_0(\epsilon) = \frac{1}{h^d} \int d^d q d^d p \; \delta[\epsilon - \mathcal{H}(\mathbf{q}, \mathbf{p})],$$

[4] A generalization of the diagonal approximation which imposes the discreteness of the quantum spectrum gives, however, similar results to those presented here (Bogomolny and Keating 1996).

where $\mathbf{q}$ and $\mathbf{p}$ are the coordinate and the momentum vectors respectively, while $\mathcal{H}(\mathbf{q},\mathbf{p})$ is the Hamiltonian. The last two terms in (36) are contributions of the periodic orbits. The first, $\rho_{po}^{(i)}$, is the contribution from orbits that are not scattered by impurities while the second term, $\rho_{po}^{(c)}$, results from the scattered orbits and is given by the Gutzwiller trace formula (Gutzwiller 1990). Each of the last two terms in (36) takes the form

$$\rho_{p.o.}(\epsilon) = \sum_p A_p \, e^{\frac{i}{\hbar} S_p}$$

where A_p is the amplitude of the p-th orbit, while S_p is its action. The amplitudes A_p for the integrable and chaotic contributions were calculated in (Berry and Tabor 1977b) and (Gutzwiller 1990).

Inserting the semiclassical expression for the density of states (36) in (2), and keeping only the diagonal part of the double sum over periodic orbits, leads to the diagonal approximation (Berry 1991) of the form factor,

$$\hat{K}_P(t) = \hat{K}_P^{(i)}(t) + \hat{K}_P^{(c)}(t).$$

Here $K_P^{(i)}$ and $K_P^{(c)}$ are the contributions of periodic orbits that are not scattered and that are scattered by impurities respectively. In this approximation the products of terms belonging to different periodic orbits, that are not related by symmetry, are assumed to be averaged out to zero. For metallic grains it was found to give results identical to those of diagrammatic perturbation theory (Argaman, Imry, and Smilansky 1993).

3.1 The Contribution of Trajectories That Are Not Scattered by Impurities

The contribution of each trajectory that is not scattered by an impurity is identical to the corresponding contribution for an integrable system. The periodic orbits contribution to the density of states for integrable systems was calculated in (Berry and Tabor 1977b). It has the form of a sum over resonant tori,

$$\rho_{p.o.}^{(integrable)} = 2\mathrm{Re} \sum_{\mathbf{m}}{}' A_{\mathbf{m}} \exp\left\{ i\, 2\pi \mathbf{m} \left(\frac{\mathbf{I_m}}{\hbar} - \frac{\alpha_{\mathbf{m}}}{4} \right) \right\},$$

where the integer vector $\mathbf{m}$ defines the topology of the various tori. Each orbit on the $\mathbf{m}$-th torus is periodic, it winds m_1 times along θ_1, m_2 times along θ_2 etc. where θ denote the angle variables on the torus. $\mathbf{I_m}$ and $\alpha_{\mathbf{m}}$ are vectors of actions and Maslov indices of the orbit of topology $\mathbf{m}$. The prime over the sum denotes exclusion of terms with negative m_i as well as of the term $\mathbf{m} = 0$. The amplitudes $A_{\mathbf{m}}$ were calculated in (Berry and Tabor 1977b) (see Eq. 18 and Sec. 4 there).

For a system with impurities, only a fraction of the families of orbits survive no scattering at all. Denoting by $f_{\mathbf{m}}$ the fraction of the trajectories with the topology $\mathbf{m}$ that are not scattered by impurities, the resulting contribution is

$$\rho^{(i)}_{p.o.} = 2\mathrm{Re} \sum_{\mathbf{m}}{}' f_{\mathbf{m}} A_{\mathbf{m}} \exp\left\{ i\, 2\pi \mathbf{m} \left(\frac{\mathbf{I_m}}{\hbar} - \frac{\alpha_{\mathbf{m}}}{4} \right) \right\}.$$

Thus the contribution to the perturbative (namely diagonal) part of the form factor (3) is

$$\hat{K}^{(i)}_P(t) = 2\pi\beta \sum_{\mathbf{m}}{}' \delta(t - t_{\mathbf{m}}) |A_{\mathbf{m}}|^2 f^2_{\mathbf{m}},$$

where $t_{\mathbf{m}}$ is the period of an orbit of the topology $\mathbf{m}$ while $\beta = 2$ if the system is time reversible and $\beta = 1$ if it is not. Notice that time here is dimensionless (measured in units of $\hbar/\Delta$). This contribution, when averaged over realizations, is

$$\langle \hat{K}^{(i)}_P(t) \rangle = 2\pi\beta \sum_{\mathbf{m}}{}' \delta(t - t_{\mathbf{m}}) |A_{\mathbf{m}}|^2 \langle f^2_{\mathbf{m}} \rangle.$$

The fractions $f_{\mathbf{m}}$ are random variables that depend on the realization of the configuration of impurities. The average $\langle f^2_{\mathbf{m}} \rangle$ is a smooth function of the length of the orbit, or its period. It will be therefore denoted by

$$\varphi(t) = \langle f^2_{\mathbf{m}} \rangle,$$

where $t = t_{\mathbf{m}}$. Using Hannay and Ozorio de-Almeida sum rule (Hannay and Ozorio de Almeida 1984) for integrable systems, $2\pi \sum_{\mathbf{m}}{}' \delta(t - t_{\mathbf{m}}) |A_{\mathbf{m}}|^2 = 1$, one obtains

$$\langle \hat{K}^{(i)}_P(t) \rangle = \beta \varphi(t). \tag{37}$$

To estimate the function $\varphi(t)$, consider the limit in which the scatterers are small and far apart so that the cross section of each scatterer is small, namely $\sigma \to 0$ and the density of scatterers n is large ($n \to \infty$) but the elastic mean free path $l = 1/n\sigma$ is finite. At this limit and for sufficiently long orbits, the fraction $f_{\mathbf{m}}$ equals to the probability that an orbit associated with the topology $\mathbf{m}$ is not scattered by the impurities. This probability is approximately $e^{-t/\tau}$ where t is the orbit period and $\tau = l/v$ is the elastic mean free time (v being the particle velocity). Thus $\varphi(t) = e^{-2t/\tau}$, and the resulting contribution of the orbits, that are not scattered by impurities, to the form factor given by (37) is

$$\langle \hat{K}^{(i)}_P(t) \rangle = \beta e^{-2|t|/\tau}\,. \tag{38}$$

Here primitive orbits of period that is longer than τ were ignored since their contribution is negligible compared to the one of the chaotic orbits that will be calculated in the next subsection. The corresponding contribution to the two point density correlator, i.e. the inverse Fourier transform of $\hat{K}_P(t)$ is

$$\langle K^{(i)}_P(s) \rangle = \frac{2\beta}{\pi} \frac{\tau}{4 + (s\tau)^2}. \tag{39}$$

3.2 The Contribution of Orbits Which are Scattered by Impurities

The perturbative part of the two point density correlator, $K_P^{(c)}(s)$, associated with the periodic orbits which are scattered by impurities emerges from the contribution of isolated orbits to the (dimensionless) density of states, given by Gutzwiller's trace formula:

$$\rho_{po}(s) = \mathrm{Re}\,\frac{1}{\pi}\sum_p t_p \sum_r \frac{e^{\frac{i}{\hbar}S_p(\epsilon)r - i\nu_p r}}{|\det(M_p^r - I)|^{1/2}},$$

where p labels a primitive orbit that is characterized by a period t_p, action $S_p(\epsilon)$, and Maslov phase ν_p. r stands for the number of the repetitions of this orbit, and M_p is the monodromy matrix associated with the linearized dynamics on the Poincarè section perpendicular to the orbit.

Our working assumption is that the periodic orbits can be constructed from sets of segments connecting the impurities, as illustrated for example by the solid line in Fig. 3. Thus among any ordered set of impurities one can construct a periodic orbit unless a segment between two impurities hits another impurity. This situation is illustrated by the dashed line in Fig. 3. It is thus assumed that a particle can be scattered from an impurity in any direction. This approximation is valid as long as the radius of the spherical impurities is small compared to the average distance between nearby scatterers.

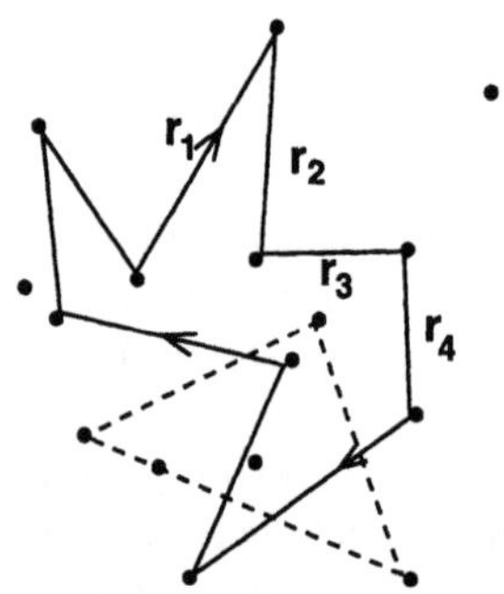

Fig. 3. An illustration of the construction of orbits that are scattered by impurities. The black dots represent the impurities, and the solid line is a periodic orbit which is scattered from them. The dashed line illustrates a forbidden orbit.

We shall construct $K_P^{(c)}(s)$ as a sum

$$K_P^{(c)}(s) = \sum_{j=1}^{\infty} K_j(s), \tag{40}$$

where $K_j(s)$ denotes the contribution from orbits which are scattered j times from impurities. Within the diagonal approximation the semiclassical formula for $K_j(s)$ is given as a sum over orbits,

$$K_j(s) = \frac{-\beta}{2\pi^2}\frac{\partial^2}{\partial s^2}\mathrm{Re}\sum_{p(j)}\left|\det\left(M_p - \mathbf{1}\right)\right|^{-1} e^{ist_p} \tag{41}$$

where $p = p(j)$ denotes all orbits which are scattered from j impurities, t_p is the period of the p-th orbit, M_p is the corresponding monodromy matrix, and $\mathbf{1}$ is the unit matrix. In the above formula we have neglected the contribution from orbits which are not primitive orbits, namely those which are repetitions of shorter ones. It is justified due to the exponential proliferation in the number of periodic orbits as their length increases. Formula (41) corresponds to a specific realization of the configuration of impurities.

Let us begin by estimating the number of periodic orbits which are scattered from j impurities. We denote by the index i the i-th segment of such an orbit. We can start the construction of an orbit from any of the n impurities which exist in the system. The number of possibilities in going to the next impurity near $\mathbf{r}_1$ in the vicinity $d^d r_1$ is given by $nr_1^{d-1}dr_1 d\Omega_1 e^{-r_1/l}$ where n is the number of the impurities which in our units is also their density; $d\Omega_1$ is an angle element of a d dimensional hypersphere, and $e^{-r_1/l}$ is the probability that the orbit does not hit any impurity on the way. Here l is the classical elastic mean free path. Thus the total number of orbits in the vicinity of the sequence of segments $(\mathbf{r}_1, \mathbf{r}_2, \cdots \mathbf{r}_j)$ is

$$n^j \prod_{i=1}^{j} e^{-r_i/l} r_i^{d-1} dr_i d\Omega_i \tag{42}$$

Notice that here we still do not assume that the orbit is periodic.

The amplitude of an orbit which undergoes the sequence of scattering represented by $p = (\mathbf{r}_1, \mathbf{r}_2, \cdots \mathbf{r}_j)$ can be estimated to be (Agam and Fishman 1996b),

$$\left|\det(M_p - \mathbf{1})\right|^{-1} = \prod_{i=1}^{j} \frac{\sigma}{r_i^{d-1}}, \tag{43}$$

where σ is the cross–section of the spherical impurities from which the particle is scattered.

Using expressions (42) and (43) one can write (41) in the form:

$$\langle K_j(s)\rangle = \frac{-\beta}{2\pi^2 j l^j}\frac{\partial^2}{\partial s^2}\mathrm{Re}\int\cdots\int\prod_{i=1}^{j}\left(dr_i d\Omega_i\, e^{-r_i/l}\right)\hat{\delta}\left(\sum_i \mathbf{r}_i\right) e^{is\sum_i r_i/v},$$

where we have used the relation $l = 1/n\sigma$ which relates the classical mean free path to the density of the impurities and their cross section. In this expression, the constraint that the orbit is periodic is imposed by the periodic delta function $\hat{\delta}(\mathbf{r})$. It is defined on the torus as

$$\hat{\delta}(\mathbf{r}) = \sum_{\mathbf{n}'} \delta(\mathbf{r} - \mathbf{n}'),$$

where $\mathbf{n}'$ is a d dimensional integer vector, and the sum runs over all its values. The prefactor $1/j$ was introduced to avoid multiple counting of the same periodic orbit (in the construction explained above each periodic orbit may start at any of the j impurities which it is visiting hence the same periodic orbit is constructed j times). The time t_p in (41) is equal to $\sum_i r_i/v$, where v is the magnitude of the velocity of the particle.

Denoting by $\langle \cdots \rangle_\Omega$ an average over all angles Ω_i,

$$\langle \cdots \rangle_\Omega = \int \cdots \int \prod_{i=1}^{j} \frac{d\Omega_i}{\mathcal{V}_d d} (\cdots),$$

where $\mathcal{V}_d = \pi^{d/2}/\Gamma(1 + d/2)$ is the volume of a d dimensional unit sphere, we obtain

$$\langle K_j(s) \rangle \simeq \frac{-\beta}{2\pi^2 j l^j} \frac{\partial^2}{\partial s^2} \mathrm{Re} \int \cdots \int \prod_{i=1}^{j} \left(dr_i \, e^{-r_i/l + i r_i s/v} \right) \left\langle \hat{\delta} \left(\sum_i \mathbf{r}_i \right) \right\rangle_\Omega . \quad (44)$$

The average of the periodic delta function can be calculated using Poisson summation formula, and the integral $\int_0^\pi e^{iz\cos\theta} \sin^{2\nu}\theta d\theta = \sqrt{\pi}\Gamma(\nu + 1/2)(2/z)^\nu J_\nu(z)$, where J_ν is the Bessel function. The result is

$$\left\langle \hat{\delta} \left(\sum_{i=1}^{j} \mathbf{r}_i \right) \right\rangle_\Omega = \sum_{\mathbf{m}} \prod_{i=1}^{j} \mathcal{Y}_d(2\pi|\mathbf{m}|r_i)$$

where $\mathbf{m}$ is an integer vector and $|\mathbf{m}|$ is its modulus, namely the square root of the sum of its d components squared. The function $\mathcal{Y}_d(z)$ is given by

$$\mathcal{Y}_d(z) = \Gamma\left(\frac{d}{2}\right) \left(\frac{2}{z}\right)^{\frac{d}{2}-1} J_{\frac{d}{2}-1}(z).$$

It is normalized so that $\mathcal{Y}_d(0) = 1$ for all d. This follows, e.g., from the asymptotic expansion of the Bessel functions at small arguments. For instance, in the two dimensional case $\mathcal{Y}_2(z) = J_0(z)$, while for $d = 3$ it is $\mathcal{Y}_3(z) = \sin(z)/z$.

Our next step is to perform the sum over the index j which counts the number of impurities the orbit is bounced from. After changing variables in (44) to $z = r_i/l$, $\langle K_j(s) \rangle$ takes the form

$$\langle K_j(s) \rangle = \frac{-\beta}{2\pi^2} \frac{\partial^2}{\partial s^2} \mathrm{Re} \sum_{\mathbf{m}} \frac{1}{j} \left[\int_0^\infty dz e^{-z(1-i\tau s)} \mathcal{Y}_d(2\pi|\mathbf{m}|lz) \right]^j ,$$

where $\tau = l/v$ is the elastic mean free time. The sum over j in (40) can be performed now. Notice that the first non–vanishing term in this sum is $j = 2$.

The reason is that on the torus geometry, where there are no boundaries, there are no closed orbits which are scattered only once from an impurity. Thus, the sum over j yields

$$\langle K_P^{(c)}(s)\rangle = \tag{45}$$
$$\frac{\beta}{2\pi^2}\frac{\partial^2}{\partial s^2}\mathrm{Re}\sum_{\mathbf{m}}\left\{\ln\left[1-\hat{\mathcal{Y}}_d(2\pi|\mathbf{m}|l,1-i\tau s)\right]+\hat{\mathcal{Y}}_d(2\pi|\mathbf{m}|l,1-i\tau s)\right\}$$

where $\hat{\mathcal{Y}}_d(x,y)$ is the Laplace transform of $\mathcal{Y}_d(xz)$, namely

$$\hat{\mathcal{Y}}_d(x,y) = \int_0^\infty dz e^{-zy}\mathcal{Y}_d(xz) =$$
$$= \frac{1}{\sqrt{x^2+y^2}}F\left(\frac{1}{2},\frac{d}{2}-1;\frac{d}{2};\frac{x^2}{x^2+y^2}\right),$$

where $F(\alpha,\beta;\gamma;\delta)$ is the hypergeometric function. The function $\hat{\mathcal{Y}}_d(x,y)$ is usually non-analytic and multivalued. The correct branch in the complex plane should be chosen such that it satisfies the relation $\hat{\mathcal{Y}}_d(0,y)=1/y$. For some specific cases:

$$\hat{\mathcal{Y}}_d(x,y) = \begin{cases} y/(x^2+y^2) & d=1 \\ 1\Big/\sqrt{x^2+y^2} & d=2 \\ \frac{1}{x}\tan^{-1}(x/y) & d=3 \\ \frac{2}{x^2}\left(\sqrt{x^2+y^2}-y\right) & d=4 \end{cases}$$

where the case $d=1$ should be understood as the quasi one dimensional case. By quasi one dimensional geometry we refer to a d dimensional box which is extended in one direction and has length 1, while in all other perpendicular directions its size is L, where $L \ll 1$. In addition, it is assumed that the system is observed over a time which is much longer than the time that takes for the particle to move a distance L. The scatterers are still assumed to be small, namely $R \ll L$.

4 The Dynamical Zeta Function $1/Z(z)$

It was shown in section 2 that the essential quantity which determines the quantal spectral properties of chaotic systems in the semiclassical limit is the dynamical zeta function $1/Z(z)=\det(z-\hat{\mathcal{L}})$ (Ruelle 1978). Its zeros, γ_μ, are the eigenvalues corresponding to the modes by which a disturbance in the particle density decays into its invariant distribution. For Anosov systems this decay is exponential in time, namely $e^{-\gamma_\mu t}$. Closed Hamiltonian systems always contain the eigenvalue $\gamma_0 = 0$ which corresponds to conservation of probability. The probability density mode associated with this eigenvalue is the invariant distribution. The normalization of $1/Z(z)$ is chosen to satisfy $\lim_{z\to 0} zZ(z) = 1$.

The relation (5) together with (35) enables us to identify the dynamical zeta function for our system. Since $K_P(s)$ is a sum of the "integrable" part (39) and the chaotic part (45) it follows that

$$Z(z) = Z_i(z)Z_c(z),$$

where $Z_i(z)$ is associated with the "integrable" dynamics and is given by

$$1/Z_i(z) = \left(\frac{2}{e}\right)^{\frac{4\pi}{\tau}} \exp\left\{-\frac{2\pi}{\tau}(2-z\tau)[\ln(2-z\tau)-1]\right\}. \tag{46}$$

It is normalized such that $Z_i(0) = 1$. The second term $Z_c(z)$, coming from orbits which are scattered by impurities, is

$$1/Z_c(z) = \prod_{\mathbf{m}} \mathcal{N}_{\mathbf{m}} \left[1 - \hat{\mathcal{Y}}_d(2\pi|\mathbf{m}|l, 1-\tau z)\right] \exp\left[\hat{\mathcal{Y}}_d(2\pi|\mathbf{m}|l, 1-\tau z)\right]. \tag{47}$$

Here $\mathcal{N}_{\mathbf{m}}$ are regularization factors which are introduced to make the product converge, and to satisfy the normalization property $\lim_{z\to 0} zZ(z) = 1$.

In the quasi one dimensional case, the dynamical zeta function can be expressed in terms of simple trigonometric functions. Using the relations $\sin x = x\prod_{m=1}^{\infty}(1-x^2/\pi^2m^2)$ and $\coth x = \sum_{m=-\infty}^{\infty} x/(x^2+\pi^2m^2)$, one obtains

$$1/Z_c(z) = \mathcal{N} \exp\left\{\frac{1}{2l}\coth\left(\frac{1-z\tau}{2l}\right)\right\}\left[\frac{\sin\left(\frac{\sqrt{z\tau-z^2\tau^2}}{2l}\right)}{\sinh\left(\frac{1-z\tau}{2l}\right)}\right]^2, \tag{48}$$

where

$$\mathcal{N} = \frac{1}{\tau}\exp\left\{-\frac{1}{2l}\coth\left(\frac{1}{2l}\right)\right\}\left[2l\sinh\left(\frac{1}{2l}\right)\right]^2.$$

The analytic structure of this function is depicted in Fig. 4 (a). It has essential singularities at $z_n = (1+i2\pi ln)/\tau$, double poles at $z_j = (1+i2\pi lj)/\tau$, double zeros at $\gamma_m^{(\pm)} = [1/2 \pm \sqrt{1/4-(2\pi lm)^2}]/\tau$, simple zero at the origin, and simple pole at $z = 1/\tau$. Here j, m, and n are integers, where $j \neq 0$ and $m > 0$. When $l > 1/4\pi$ all the zeros, except the one at the origin, are located along the line $\mathrm{Re}\, z = 1/2\tau$. In the other limit, $l \ll 1/4\pi$, there are many real zeros near the origin $\gamma_m^{(-)} \approx (2\pi lm)^2/\tau$. These zeros are associated with the diffusion modes of the system. In order to show that, consider the perturbative part of the form factor in the quasi one dimensional case. Calculating it from the Fourier integral of expression (5) one can show that it is approximately given by

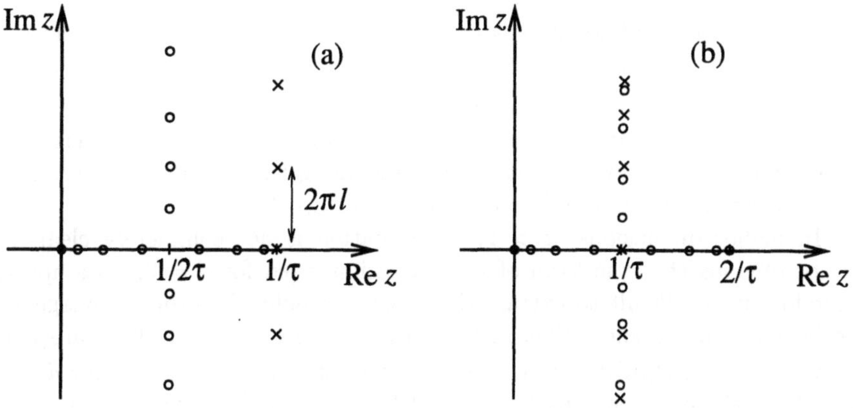

Fig. 4. The analytic structure of $1/Z_c(z)$ for (a) quasi one dimensional case (b) two dimensional case. $\circ$ and $\times$ represent zeros and singularities respectively. In the two dimensional case there are also branch cuts with branch points at the singularities.

$$\langle K_P(t)\rangle \simeq \frac{\beta}{2\pi}|t| \sum_{m=-\infty}^{\infty} e^{-\gamma_m^{(-)}|t|},$$

where the contributions from the poles and the essential singularities are neglected, since they are located far from the real s axis when $\tau \to 0$. Substituting $\gamma_m \approx (2\pi l m)^2/\tau$ and approximating the sum over m by an integral from $-\infty$ to $+\infty$ yields

$$\langle K_P(t)\rangle \simeq \frac{\beta}{2\pi}\sqrt{\frac{\tau|t|}{4\pi l^2}}. \tag{49}$$

This is the result of Altshuler and Shklovskii for the form factor of quasi one dimensional diffusive systems (Altshuler and Shklovskii 1986) written here in terms of dimensionless quantities. The influence of the complex zeros and poles is manifested only on time scale shorter or of order τ which here is assumed to be very small.

In the two dimensional case the regularization factors $\mathcal{N}_{\mathbf{m}}$ may be chosen to be

$$\mathcal{N}_{\mathbf{m}}^{-1} = \left[1 - \hat{\mathcal{Y}}_2(2\pi l|\mathbf{m}|, 1)\right] \exp\left[\hat{\mathcal{Y}}_2(2\pi|\mathbf{m}|l, 1)\right], \tag{50}$$

where $\mathbf{m} \neq \mathbf{0}$, while $\mathcal{N}_{\mathbf{0}} = 1$. It is easy to verify that with this choice, the product (47) converges. Therefore the zeros of $1/Z(z)$ are those of the various terms in the product, and they are located at $\gamma_{\mathbf{m}}^{(\pm)} = \left[1 \pm \sqrt{1-(2\pi l|\mathbf{m}|)^2}\right]/\tau$. When $l > 1/2\pi$ all of them, except the one at the origin, are complex and located along the line $\mathrm{Re}\, z = 1/\tau$. In the other limit, $l \ll 1/2\pi$, there are many real zeros associated with diffusive modes of the system, $\gamma_{\mathbf{m}}^{(-)} \approx (2\pi l\mathbf{m})^2/2\tau$. A straightforward calculation, similar to the one which led to (49), shows that these zeros lead to the Altshuler–Shklovskii result for the form factor in two dimensions, $\langle K_P(t)\rangle \simeq \beta\tau/4\pi^2 l^2$. The poles are located at $z_{\mathbf{j}}^{\pm} = [1 \pm i2\pi l|\mathbf{j}|]/\tau$, and there are branch cuts connecting $z_{\mathbf{j}}^{(+)}$ and $z_{\mathbf{j}}^{(-)}$. The essential singularities are at $[1 \pm i2\pi l|\mathbf{n}|]/\tau$. In difference from the quasi one dimensional case, they are not equidistant along the line $\mathrm{Re}\, z = 1/\tau$. The analytic structure described here is shown in Fig. 4(b).

In higher dimensions, $d > 2$, the situation is not completely clear. It is not obvious that the form of regularization, used for $d = 2$, does apply, therefore it is difficult to extract the zeros and poles. Ignoring convergence problems, one can show that in the diffusive regime, $l \ll 1$ there are real zeros, $\gamma_{\mathbf{m}}^{(-)} \approx (2\pi l|\mathbf{m}|)^2/d\tau$, in the vicinity of the origin, and that for $d = 3$ and $d = 4$ there are no complex zeros of $1 - \hat{\mathcal{Y}}_d(2\pi|\mathbf{m}|l, 1 - \tau z)$ for any value of l.

The formula (4) for the spectral density correlator is applicable in the limit where the real part, γ_1', of the first non-vanishing zero of $1/Z(z)$ is sufficiently large compared to unity. Near the diffusive regime, where $l \ll 1$, the smallest non-vanishing zero is $\gamma_1 = (2\pi l)^2/d\tau$. Identifying $l^2/d\tau$ with the diffusion constant of the particle in the system, it follows that $\gamma_1 = 4\pi^2/\tau_c$, where τ_c is the Thouless time measured in units of $\hbar/\Delta$. Therefore, up to a constant of $4\pi^2$, γ_1 is the dimensionless conductance of the system, g. When $g \to \infty$, $\mathcal{D}(s)$ reduces to $1/s^2$ and the RMT result for the spectral density correlator, namely $\langle K(s)\rangle = -\sin^2(\pi s)/\pi^2 s^2 \ \ (s \neq 0)$, is recovered. The reason is that in this limit all the zeros and poles of $\mathcal{D}(s)$ except the zero at the origin, are far from the real s axis, therefore their contribution becomes small. Formula (4) with the above form of the spectral determinant $\mathcal{D}(s)$ enables to calculate the deviations from the RMT universal result when g is finite but sufficiently large.

5 The Two-Dimensional Case, d=2

In this section we shall study in more detail the two dimensional case ($d = 2$), and in particular systems which belong to the unitary ensemble. Here we prefer to present most of the results in terms the form factor and to compare them with the RMT universal function which in the unitary case takes the form (Mehta 1991):

$$K(t) = \begin{cases} \frac{|t|}{2\pi} & |t| \leq 2\pi \\ 1 & \text{otherwise.} \end{cases} \tag{51}$$

The nature of the classical dynamics of the system is determined by the elastic mean free path l. When l is larger than unity the system is in the ballistic regime, while when $l \ll 1/2\pi$ the dynamics is diffusive. In terms of the dynamical zeta function, these two cases correspond, respectively, to the situations where $1/Z(z)$ has only complex zeros besides the one at the origin, or also many real zeros near the origin. As will be shown this is reflected in the form of the non-universal features which decorate the universal result (51). In what follows we shall study the full range of the parameter l.

Let us start with the ballistic regime $l \gg 1$. It is convenient here to split the contribution of the term $\mathbf{m} = \mathbf{0}$ in (45) from the others, and to write the perturbative part of the density correlator, associated with orbits which are scattered from impurities, as

$$\langle K_P^{(c)}(s)\rangle = \frac{\beta}{4\pi^2}\frac{\partial^2}{\partial s^2}\left[\mathcal{F}_0 + \mathcal{F} + C.C.\right], \tag{52}$$

where the contribution of the $\mathbf{m} = \mathbf{0}$ term is

$$\mathcal{F}_0 = \ln\left(1 - \frac{1}{1 - is\tau}\right) + \frac{1}{1 - is\tau},$$

while $\mathcal{F}$ is the contribution from $\mathbf{m} = (m_1, m_2) \neq (0,0)$. A crude estimate for $\mathcal{F}$ can be obtained replacing the sum over $\mathbf{m}$ by an integral. The result is

$$\mathcal{F}(x) \approx \frac{1}{4\pi l^2}\left[(x^2 - 1)\ln(x - 1) - x^2 \ln x + x\right]$$

where

$$x = \sqrt{(2\pi l)^2 + (1 - is\tau)^2}.$$

Substituting this result in (52) one can identify three domains of the parameter s where the two point correlation function reduces to

$$\langle K_P(s)\rangle \sim \frac{\beta}{2\pi^2}\begin{cases} -\frac{1}{s^2} & \tau \ll \frac{1}{s} \\ \frac{4\pi}{s^2\tau} + \frac{3}{\tau^2 s^4} & \tau_f \ll \frac{1}{s} \ll \tau \\ \frac{-1}{4\pi l^2 s^2} & \frac{1}{s} \ll \tau_f \end{cases} \qquad l \gg 1 \tag{53}$$

These three domains are referred as B1, B2, and B3 in (Altland and Gefen 1995). The first of them corresponds to long time scales over which the motion of the particle is already ergodic. The result is therefore the universal one (RMT). In the second regime, the time is smaller than the elastic mean free time but longer than the time of flight. Here, depending on the actual value of the parameter τ, the contribution from orbits which do not scatter from impurities may be the dominant one. The last s domain corresponds to time scales that are smaller than the time required to cross the sample. Here the contribution to the density correlator comes only from short orbits which are scattered by impurities, and the result is only a crude estimate.

We shall present now the behavior of the form factor of a system belonging to the unitary ensemble for some representative values of the parameters. Here the form factor is calculated from the Fourier transform (3) of the formula (4) for the spectral density correlator. The spectral determinant $\mathcal{D}(s)$ given by (35), (46) and the product (47) that is numerically calculated by including the terms with $|m_i| \leq 1000$ $(i = 1, 2)$, and using the regularization (50).

In Fig. 5 we present the results for the form factor in the ballistic regime where the elastic mean free path $l = 10$ is by an order of magnitude larger than the size of the system. Here, the contribution from the orbits which are not scattered by impurities is large, while the effect of $\mathcal{F}$ is negligible.

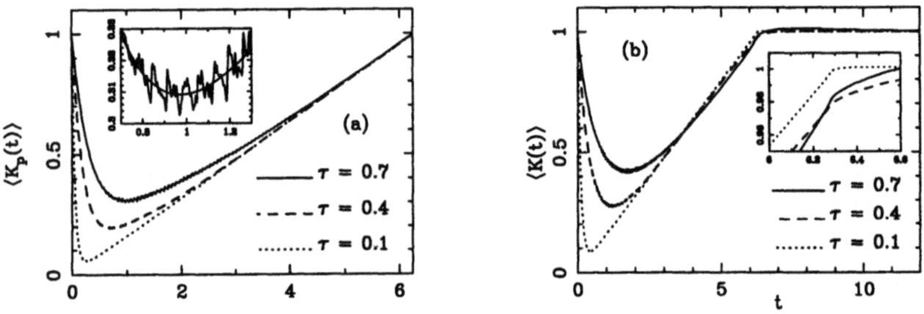

Fig. 5. (a) The perturbative part of the form factor for $l = 10$. The inset is a magnification of the region $0.7 \leq t \leq 1.3$ for the case $\tau = 0.7$ and $l = 10$. (b) The full form factor for the same parameters as in (a). The inset is a magnification of the vicinity of the Heisenberg time.

In Fig. 5(a) we present the perturbative part of the form factor for various values of τ. A magnification of the domain $0.7 \leq t \leq 1.3$ is the inserted for the case $\tau = 0.7$ where the time of flight is $\tau_f = 0.07$. The smooth line corresponds to a calculation in which the $\mathcal{F}$ term has been omitted. Several conclusions follow from this figure: (i) the contribution from orbits which are not scattered by impurities can be important, depending on the values of τ, (ii) the effect of $\mathcal{F}$, i.e. the contribution from nonvanishing $\mathbf{m}$ terms, is small in the ballistic regime, and (iii) there is a minimum of the form factor at t of order τ. In Fig 5(b) the full form factor is depicted. It was calculated under the assumption that formula (8) for the non-perturbative part of the density–density correlator can be still used with the approximation (46) for the part of the spectral determinant associated with orbits which are not scattered by impurities. The results of Fig. 5 are therefore valid only when $t \gg \tau_f$ and $|t - 2\pi| \gg \tau_f$. This figure shows that the singularity of the RMT universal form factor (51), at the Heisenberg time, is smoothed out.

The form factor in an intermediate regime where $l = 0.5$ is depicted in Fig. 6. Here $\tau = 0.25$, and hence $\tau_f = 0.5$. Inserted is the result for $l = 0.5, \tau = 0.15$. The contribution of $\langle K^{(i)}(t)\rangle$ has been ignored since it is meaningless when $\tau < \tau_f$. The deviations from universality present themselves in two time

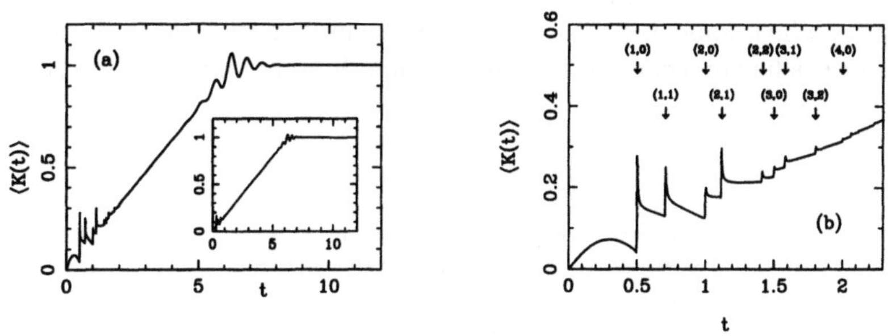

Fig. 6. (a) The form factor for the parameters $l = 0.5$ and $\tau = 0.25$. The inset is the form factor for $l = 0.5$ and $\tau = 0.15$ (b) A magnification of the domain $t \leq 2.3$. The times indicated by arrows are the periods of the orbits which are not scattered by impurities. The pair (m_x, m_y) above each arrow represents the winding numbers, namely the topology.

domains. Near the origin where $t \approx \tau_f$, and in the vicinity of the Heisenberg time $t \approx 2\pi$. Near the origin $\langle K(t) \rangle$ shows a singular behavior. The spikes which appear in $\langle K(t) \rangle$ result from the underlying torus geometry. They can be associated with short orbits that are scattered from a very small number of impurities, and therefore still preserve the topology of the orbits of the clean system. The positions of the spikes are at the periods of the periodic orbits which are not scattered by impurities. Their amplitudes decay as $e^{-t/\tau}$ since the probability for the existence of such orbits decays exponentially with time. In Fig. 6(b) a magnification of Fig. 6(a) for the domain $0 \leq t \leq 2.3$, $(\tau = 0.25)$, is presented. The times indicated by arrows are the periods of the orbits which are not scattered by impurities, and the corresponding winding numbers (m_x, m_y) are shown above the respective arrows.

Near the Heisenberg time, the universal RMT behavior (51) is completely changed. The singularity disappears and is replaced by oscillations with a period approximately equal to the time of flight τ_f. The parameters in Fig. 6 were chosen such that $1/Z(z)$ has no real zeros except the one at the origin. This fact reflects itself in the oscillatory behavior of the form factor near the Heisenberg time. The real part of the first non-vanishing zero is $\gamma_1' = 1/\tau = 4$. The universal RMT limit (51) of $\langle K(t) \rangle$ is expected to be reached when τ decreases. This is demonstrated in the inserted figure where $\gamma_1' = 1/\tau = 20/3$. It shows that as τ decreases, the time intervals over which the non-universal features are large shrink, and their amplitude decreases.

The oscillatory behavior of the form factor near the Heisenberg time changes when $1/Z(z)$ has real zeros in addition to the one at origin and the complex ones. In Fig. 7 the form factor calculated for the parameters $l = 2\tau = \sqrt{5}/6\pi \approx 0.12$, $\tau_f = 1/2$ is presented. For these parameters the first non-vanishing zero is $\gamma_1 = 4/3l \simeq 11.23$, the second one is at $8/3l$, and all the others are complex. Comparing Figs. 6 and 7, one sees that the singularities, which appear in Fig. 6 near the origin, almost disappear, and the oscillatory behavior near the Heisenberg time becomes smooth. Inserted in

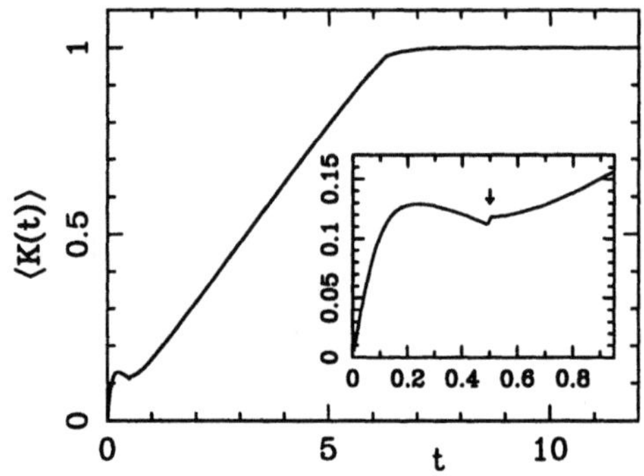

Fig. 7. The form factor for the parameters $l = 5^{1/2}/6\pi$ and $\tau = l/2$. The inset is for the time domain $0 < t < 0.9$.

Fig. 7 is a magnification of the time domain $0 < t < 0.9$. The jump indicated by the arrow results from the toroidal topology of the system. It appears at the period of the shortest orbit of the clean system $[(m_x, m_y) = (1,0)$ and $(0,1)]$.

The structure of the non-universal features does not change much as l is further decreased. In Fig. 8 we present the form factor for $l = 1/40$ that is much deeper in the diffusive regime. τ is set to be equal 1/200 so that the time of flight is $\tau_f = 1/5$. For these parameters, the smallest real zero is $\gamma_1 \approx 2.483$ and there are 127 real zeros in the interval $[0, 1/\tau)$. A similar number of zeros exist in the interval $(1/\tau, 2/\tau]$, and all the other ones are complex.

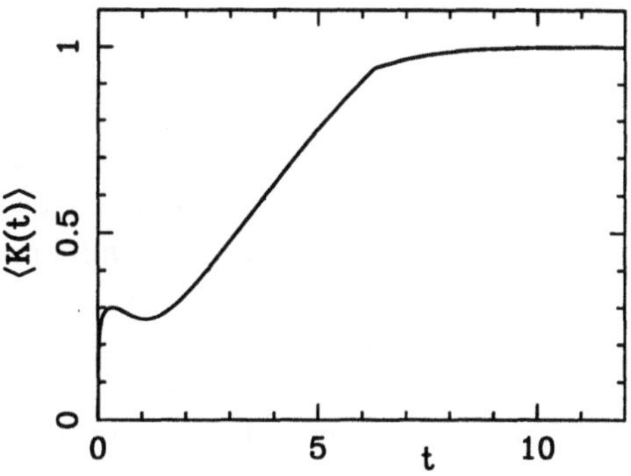

Fig. 8. The form factor in the diffusive regime for the parameters $l = 1/40$ $\tau = 1/200$ leading to $\tau_f = 1/5$

6 Summary and Discussion

In this review the spectral density correlator and the form factor of a system in which the amount of disorder is varied was analyzed with the help of periodic orbit theory and the supersymmetric σ-model. Results which are beyond the traditional diagonal approximation and the random matrix theory were presented. It was demonstrated how the eigenvalues of the Perron–Frobenius (PF) operator are transformed into those of the diffusion operator as the disorder is increased. It was also shown how the form factor $\hat{K}(t)$ varies with increasing disorder from the ballistic to the diffusive regime. Two parameters govern the behavior of the results: l the elastic mean free path, and τ the elastic mean free time in units of the Heisenberg time τ_H. The elastic mean free path measures the amount of disorder in the system. When $l > 1$, i.e. when the elastic mean free path is larger than the system size, the dynamics is ballistic, while it is diffusive when $l \ll 1$ (Altland and Gefen 1995), (Kravtsov and Lerner 1995). The elastic mean free time determines the classical time scales of the system, since $v = l/\tau$ where v is the dimensionless velocity. As $\tau \to 0$ other time scales effectively increase relative to τ and the RMT universal behavior is recovered (this will be discussed in details in what follows).

The non–universal features, that appear for finite small values of τ, decorate the RMT result of the form factor (51) mainly in two regions: near the origin $t = 0$, and in the vicinity of the Heisenberg time $t = 2\pi$. They are appreciable over small intervals that scale with τ. Thus as $\tau \to 0$ these time domains shrink and the universal result is reached. The nature of the non–universal behavior of the system is determined by the elastic mean free path l. Detailed analysis of this behavior, that was carried out for a system with two degrees of freedom belonging to the unitary ensemble, will now be summarized.

The typical structure of the form factor in the ballistic regime, $l \gg 1$ is depicted in Fig. 5. For $t < \tau$ it is dominated by the contribution, $e^{-2t/\tau}$, of periodic orbits that are not scattered by impurities (38). Near the Heisenberg time $|t-2\pi| < \tau$, the RMT singularity is smoothed out. The typical line shape of the form factor in this regime is characterized by a minimum in the vicinity of $t = \tau$.

The form factor, in the intermediate regime $l = 0.5$, is depicted in Fig. 6. Here the contribution from periodic orbits that are not scattered by impurities is negligible, and the behavior is determined by orbits that do scatter. Near the origin, [see Fig. 6(b)] the form factor exhibits a singular behavior. These singularities can be associated with short orbits that are scattered from a very small number of impurities, and therefore still preserve the topology of the orbits of the clean system. The amplitudes of the singularities decay exponentially as $e^{-t/\tau}$ since the probability for the existence of such orbits decreases exponentially with period. The behavior near the Heisenberg time is oscillatory with a period of the order of the time of flight across the system τ_f.

Moving towards the diffusive regime where $l < 1/2\pi$, the non–universal features change their character as presented in Fig. 7, for $l = \sqrt{5}/6\pi \simeq 0.12$. The singular behavior near the origin almost disappears, and the oscillations near the Heisenberg time transform into an overall smooth curve. The situation does not change much in the diffusive regime, as can be seen from Fig. 8 where $l = 1/40$.

An understanding of the scenario described above was obtained by analyzing the classical dynamical zeta function of the system, $1/Z(z)$. The zeros of this function are the eigenvalues of the Perron–Frobenius operator of the system (Ruelle 1986). They are thus associated with the decaying modes of the probability distribution of particles towards the invariant distribution. The dynamical zeta function, in an arbitrary dimension, has been calculated in our model, and was expressed in terms of an infinite product (47). For the quasi one dimensional case a simple formula has been obtained in terms of simple trigonometric functions (48). Its structure in the complex z plane is shown in Fig. 4(a).

A crucial role is played by the singularities and the zeros of the dynamical zeta function, since, up to constants, the perturbative part of the form factor, $\langle \hat{K}_P(t) \rangle$, and nonperturbative part, $\langle \hat{K}_{NP}(t) \rangle$, are the Fourier transforms (in $s = -iz$) of $\frac{\partial^2}{\partial z^2}[\ln Z(z) + \ln Z(-z)]$ and $Z(z)Z(-z)\cos(-i2\pi z)$ respectively. Therefore, the zeros and singularities that are closest to the imaginary z axis dominate the behavior of the form factor.

The analysis of $1/Z(z)$ in the two dimensional case shows that its zeros associated with orbits which are scattered from impurities appear in the interval $[0, 2/\tau]$ with $\mathrm{Im}\, z = 0$, and on the line $\mathrm{Re}\, z = 1/\tau$ [see Fig. 4(b)]. When $l > 1/2\pi$ all of them except two zeros, one at the origin and the second at $z = 2/\tau$, lie on the line $\mathrm{Re}\, z = 1/\tau$. The singularities are located only along this line, $\mathrm{Re}\, z = 1/\tau$. The part of $1/Z(z)$ associated with orbits that are not scattered has a cut and a branch point at $z = 2/\tau$. Since except the zero at the origin all these zeros and singularities scale as $1/\tau$, their contribution to $\langle \hat{K}(t) \rangle$ is expected to fall off exponentially with an exponent that scales as $1/\tau$. It therefore vanishes rapidly in the limit $\tau \to 0$, and only the RMT term associated with the zero at the origin survives.

In the ballistic regime $l \gg 1$, the contribution from all zeros and singularities that lie at $\mathrm{Re}\, z = 1/\tau$ is negligible. Thus for $|t| < \tau$, $\langle \hat{K}(t) \rangle$ is dominated by the contribution (38) of periodic orbits that are not scattered by impurities.

As l becomes of order unity, the contribution of non-scattered orbits becomes negligible and the behavior is governed by orbits which do scatter. If also $l > 1/2\pi$, then all the zeros z_n of $1/Z(z)$, except those at the origin and at $2/\tau$, are complex and located along the line $\mathrm{Re}\, z = 1/\tau$. The non–universal features are therefore oscillatory, because the contribution of a zero, γ_n, to $\langle \hat{K}_P(t) \rangle$ and $\langle \hat{K}_{NP}(t) \rangle$ is proportional to $\exp(-\gamma_n t)$ and $\exp(-\gamma_n |t - 2\pi|)$ respectively.

Zeros of $1/Z(z)$ start to appear on the real z axis, when l becomes smaller than $1/2\pi$. Some of them dominate the behavior of the form factor since they are closer to the imaginary z axis than the complex zeros. In particular the non–universal features of the form factor become non–oscillatory. This change of behavior is demonstrated in Figs. 6, 7 and 8.

The diffusive limit is reached when $l \ll 1/2\pi$. The real zeros in this limit concentrate near the origin, and coincide with the eigenvalues of the diffusion operator. By summing over their contributions one recovers the Altshuler and Shklovskii result for the perturbative part of the form factor (Altshuler and Shklovskii 1986) which applies for times smaller than $\tau_c/4\pi^2$ where $\tau_c = d\tau/l^2$ is the Thouless time.

In this work the semiclassical approximation was used. It requires the wavelength to be much smaller than the radius of the spheres R, that is much smaller than the size of the system. This leads to a constraint on the time of flight, $\tau_f \ll 1$. Nevertheless, the contribution of the scattered trajectories to is identical to the one found in (Altland and Gefen 1995) for δ scatterers. The contribution of trajectories that are not scattered by impurities is, however, different for these models.

Throughout this review the classical elastic mean free path l was used. In the semiclassical limit, for hard spheres, the quantum elastic mean free path is $l/2$. This is a well know result in three dimensions (Messiah 1970). It can be easily extended to two dimensions (Averbuch 1986).

Several important questions require a further study. The most important one concerns the better understanding of the Perron–Frobenius operator and in particular its renormalization due to quantum interference effects. In the diffusive regime this is manifested by renormalization of the diffusion constant (Kravtsov and Lerner 1995) which leads to weak localization effects. These corrections, therefore, are of higher order in $1/g$ where g is the dimensionless Thouless conductance. A deeper understanding of the analytical structure of the dynamical zeta function is desirable for higher number of degrees of freedom $d > 2$. Another problem concerns the generalization of this formalism to the situation of small angle scattering, which is the common scattering event an electron in a quantum dot experiences. Finally, effects of diffraction (Vattay, Wirzba, and Rosenquist 1994), tunneling and ghost orbits (Haake, Kuś, and Delande 1993), (Kuś, Haake, and Eckhardt 1993) were not taken into account for this system and their exploration should be subject of further studies.

Acknowledgments

It is our great pleasure to thank A. Altland, B. L. Altshuler, A. V. Andreev, E. Bogomolny, Y. Gefen, J. P. Keating, B. D. Simons, and M. L. Zirnbauer for informative and stimulating discussions. This work was supported in part by the US-Israel Binational Science Foundation (BSF), by the National Science Foundation under Grant No. PHY94-07194 and by the Fund for Promotion of Research at the Technion. O. A. acknowledges the support of the Rothschild Fellowship.

References

Agam, O., B. L. Altshuler, and A. V. Andreev (1995): *Phys. Rev. Lett.* **75**, 4389.
Agam, O. and S. Fishman (1996a): *Phys. Rev. Lett.* **76**, 726.
Agam, O. and S. Fishman (1996b): *J. Phys. A: Math. Gen.* **29**, 2013.
Altland, A. and Y. Gefen (1993): *Phys. Rev. Lett.* **71**, 3339.
Altland, A. and Y. Gefen (1995): *Phys. Rev. B* **51**, 10671.
Altshuler, B. L. and B. I. Shklovskii (1986): *Sov. Phys. JETP* **64**, 127.
Andreev, A. V., O. Agam, B. D. Simons, and B. L. Altshuler (1996): *Phys. Rev. Lett.* **76**, 3947.
Andreev, A. V. and B. L. Altshuler (1995): *Phys. Rev. Lett.* **75**, 902.
Argaman, N., Y. Imry, and U. Smilansky (1993): *Phys. Rev. B* **47**, 4440.
Averbuch, P. G. (1986): *J. Phys. A: Math. Gen* **9**, 2325.
Baranger, H. U., R. A. Jalabert, and A. D. Stone (1993): *Chaos* **3**, 665.
Berry, M. V. (1985): *Proc. R. Soc. Lond. A* **400**, 229.
Berry, M. V. (1991): Some quantum–to–classical asymptotics. In M. J. Giannoni, A. Voros, and J. Zinn-Justin (Eds.), *Chaos and Quantum Physics,* Proc. of Les-Houches Summer School, *Session LII, 1989.* Elsevier.
Berry, M. V. and M. Tabor (1977a): *Proc. Roy. Soc. Lond.* **A 356**, 375.
Berry, M. V. and M. Tabor (1977b): *J. Phys. A: Math. Gen.* **10**, 371.
Bogomolny, E. and J. P. Keating (1996): *preprint* .

Chang, A. M., H. U. Baranger, L. N. Pfeiffer, and K. W. West (1994): *Phys. Rev. Lett.* **73**, 2111.
Cvitanoviċ, P. and B. Eckhardt (1991): *J. Phys. A: Math. Gen.* **24**, L237.
Doron, E. and U. Smilansky (1990): *Phys. Rev. Lett.* **65**, 3072.
Doron, E. and U. Smilansky (1991): *Physica D* **50**, 367.
Efetov, K. B. (1983): *Adv. Phys.* **32**, 53.
Fleischmann, R., T. Geisel, and R. Ketzmerick (1992): *Phys. Rev. Lett.* **68**, 1367.
Fromhold, T. M., L. Eaves, F. W. Sheard, T. J. Foster, M. L. Leadbeater, and P. C. Main (1994): *Physica B* **201**, 367; *Phys. Rev. Lett.* **72**, 2608 (1994).
Gefen, Y., D. Braun, and G. Montambaux (1994): *Phys. Rev. Lett.* **73**, 154; *Ann. Physik* **3**, 467 (1994).
Gutzwiller, M. C. (1990): *Chaos in Classical and Quantum Mechanics.* New York: Springer.
Haake, F., M. Kuś, and D. Delande (1993): *Phys. Rev. Lett.* **71**, 2167.
Hannay, J. H. and A. M. Ozorio de Almeida (1984): *J. Phys. A: Math. Gen.* **17**, 3429.
Hauge, E. H. (1974): In G. Kirczenow and J. Marro (Eds.), *Transport Phenomena* Lecture Notes in Physics Vol. 31, Berlin, pp. 337. Springer-Verlag.
Jalabert, R. A., H. U. Baranger, and A. D. Stone (1990): *Phys. Rev. Lett.* **65**, 2442.
Kravtsov, V. E. and I. V. Lerner (1995): *Phys. Rev. Lett.* **74**, 2563.
Kudrolli, A., V. Kidambi, and S. Sridhar (1995): *Phys. Rev. Lett.* **75**, 822.
Kuś, M., F. Haake, and B. Eckhardt (1993): *Z Phys B* **92**, 221.
Levy, L. P., L. Reich, L. Pfeiffer, and K. West (1993): *Physica B* **189**, 204.
Marcus, C. M., A. J. Rimberg, R. M. Westervelt, P. F. Hopkins, and A. C. Gossard (1992): *Phys. Rev. Lett.* **69**, 509.
Mehta, M. L. (1991): *Random Matrices.* New York: Academic Press.
Messiah, A. (1970): *Quantum Mechnics.* Amsterdam: North-Holland.
Moyal, J. E. (1949): *Proc. Camb. Phil. Soc. Math. Phys. Sci.* **45**, 99.
Muzykantskii, B. A. and D. E. Khmelnitskii (1995): *JETP Lett.* **62**, 76.
Pollicott, M. (1990): *Ann. Math* **131**, 33.
Ruelle, D. (1978): *Statistical Mechanics, Thermodynamic Formalism.* MA: Addison-Wesley: Reading.
Ruelle, D. (1986): *Phys. Rev. Lett.* **56**, 405.
Sridhar, S. and E. J. Heller (1992): *Phys. Rev. A* **46**, 1728.
Stein, J. and H. J. Stöckman (1992): *Phys. Rev. Lett.* **68**, 2867.
Ullmo, D., K. Richter, and R. A. Jalabert (1995): *Phys. Rev. Lett.* **74**, 383.
Vattay, G., A. Wirzba, and P. R. Rosenquist (1994): *Phys. Rev. Lett.* **73**, 2304.
Weiss, D., M. L. Roukes, A. Menschig, P. Grambow, K. von Klitzing, and G. Weimann (1991): *Phys. Rev. Lett.* **66**, 2790.
Zirnbauer, M. L. (1986): *Nucl. Phys.* **B 265**, 375.
Zirnbauer, M. L. : *private communication* .

Microwave Ionization of Hydrogen Rydberg Atoms: Chaos, Symmetry and Dynamical Localization

Reinhold Blümel

Albert-Ludwigs-Universität, Hermann-Herder-Str. 3, D-79104 Freiburg, Germany

Abstract. Together with the hydrogen atom in a strong magnetic field the hydrogen atom in a strong microwave field is one of the best studied atomic physics examples of a quantum Hamiltonian system chaotic in its classical limit. The chaos link is strong. Excitation, ionization and dynamical localization are most easily understood in the framework of chaos. This paper focusses on dynamical localization in driven hydrogen Rydberg atoms and suggests to study the effect of symmetry and symmetry breaking on the dynamical localization length in strongly driven Rydberg atoms.

1 Introduction

There are two atomic physics systems which have received considerable attention in connection with chaos in atomic physics: The hydrogen atom in a strong magnetic field and the hydrogen atom in a strong microwave field. The hydrogen atom in a strong magnetic field is an example of a time independent atomic physics system chaotic in the classical limit. A major contribution to the field was a kind of inverse Gutzwiller formula, i.e., the extraction of periodic orbit information from quantum spectra (Wintgen (1987)). This method is generally applicable to all scaling quantum systems, regular, mixed or chaotic. The chaotic properties of the hydrogen atom in a strong magnetic field were reviewed by Friedrich and Wintgen in 1989. Some of the recent developments in the field, experimental as well as theoretical, are reviewed in a collection of papers edited by J.-C. Gay in 1992. The hydrogen atom in a strong magnetic field continues to play its role as one of the foremost paradigms in quantum chaos (see, e.g., the text book by Friedrich (1991), the book on quantum chaos edited by Casati and Chirikov (1995) and the article by Jörg Main constituting a later chapter of this volume).

In this chapter we focus on the hydrogen atom in a strong microwave field. Major reviews exist that cover many of the developments in the field that occurred from its beginnings (Bayfield and Koch (1974)) up to very recent times. We mention the reviews by Casati *et al.* (1987), Jensen *et al.* (1991), Moorman and Koch (1992), Koch and van Leeuwen (1995). Here we focus on a particular aspect of microwave excitation: dynamical localization. In particular, we investigate the question of whether the "weak localization signature", a factor two enhancement of the staying probability (Dittrich and Smilansky (1991)) can be observed in driven hydrogen Rydberg atoms. Far from being able to answer this question for fully three-dimensional microwave-perturbed hydrogen Rydberg atoms, we concentrate here on the investigation of a simple model atom perturbed by narrow microwave pulses. Since we draw considerably on previous experience with the classical and quantum kicked rotor, this system is also briefly discussed. We shall present indications for the existence of dynamical localization and quantum enhancement of the staying probability for the kicked hydrogen model. A final clarification of the question has to await further detailed study, especially on the structure of periodic and returning orbits of the kicked hydrogen atom.

2 The Kicked Rotor

The kicked rotor was the first quantum system sytematically studied in the context of quantum chaos (Casati *et al.* (1979)). A host of new concepts was found and explored in connection with the kicked rotor. One of the most important concepts is dynamical localization. It was first discussed by Fishman *et al.* in 1982 in order to explain the phenomenon of quantum diffusion limitation found earlier by Casati *et al.* in 1979. Excellent review articles on the kicked rotor exist in the physics literature. We mention, e.g., the review article by Izrailev which covers the progress of the subject up to 1990.

Dynamical localization and quasienergy spectral correlations were studied in detail by Dittrich and Smilansky in two papers in 1991. The effect of symmetry breaking on the dynamical localization length of the kicked rotor was investigated by Blümel and Smilansky in 1992. An extension to the localization properties of the symplectic kicked rotor was reported by Thaha *et al.* in 1993.

2.1 Kicked Rotor Basics

The Hamiltonian of a dipole in an impulsively switched homogeneous field constrained to rotate in a plane is given by

$$\hat{H} = \frac{\hat{L}^2}{2I} + \mu\mathcal{E}\cos(\theta)\sum_{j=-\infty}^{\infty}\delta(t/T-j) \ , \qquad (2.1)$$

where θ is the rotation angle, $\hat{L} = -i\hbar\partial/\partial\theta$ is the angular momentum operator, I is the moment of inertia, μ is the dipole moment, $\mathcal{E}$ is the field amplitude and T is the period of the kicks. The quantum dynamics of (2.1) may be represented in the form of a quantum mapping

$$|\,\psi_{n+1}\rangle = \hat{U}\,|\,\psi_n\rangle \ , \qquad (2.2)$$

where $\mid \psi_n\rangle$ is the wave function immediately before kick number n. The mapping operator $\hat{U}$, also referred to as the one-step propagator, is given by

$$\hat{U} = \exp\left[-\frac{i}{\hbar}\frac{\hat{L}^2}{2I}T\right] \exp\left[-\frac{i}{\hbar}\mu\mathcal{E}T\cos(\theta)\right] . \tag{2.3}$$

If we measure $\hat{L}$ in units of $\hbar$, $\hat{L} = \hbar\hat{l}$, and introduce the dimensionless control parameters

$$\tau = \hbar T/I \ , \quad k = \mu\mathcal{E}T/\hbar \ , \tag{2.4}$$

the mapping operator $\hat{U}$ can be written as

$$\hat{U} = \exp\left[-i\frac{\tau}{2}\hat{l}^2\right] \exp\left[-ik\cos(\theta)\right] . \tag{2.5}$$

The classical mapping corresponding to (2.5) is obtained as the solution of the classical canonical equations of (2.1):

$$\dot{L} = -\frac{\partial H}{\partial \theta} = \mu\mathcal{E}\sin(\theta) \sum_{j=-\infty}^{\infty} \delta(t/T - j) \ ,$$

$$\dot{\theta} = \frac{\partial H}{\partial L} = L/I \ . \tag{2.6}$$

The solution of (2.6) is the mapping

$$l_{n+1} = l_n + k\sin(\theta) \ ,$$

$$\theta_{n+1} = \theta_n + \tau\, l_{n+1} \ , \tag{2.7}$$

where (θ_n, l_n) refers to the values of (θ, l) immediately before kick number n. The mapping (2.7) can be rescaled. Defining

$$\lambda_n = \tau\, l_n \ , \tag{2.8}$$

we obtain the standard mapping (Chirikov (1979))

$$\lambda_{n+1} = \lambda_n + K\sin(\theta_n) \ ,$$

$$\theta_{n+1} = \theta_n + \lambda_{n+1} \ , \tag{2.9}$$

where

$$K = \tau k \ . \tag{2.10}$$

The mapping (2.9) depends on the single control parameter K only. This is not the case for the quantum rotor mapping induced by the mapping operator (2.5) which depends nontrivially on the two control parameters k and τ. For $K > K_c \approx 1$ the mapping (2.9) is chaotic with a mixed phase space. In this regime the angular momentum λ is said to diffuse. In quasi-linear approximation, neglecting all correlations between the angular momentum and the angle,

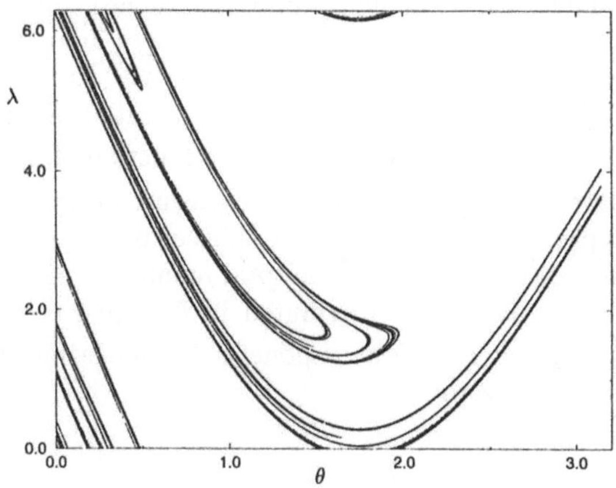

Fig. 1. The set $F_{10}^{(+)}$ of generalized accelerator modes.

$$\langle \lambda_n^2 \rangle \approx \frac{1}{2} K^2 n \ . \tag{2.11}$$

The diffusive behavior of (2.9) is important in the context of dynamical localization where the classical diffusion constant is used to normalize the localization length.

Complications arise due to regions in phase space which are always accelerating, i.e., $\lambda_{n+1} - \lambda_n > 0$ for all n, or $\lambda_{n+1} - \lambda_n < 0$ for all n. Special cases of these accelerator modes were already studied by Chirikov (1979). Here we generalize the concept of accelerator modes and prove that the phase space may contain accelerating regions of finite area.

2.2 Generalized Accelerator Modes

Generalized accelerator modes and their importance for plasma heating were recently investigated by Fan (1994). We define accelerator modes of order M as subsets of phase space with

$$\begin{aligned} F_M^{(+)} &= \{(\theta \equiv \theta_0, \lambda \equiv \lambda_0) \mid \lambda_{n+1} - \lambda_n > 0 \quad \text{for all } n \le M\} \ , \\ F_M^{(-)} &= \{(\theta \equiv \theta_0, \lambda \equiv \lambda_0) \mid \lambda_{n+1} - \lambda_n < 0 \quad \text{for all } n \le M\} \ . \end{aligned} \tag{2.12}$$

We also define the limiting sets for $M \to \infty$ as

$$F^{(\pm)} = \lim_{M \to \infty} F_M^{(\pm)} \ . \tag{2.13}$$

Figure 1 shows $F_{10}^{(+)}$ for the mapping (2.9) with $K = 5$ starting at $n = 0$ with 10^8 points equi-distributed in $(\theta, \lambda) \in [0, 2\pi] \times [0, 2\pi]$. $F_{10}^{(+)}$ resembles a fractal in the first stages of its construction by removing successive sets of finite size. Thus, $F_{10}^{(+)}$ can be interpreted as $F^{(+)}$ in stage number 10 of its construction. It still contains regions of finite area. These regions are most probably broken up for larger M, leaving the fractal set $F^{(+)}$ in the limit $M \to \infty$. The limiting fractal may be a fractal of zero measure. This is motivated by Fig. 2 which shows $\mu(F_n^{(+)})$ as a function of n, i.e., the measure of the accelerator set $F_n^{(+)}$ of order n. This measure, normalized to 1 at $n = 0$, is computed approximately as the ratio of phase-space points accelerating up to and including kick number n to the total number of 25×10^6 phase-space points started at $n = 0$ equi-distributed in $\theta \in [0, 2\pi]$, $\lambda \in [0, 2\pi]$. The measure of $F_n^{(+)}$ appears to decay exponentially indicating that the limiting set is a set of measure zero.

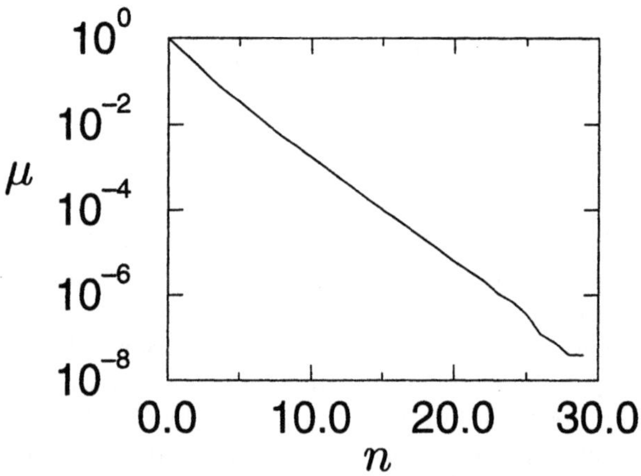

Fig. 2. Normalized measure of $F_n^{(+)}$ after n applications of the standard mapping with $K = 5$.

In the theory of symmetry breaking and localization generalizations of the rotor mapping (2.9) are investigated. We define

$$\lambda_{n+1} = \lambda_n + K[\cos(\pi q/2)\sin(\theta_n) - \sin(\pi q/2)\cos(2\theta_n)] \ ,$$
$$\theta_{n+1} = \theta_n + \lambda_{n+1} \ , \qquad (2.14)$$

where $q \in [0, 1]$ is an additional control parameter. It is also referred to as a symmetry breaking parameter since $q \neq 0$ destroys the conjugation symmetry (Dittrich and Smilansky (1991)) of (2.9).

Fig. 3 shows $F_{10}^{(+)}$ for the mapping (2.14) with $K = 5$ and $q = 0.25$. Again the set looks fractal. This time, however, the finite phase-space regions remaining after 10 kicks appear to be much larger in area than in the corresponding case of Fig. 1. This is corroborated in Fig. 4 which shows a magnification of a detail of Fig. 3 around $\lambda = 2\pi$ and $\theta = \pi/2$. A major

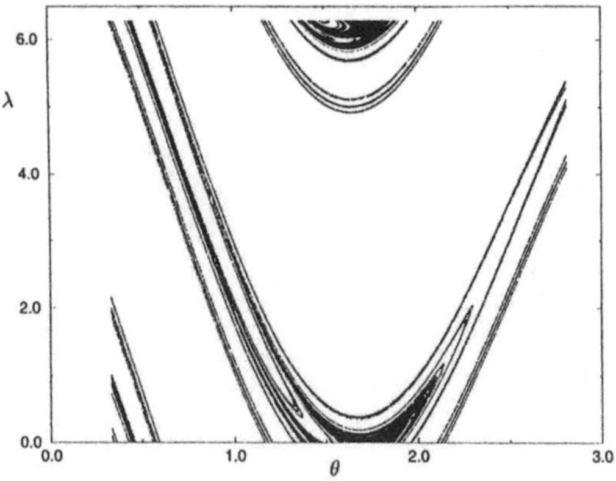

Fig. 3. The set of accelerator modes $F_{10}^{(+)}$ for the modified rotor mapping with $K = 5$ and $q = 0.25$.

portion of the regions of finite measure shown in Fig. 4 may survive in the limit $M \to \infty$. This is motivated in Fig. 5 which shows $F_{100}^{(+)}$ on the same scale as Fig. 4.

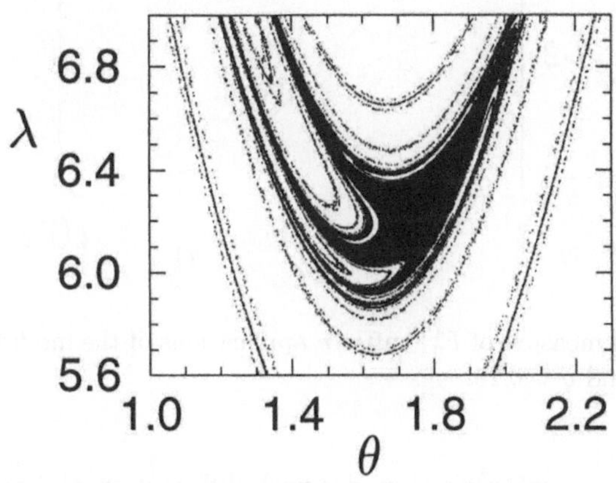

Fig. 4. Magnification of a detail of Fig. 3 in the region around $\lambda = 2\pi$, $\theta = \pi/2$.

That the finite region in Fig. 5 really survives in the limit $M \to \infty$ is corroborated in Fig. 6. It shows the fraction of accelerator points as a function of n starting with 25×10^6 points equi-distributed in $\theta \in [0, 2\pi]$, $\lambda \in [0, 2\pi]$. Apparently the measure $\mu(F_n^{(+)})$ reaches a plateau indicating that a finite phase-space area is accelerated forever. Finite accelerator regions may find important applications in plasma heating where the mapping (2.14) may be

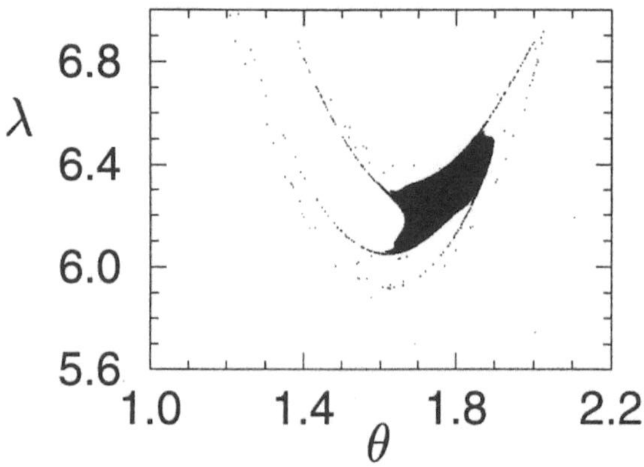

Fig. 5. The set of generalized accelerator modes $F_{100}^{(+)}$ for the modified rotor mapping with $K = 5$ and $q = 0.25$.

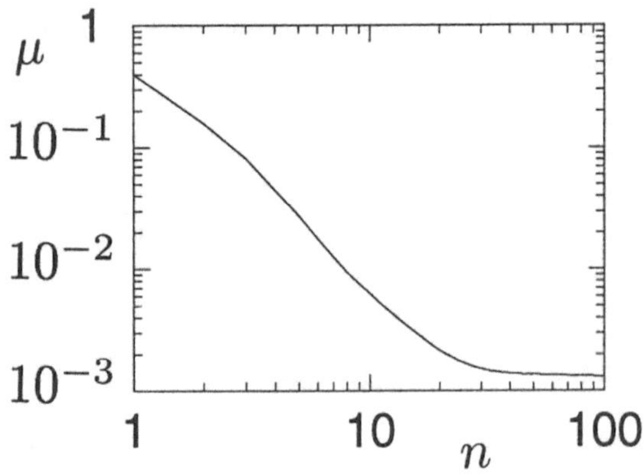

Fig. 6. Normalized measure of $F_n^{(+)}$ after n applications of the modified rotor mapping with $K = 5$ and $q = 0.25$.

realized by combining microwave modes with different spatial mode structure (Fan (1994)). Accelerator regions, fractal or nonfractal, of zero or of finite measure complicate investigations of dynamical localization considerably.

2.3 Symmetry and Localization

A useful concept in the theory of dynamical localization is the staying probability. It is defined as the probability to stay in (return to) the initial state after n kicks:

$$P^{(Q)}(n) = |\langle\psi_0 \mid \psi_n\rangle|^2 = |\langle\psi_0 \mid \hat{U}^n \mid \psi_0\rangle|^2 \quad . \tag{2.15}$$

In the rotor basis

$$\langle \theta \mid l \rangle \;=\; \frac{1}{\sqrt{2\pi}} \exp(il\theta) \tag{2.16}$$

the matrix elements of the one-cycle propagator can be computed explicitly. They are given by

$$U_{lm} \;=\; \langle l \mid \hat{U} \mid m \rangle \;=\; i^{l-m} \exp\left[-i\frac{T}{2}l^2\right] J_{|l-m|}(k) \;, \tag{2.17}$$

where $J_p(x)$ are Bessel functions of the first kind. The explicit formula for U_{lm} allows us to compute the quantum staying probability after one kick

$$P_l^{(Q)}(1) \;=\; J_0^2(k) \;. \tag{2.18}$$

The classical staying probability is given by

$$P_l^{(C)}(1) \;=\; 1/\pi k \;. \tag{2.19}$$

Apparently, neither $P_l^{(Q)}(1)$ nor $P_l^{(C)}(1)$ depend on the initial state $\mid l\rangle$. This can be understood on the basis of the structure of U_{lm}. The transition probabilities $\mid U_{lm} \mid^2$ depend only on the difference of l and m. Thus, average properties of the quantum dynamics are expected to be independent of the starting state $\mid l\rangle$. This motivates the definition of the average staying probability

$$P_s^{(Q,C)}(n) \;=\; \langle P_l^{(Q,C)}(n)\rangle_l \;, \tag{2.20}$$

where the average is on a large set of initial angular momenta l. We define the eigenenergies and eigenstates of $\hat{U}$, referred to as the quasienergies and quasienergy states, respectively, as

$$\hat{U} \mid \alpha \rangle \;=\; e^{i\alpha} \mid \alpha \rangle \;. \tag{2.21}$$

With the help of the quasienergy states we define the inverse participation ratio

$$\frac{1}{\xi} \;=\; \langle \sum_{\alpha} \mid \langle l \mid \alpha \rangle \mid^4 \rangle_l \;. \tag{2.22}$$

For large n, and in the localized case the time averaged staying probability is given by:

$$\begin{aligned} \langle P_s^{(Q)}(n)\rangle_n \;&=\; \langle \mid \langle l \mid \hat{U}^n \mid l \rangle \mid^2 \rangle_{l,n} \\ &=\; \langle \mid \sum_{\alpha} \langle l \mid \alpha \rangle \, e^{in\alpha} \, \langle \alpha \mid l \rangle \mid^2 \rangle_{l,n} \;=\; 1/\xi \;. \end{aligned} \tag{2.23}$$

Dittrich and Smilansky (1991a,b) conjectured that $P_s^{(Q)}(n)$ is a scaling function which can be related to a universal function f by

$$P_s^{(Q)}(n) \;=\; \frac{1}{\xi} f(n/\xi) \;. \tag{2.24}$$

The function f was studied in great detail for various symmetry classes of the kicked rotor (Dittrich and Smilansky (1991a,b), Thaha *et al.* (1993)). Because of (2.23) $f(x) \to 1$ for $x \to \infty$. We define a function

$$\rho(n) \;=\; P_s^{(Q)}(n)\,/\,P_s^{(C)}(n) \;. \tag{2.25}$$

There is by now ample numerical evidence that for the standard kicked rotor there exists an interval of n values for which $\rho(n)$ is constant and equal to 2. Breaking all antiunitary symmetries of the kicked rotor reduces the plateau-value of ρ to $\rho = 1$.

For the standard kicked rotor we can compute $\rho(1)$ explicitly. According to (2.18) and (2.19) it is given by

$$\rho(1) \;=\; \pi k\, J_0^2(k) \to 2\cos^2(k-\pi/4) \quad \text{for large } k \;. \tag{2.26}$$

The ratio $\rho(1)$ is constant and equal to 1 on average, but can reach the values 2 and 0 depending on k.

3 The Kicked Hydrogen Atom

So far the factor 2 enhancement in the staying probability of time reversal conserving dynamically localizing systems was seen only in the kicked rotor. A first step towards establishing the existence (or non-existence) of the factor 2 in the context of dynamically localizing atomic physics systems is the investigation of the kicked hydrogen atom.

3.1 The Model

In atomic units the Hamiltonian of the kicked hydrogen atom is defined as

$$\hat{H} \;=\; \hat{H}_0 \;-\; \mathcal{E}x \sum_m \delta(t/T - m) \;, \tag{3.1}$$

where

$$\hat{H}_0 \;=\; \frac{\hat{p}^2}{2} - \frac{1}{x} \quad \text{for } x>0, \quad \hat{H}_0 \;=\; \infty \quad \text{for } x<0 \;. \tag{3.2}$$

The classical and quantum dynamics of (3.1) were studied by Blümel and Smilansky (1984a,b) in the context of dynamical localization. We assume in the following that $\mathcal{E} > 0$ since it was established (Hillermeier *et al.* (1992), Blümel (1993)) that in this case there are no stable islands in phase space which could possibly complicate the investigation. The bound spectrum of $\hat{H}_0$ is given by

$$\hat{H}_0 \;|\, l\rangle \;=\; -\frac{1}{2l^2}\;|\, l\rangle \;, \quad l = 1,2,\ldots \;. \tag{3.3}$$

We are only interested in the localization properties of the bound space. Therefore we neglect the continuum part of the spectrum altogether and confine our attention exclusively to the Hilbert space spanned by the states (3.3). The quantum mapping induced by (3.1) is given by

$$| \psi_{n+1}\rangle = \hat{U} \mid \psi_n\rangle = e^{i\tau/2l^2} e^{ikx} \mid \psi_n\rangle , \tag{3.4}$$

where, in analogy to the kicked rotor, we introduced the dimensionless control parameters

$$\tau = T, \quad k = \mathcal{E}T . \tag{3.5}$$

3.2 Chaos and Diffusion

The classical mapping of (3.1) is constructed in the following way. First, we define the canonical transformation to action and angle variables according to

$$p = \frac{1}{l} \cot(\eta), \quad x = 2l^2 \sin^2(\eta) ,$$
$$\theta = 2\eta - \sin(2\eta), \quad 0 \le \eta \le \pi . \tag{3.6}$$

Then, we observe that at a kick the position x is not changed. We obtain

$$l \sin(\eta) = l' \sin(\eta') . \tag{3.7}$$

The kick changes the momentum of the electron according to

$$p' = p + k , \tag{3.8}$$

from which we get

$$\frac{1}{l'} \cot(\eta') = \frac{1}{l} \cot(\eta) + k , \tag{3.9}$$

or

$$\sin(\eta') = \sin(\eta) \left[1 - 2kl \cot(\eta) - (lk)^2\right]^{1/2} . \tag{3.10}$$

The angle η' depends on the sign of p' which can be computed from (3.8). During the free motion part of (3.1) the angle θ changes linearly according to $\theta \to \theta + \tau/l'^3$. The final result for the kicked hydrogen mapping is now given by

$$l_{n+1} = l_n / \left[1 - 2kl_n \cot(\eta_n) - (kl_n)^2\right]^{1/2} ,$$
$$\eta' = \arcsin[l_n \sin(\eta_n)/l_{n+1}], \quad p' > 0 ,$$
$$\eta' = \pi - \arcsin[l_n \sin(\eta_n)/l_{n+1}], \quad p' < 0 ,$$
$$\theta_{n+1} = 2\eta'_n - \sin(2\eta'_n) + \tau/l^3_{n+1} . \tag{3.11}$$

The mapping (3.11) is considerably more complicated than the mapping (2.7). There also exists a scaled version of (3.11) (see Hillermeier *et al.* (1992)) analogous to the standard mapping (2.9).

The fundamental difference between "positive kicks" and "negative kicks" is best illustrated with the help of Poincaré sections. Fig. 7 shows the 200 iterates $(\theta_n^{(j)}, l_n^{(j)})$, $n = 1, 2, \ldots, 200$ of 100 phase-space points started at $(\theta_0^{(j)}, l_0^{(j)})$

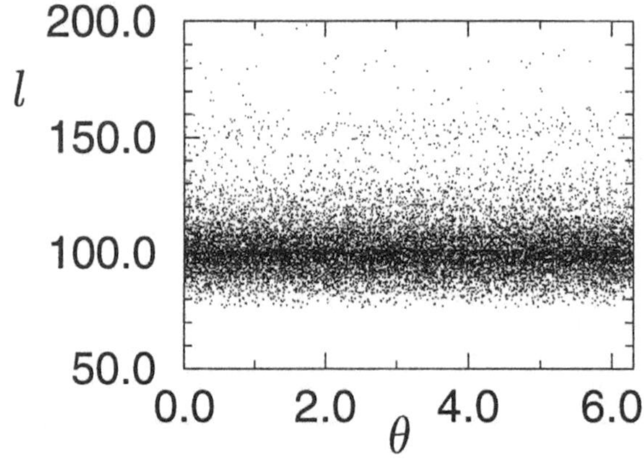

Fig. 7. Poincaré section for the positively kicked hydrogen atom.

where $l_0^{(j)} = 100$, $\theta_0^{(j)} = j\Delta\theta$, $\Delta\theta = 2\pi/101$, $j = 1, 2, .., 100$ for $k = 10^{-4}$ and $\tau = 10^8$. For this choice of parameters no ionization was observed for the first 500 kicks. The τ value was chosen to be such a large number because according to (3.11) at $l = 100$ we get significant rotation in angle only for $\tau > 10^6$. In Fig. 7 we see no indications of regular behavior. This is consistent with the claim that the positively kicked hydrogen atom has no stable islands of any order. Fig. 8 is a magnification of a part of phase space shown in Fig. 7. Fig. 8 confirms that apart from a straight line of points at $l = 100$ (the starting points) and a slanted line of points (the first iteration of the starting points) there are no discernable regular features in phase space.

Flipping the sign of k we switch to the negatively kicked hydrogen atom and obtain the phase-space portrait shown in Fig. 9. We see an elliptic island surrounded by a chain of satellites. We also see the straight line of initial conditions chosen to produce this portrait.

Next we address the question of whether there is diffusion in the kicked hydrogen atom. This is important for establishing an analogy with the kicked rotor. It is also important for localization considerations. For diffusive dynamics, the localization length of the kicked hydrogen atom can be normalized to the classical diffusion constant which allows a close comparison with the results on localization, and the breaking thereof, in the case of the kicked rotor. We choose the same parameters as in Fig. 7 and compute the staying probability of the classical kicked hydrogen atom i.e., the probability to be mapped into the interval $l = 100 \pm 1/2$ at kick number n. The result is shown in Fig. 10. The staying probability in Fig. 10 is computed on the basis of 1000 trajectories equi-distributed in angle with $l_0 = 100$. Computed is $P_s^{(C)}(n)$, the ensemble average of the probability to be mapped into the interval $99.5 < l < 100.5$ at kick number n. Figure 10 shows that the staying probability decays approximately inversely proportional to $\sqrt{n}$. This result

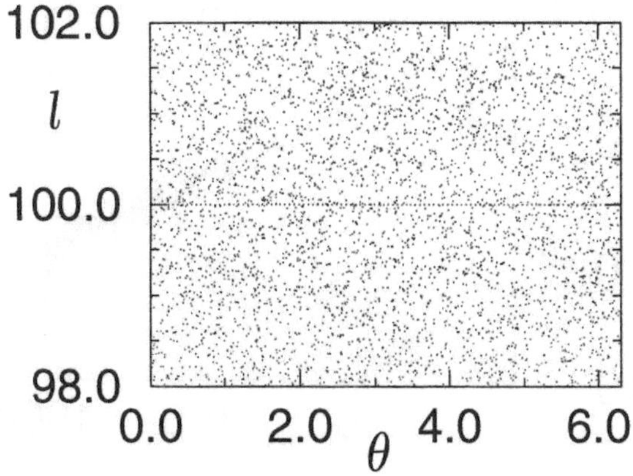

Fig. 8. Phase-space portrait of the positively kicked hydrogen atom. The parameters are the same as in Fig. 7. The portrait is a magnification of a part of phase space shown in Fig. 7. Except for trivial features that originate in the choice of initial conditions it confirms the absence of regular structure in the phase space of the positively kicked hydrogen atom.

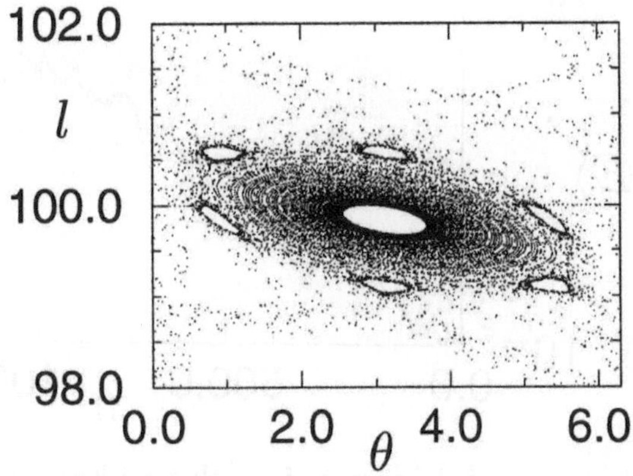

Fig. 9. Regular features in the phase-space portrait of the negatively kicked hydrogen atom. Up to the sign of the kicking strength the parameters are the same as in Fig. 7.

indicates that the phase-space dynamics is diffusive. Thus, we established that the dynamics of the kicked hydrogen atom and the dynamics of the kicked rotor are very similar.

The similarity carries even further. Fig. 11 shows the ensemble average $\langle E \rangle = \langle -1/2l^2 \rangle$. According to Fig. 11 the energy increases approximately linearly as a result of the external kicks. According to (2.11) the same result

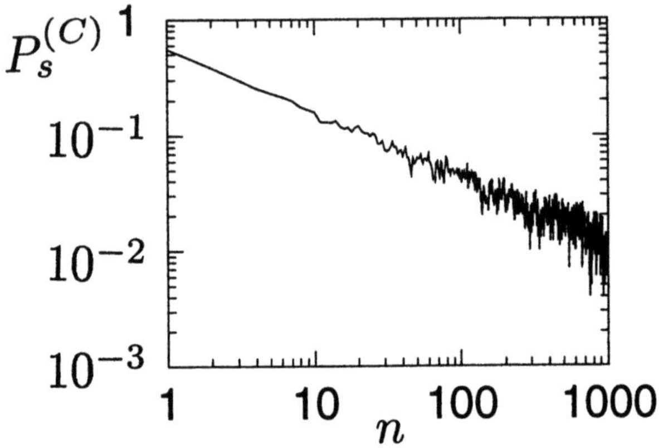

Fig. 10. Staying probability $P_s^{(C)}$ of the classical kicked hydrogen atom in the angular momentum interval $99.5 < l < 100.5$. The field parameters are the same as in Fig. 7. The staying probability decays approximately inversely proportional to the square root of n indicating diffusive behavior of phase-space populations.

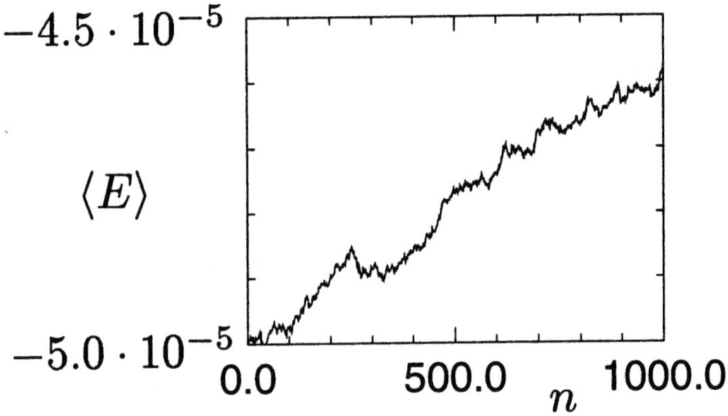

Fig. 11. Ensemble average of the energy of the classical kicked hydrogen atom. The field parameters are the same as in Fig. 7. The energy increases approximately linearly in n indicating diffusive behavior of phase-space populations.

is expected (and holds, at least approximately) for the kicked rotor in the chaotic regime.

3.3 Symmetry and Localization

In the previous sections we established that the kicked hydrogen atom is very close to the kicked rotor as far as its classical properties are concerned. For positive kick strength the kicked hydrogen atom is a strongly chaotic system.

The energy diffuses, and the staying probability decays roughly as expected from a diffusive chaotic system. We show now that even as far as the quantum mechanics is concerned the two systems are very close. In particular, we shall see that the quantum kicked hydrogen atom is a dynamically localizing system and shows enhancement of the staying probability with respect to the classical dynamics. We show these facts explicitly by constructing the one-step propagator of the kicked hydrogen atom. We choose a basis consisting of 91 of the bound states (3.3), $l_{min} = 70 \leq l \leq l_{max} = 160$. We compute the matrix elements $\langle l \mid x \mid m \rangle$ to obtain a 91×91 matrix M representing the dipole operator in our finite basis set. We obtain the kicking operator in the unitary approximation by exponentiating M. Multiplying the phases of the free propagation, we construct the one-step propagator $\hat{U}(\tau, k)$. We choose $k = 10^{-4}$, the same value as used above in the classical investigations of the kicked hydrogen atom. It is known from the work on the kicked rotor that the eigenstates of $\hat{U}$ fluctuate strongly as a function of the control parameters τ and k. This applies also to the detailed shape of a particular eigenstate which strongly depends on its localization center, i.e., the unperturbed rotor state which has maximal overlap with the chosen eigenstate of $\hat{U}$. In the case of the kicked rotor mean values are extracted by averaging over a large set of initial states $\mid l_0 \rangle$. Instead, we choose here averaging over τ. This is a more natural choice concerning possible experiments with kicked hydrogen atoms (see section 3.4). It is also indicated since contrary to the kicked rotor the absolute squares of the matrix elements of the hydrogen one-step propagator are not translationally invariant, i.e., for the kicked hydrogen atom $\mid U_{lm} \mid$ is not just a function of the difference of the indices. In the semiclassical regime, however, and up to a scaling factor, this translational invariance holds again, but only approximately. The τ averaging is done by averaging states and staying probabilities over 20 τ values chosen to be $\tau_j = 10^8 + 2\pi j g \times 10^4$, where g is the golden mean $g = (\sqrt{5} - 1)/2$. We define the amplitudes

$$a_l(n; \tau, k, l_0) = \sum_m U_{lm}(\tau, k)\, a_m(n-1; \tau, k, l_0) \; , \quad a_l(0) = \delta_{l, l_0} \quad (3.12)$$

and compute the time and τ averaged occupation probabilities

$$\bar{p}_l = \frac{1}{101 \times 20} \sum_{n=100}^{200} \sum_{j=1}^{20} \mid a_l(n; \tau_j, k = 10^{-4}, l_0 = 100) \mid^2 . \quad (3.13)$$

The result is shown in Fig. 12. The time and τ averaged occupation probability is strongly localized at $l \approx l_0 = 100$. This localization is a quantum phenomenon, since Figs. 10 and 11 show that the classical dynamics is diffusive. The transition to localization is more dramatically demonstrated in Fig. 13 which shows the classical staying probability $P_s^{(C)}(n)$ together with the τ averaged quantum staying probability $P_s^{(Q)}(n)$. After an initial decay the quantum staying probability reaches a plateau at the level $P_s^{(Q)} \approx 0.45$.

According to (2.24) together with the asymptotic behavior of f this level is expected to be close to the inverse participation ratio. We have $\xi \approx 2$. The existence and finiteness of ξ again indicates the presence of dynamical localization in the kicked hydrogen atom.

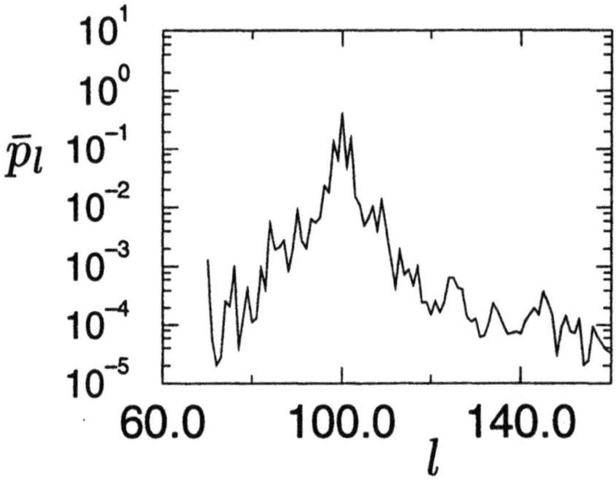

Fig. 12. Time and τ averaged occupation probability $\bar{p}_l$ of the kicked hydrogen atom. The probability distribution is sharply peaked on the initial state $| l_0 = 100\rangle$ and falls off approximately exponentially in $| l - l_0 |$ thus indicating dynamical localization.

The most interesting property of the staying probability, however, is the fact that $P_s^{(Q)}$ in Fig. 13 is systematically higher than the classical staying probability $P_s^{(C)}$. Although the difference does not quite reach a factor two expected on the basis of the kicked rotor results, the behavior of $P_s^{(Q)}$ displayed in Fig. 13 is encouraging. The enhancement of the staying probability at the starting level $| l_0\rangle$ can also be seen in Fig. 12 as a distinct spike in $\bar{p}_{100}$, much larger than the occupation probabilities of neighboring states.

For the present choice of parameters the localization length is very small (only about 2 states). One does not really see a pronounced quantum diffusive regime and the transition to quantum localization occurs very early. Moreover, the first few kicks are not representative of the dynamics at intermediate and longer time scales since for short times (the first few kicks) the dynamics does not have enough time to mix phase-space populations. This is demonstrated in Fig. 8, where the image of the first kick results in slanted lines easily recognized as regular features in the otherwise chaotic phase space. Larger k values increase the localization length of the kicked hydrogen states. Therefore, additional computations with larger values of k are needed to obtain a longer time interval of quantum diffusion in order to see the precise factor of quantum enhancement (if any) of the staying probability. The factor 2 enhancement hinges on the existence of conjugate trajectories

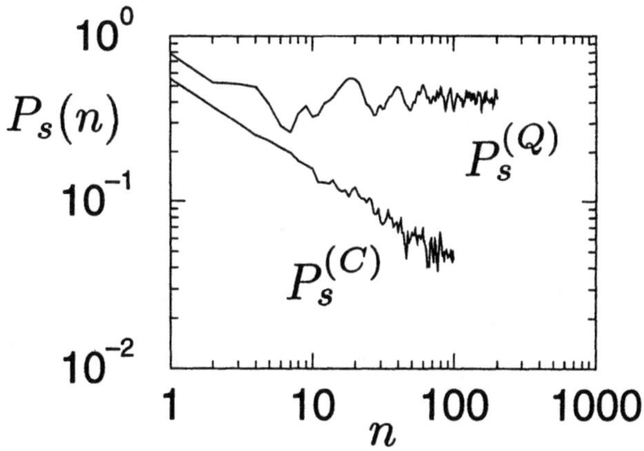

Fig. 13. Classical and quantum staying probabilities of the kicked hydrogen atom. While the classical atom shows uninhibited diffusion, quantum diffusion breaks early resulting in a stationary asymptotic behavior of the quantum staying probability.

with identical actions. It is still an open question whether such conjugate trajectories exist for the mapping (3.11). If the conjugation symmetry is only an approximate symmetry of (3.11) a reduction of quantum enhancement is expected. Larger k computations may also contribute to the solution of the question whether the scaling function f introduced in connection with the kicked rotor is indeed a universal function relevant for all dynamically localizing systems.

3.4 Experimental Implementation

We discussed the kicked hydrogen atom as a convenient model for studying dynamical localization in the atomic physics context. But there is no reason why the kicked hydrogen atom should not be implemented as a real system in the laboratory. The kicked hydrogen atom can be realized as soon as it is possible to generate short electric field pulses. This might be done using fast beam techniques in connection with a grid of electrodes, as shown in Fig. 14. This possibility of generating narrow kicks was mentioned to me in 1989 by U. Smilansky and S. Fishman. It was independently mentioned to me in 1992 by P. Koch. Short microwave pulses can also be generated directly by combining several phase-locked microwave fields as suggested some time ago in connection with kicked molecules (Blümel, Fishman and Smilansky (1986)).

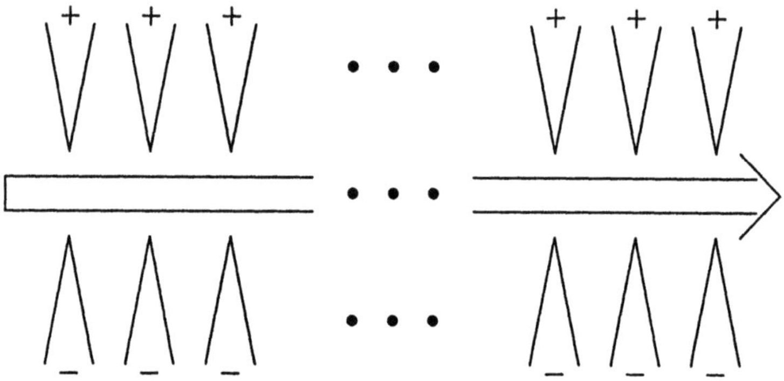

Fig. 14. Implementation of the kicked hydrogen atom as a fast beam of hydrogen Rydberg atoms traversing a regularly spaced grid of charged electrodes.

4 The Microwave Driven Hydrogen Atom: Perspectives

The continuously driven hydrogen atom is a much more complicated system than the kicked hydrogen atom discussed above. First of all it is a three-dimensional system. Second it has a mixed phase space which entails much more complicated diffusion properties than exhibited by the simple positively kicked hydrogen atom. Working with linearly polarized microwave fields it is always possible, at least in principle, to select a specific magnetic quantum number thus reducing the dimensions from three to two. For strong fields and short interaction times (< 1000 microwave cycles) the deviations from pure diffusion in the classical phase space are expected to be small. Since several laboratories are already set up to measure the quantum staying probability, i.e., the population of Rydberg atoms in the initially prepared state as a function of time, the investigation of a possible quantum enhancement of the staying probability and its destruction by symmetry breaking may be a promising subject for further experimental and theoretical investigations of dynamical localization.

This work was supported by Sonderforschungsbereich 276 of the Deutsche Forschungsgemeinschaft. I would like to thank L. Westling for stimulating discussions and valuable suggestions concerning format and contents of this paper.

References

Bayfield J.E., Koch P.M. (1974): Multiphoton Ionization of Highly Excited Hydrogen Atoms: Phys. Rev. Lett. **33**, 258–261

Blümel R., Smilansky U. (1984a): Quantum Mechanical Suppression of Classical Stochasticity in the Dynamics of Periodically Perturbed Surface-State-Electrons. Phys. Rev. Lett. **52**, 137–140

Blümel R., Smilansky U. (1984b): Suppression of Classical Stochasticity by Quantum Mechanical Effects in the Dynamics of Periodically Perturbed Surface-State Electrons. Phys. Rev. **A30**, 1040–1051

Blümel R., Fishman S., Smilansky U. (1986): Excitation of Molecular Rotation by Periodic Microwave Pulses. A Testing Ground for Anderson Localization. J. Chem. Phys. **84**, 2604–2614

Blümel R., Smilansky U. (1992): Symmetry Breaking and Localization in Quantum Chaotic Systems: Phys. Rev. Lett. **69**, 217–220

Blümel R. (1993): Exotic Fractals and Atomic Decay, in *Quantum Chaos*, ed. G. Casati, I. Guarneri and U. Smilanski (North-Holland, Amsterdam), 385–398

Casati G., Chirikov B.V., Izraelev F.M., Ford J. (1979): Stochastic Behavior of a Quantum Pendulum under a Periodic Perturbation, in *Stochastic Behavior in Classical and Quantum Hamiltonian Systems*, ed. G. Casati and J. Ford (Springer, New York)

Casati G., Chirikov B.V., Guarneri I., Shepelyansky D.L. (1987): Relevance of classical chaos in quantum mechanics: The hydrogen atom in a monochromatic field. Phys. Rep. **154**, 77–123

Casati G., Chirikov B. (editors) (1995): *Quantum Chaos* (Cambridge Univ. Press, Cambridge)

Chirikov B.V. (1979): A Universal Instability of Many-Dimensional Oscillator Systems. Phys. Rep. **52**, 263–379

Dittrich T., Smilansky U. (1991a): Spectral Properties of Systems with Dynamical Localization: I. The Local Spectrum. Nonlinearity **4**, 59–84

Dittrich T., Smilansky U. (1991b): Spectral Properties of Systems with Dynamical Localization: II. Finite Sample Approach. Nonlinearity **4**, 85–101

Fan W. (1994): Generalized Accelerator modes and their Application in Atomic Physics and Plasma Heating. (Thesis, University of Delaware)

Fishman S., Grempel D.R., Prange R.E. (1982): Chaos, Quantum Recurrences, and Anderson Localization. Phys. Rev. Lett. **49**, 509–512

Friedrich H., Wintgen D. (1989): The Hydrogen Atom in a Uniform Magnetic Field: An Example of Chaos. Phys. Rep. **183**, 37–79

Friedrich H. (1991): *Theoretical Atomic Physics.* (Springer, Berlin)

Gay J.-C. (editor) (1992): *Irregular Atomic Systems and Quantum Chaos* (Gordon and Breach, Philadelphia)

Hillermeier C.F., Blümel R., Smilansky U. (1992): Ionization of H Rydberg Atoms: Fractals and Powerlaw Decay. Phys. Rev. **A45**, 3486–3502

Izrailev F.M. (1990): Simple Models of Quantum Chaos: Spectrum and Eigenfunctions. Phys. Rep. **196**, 299–392

Jensen R.V., Susskind S.M., Sanders M.M. (1991): Chaotic Ionization of Highly Excited Hydrogen Atoms: Comparison of Classical and Quantum Theory with Experiment. Phys. Rep. **201**, 1–56

Koch P.M. (1992): Private communication.

Koch P.M., van Leeuwen K.A.H. (1995): The Importance of Resonances in Microwave "Ionization" of Excited Hydrogen Rydberg Atoms. Phys. Rep. **255**, 289–403

Main J. (1997): Semiclassical Approaches to Atoms in External Field. pp. 247 of this volume.

Moorman L., Koch P.M. (1992): Microwave Ionization of Rydberg Atoms, in *Quantum Non-Integrability*, edited by D. H. Feng and J.-M. Yuan (World Scientific, Singapore)

Smilansky U., Fishman S. (1989): Private communication.

Thaha M., Blümel R., Smilansky U. (1993): Symmetry Breaking and Localization in Quantum Chaotic Systems. Phys. Rev. **E48**, 1764–1781

Wintgen D. (1987): Connection between Long-Range Correlations in Quantum Spectra and Classical Periodic Orbits. Phys. Rev. Lett. **58**, 1589–1592

The Periodically Driven Excited Hydrogen Atom

Derek Richards

Mathematics Faculty, The Open University, Milton Keynes, UK.

Abstract. This article describes the effects of relatively strong, periodic perturbations on the classical and quantum dynamics of excited hydrogen atoms and the relation between these two dynamics. By considering a range of driving frequencies it is shown how the relative ratio of this to the unperturbed atomic frequencies affects the response and the relation between the classical and quantum dynamics. After providing an introduction to the system a brief overview of its behaviour in six different frequency regimes is given. This is followed by a detailed discussion of the response of a one-dimensional atom driven by a resonant field describing the subtle relations between the classical and quantum dynamics. This discussion is followed by a description of the more complex response of a three-dimensional hydrogen atom to a low frequency elliptically polarised field where, again, we demonstrate that the relative time scales of the system affect the dynamics in unexpected ways.

1 Introduction

In this article we discuss selected aspects of the effects of a strong periodic electric field on excited hydrogen atoms. The original impetus for these studies came from the experiment of Bayfield and Koch (1974) in which a beam of excited hydrogen atoms with initial principal quantum number $n_0 \simeq 66$ was passed through a 9.9 GHz electric field; the ionisation signal obtained was inconsistent with the then current theories. It is also interesting to note that these observations were accidental and made while studying a different problem; the article by Koch (1990a) provides an interesting short social history of this area of physics. Since then there have been many further experiments and numerous calculations, which have been reviewed in Casati (1987, 1988), Koch (1990a, b, 1992, 1995), Koch *et al.* 1992, Koch and van Leeuwen 1995 and Richards (1989, 1992b). In addition Bayfield and co-workers have performed experiments on one-dimensional hydrogen atoms, equation 3 below, Bayfield and Pinnaduwage (1985), Bayfield (1987) and Bayfield *et al.* (1995, 1996). As a consequence of this activity we now have a reasonable overall understanding of the behaviour of this type of system, although many unresolved problems remain; this behaviour is summarised in Sect. 2.

It is impossible to review all the relevant work in a short article, so I have chosen to provide a brief overall description of the dynamical behaviour in Sect. 2 and in Sects. 3 and 4 describe in some detail two aspects of the dynamics which arise as a consequence of the perturbing field being strong and nearly periodic, and the atom initially in an excited state. The aim is to provide the reader with some notion of the rich diversity of behaviour exhibited by such systems and to emphasise the importance of time scales.

It is first necessary, however, to explain why this system is worthy of study and then to describe the physical system, and equations of motion, and to consider the similarities and differences between this and other systems.

The periodically driven hydrogen atom is an interesting problem worthy of study for a variety of reasons. First, accurate experiments have been performed and these provide a serious challenge to theory, as it is necessary to develop accurate approximations to the solution of Schrödinger's equation which can cope with a large number of bound states and the continuum. Second, experiments can be performed over a sufficiently wide parameter range to produce a rich variety of dynamical behaviour which needs to be understood theoretically; this is distinct from the problem of obtaining accurate numerical solutions to the equations of motion. Third, the classical system is nonlinear and, for most frequencies, only chaotic orbits ionise: therefore there is intrinsic interest in the relation between the classical and quantal solutions besides the relation between these and experimental observations; an important and related problem is the development of semiclassical methods for relating the classical to the quantal solutions, for these time-dependent systems.

The real physical system can be considered as an excited hydrogen atom passing through a field fixed in the laboratory, so the general Hamiltonian describing the interaction of the atom with an elliptically polarised field is, [see Leopold and Richards (1991) for a derivation in the case of an linearly polarised field]

$$H_3 = \frac{1}{2\mu}\mathbf{P}^2 - \frac{e^2}{R} + \mathbf{R}\cdot\mathbf{F}_s + \frac{F\lambda(\tau)}{\mu\Omega}\left\{P_x \sin\Omega\tau + P_y \sin(\Omega\tau+\delta)\right\} \qquad (1)$$

where we take μ and e to be the mass and charge of the electron, $\lambda(\tau)$ is the field envelope produced by the passage of the atom through the oscillatory field, $\mathbf{F}_s$ represents a static field sometimes present, Ω is the field frequency and $\delta \in [0, \pi/2]$ a parameter which determines the field polarisation: if $\delta = 0$ it is linearly polarised with amplitude $F\sqrt{2}$ and if $\delta = \pi/2$ it is circularly polarised with amplitude F. The field envelope, λ, is zero for $\tau < 0$ and $\tau > T_m = 2\pi N_f/\Omega$, where N_f is generally between 100 and 300, and $\lambda(\tau)$ varies very slowly by comparison with the field oscillations, $\dot{\lambda}/\lambda \ll 2\pi/\Omega$. Until recently, all experiments on excited atoms used linearly polarised fields; experimental results using elliptically polarised fields have now been obtained, Bellermann *et al* (1996).

The Hamiltonian (1) may be put in the more familiar dipole form by making a time-dependent translation in momentum to obtain the new Hamiltonian

$$K_3 = \frac{1}{2\mu}\mathbf{P}^2 - \frac{e^2}{R} + \mathbf{R}\cdot\mathbf{F}_s - F\lambda(\tau)\left(X\cos\Omega\tau + Y\cos(\Omega\tau+\delta)\right)$$

$$-\frac{F\dot{\lambda}}{\Omega}\left(X\sin\Omega\tau + Y\sin(\Omega\tau+\delta)\right). \tag{2}$$

In this Hamiltonian the momentum $\mathbf{P}$ is not the same as in H_3. Generally $\dot{\lambda} \ll \Omega\lambda$ and the last term may be neglected. This dipole Hamiltonian is more useful at low frequencies and the original, equation (1), more useful when the scaled frequency, equation (7) below, is greater than unity.

In the experiments of Koch and co-workers the atoms are in an initial state with a known principal quantum number, n_0, with all sub-states approximately equally populated and usually $F_s = |\mathbf{F}_s| = 0$. In the experiments of Bayfield and co-workers $F_s \neq 0$ and the oscillatory field is linearly polarised, $\delta = 0$, in the direction of the static field. The atoms are initially in the extremal Stark state with known principal quantum number n_0 and $m = 0$ and $n_e = n_2 - n_1 = n_0 - 1$: since $n_0 \gg 1$ it is a good approximation to assume the electron to be confined to the field axis and the Hamiltonian which models this experiment is simpler, and can be written in the form

$$H_1 = \frac{1}{2\mu}P^2 - \frac{e^2}{Z} + ZF_s + \frac{FP}{\mu\Omega}\lambda(\tau)\sin\Omega\tau, \quad z \geq 0. \tag{3}$$

It is interesting to note that if the electron motion is initially in the direction of the field and if $F_s = 0$ then the motion does not remain aligned in the field direction; the one-dimensional character of the system is preserved only if F_s is greater than some critical value which depends upon F and Ω, Leopold and Richards (1987).

Experiments on other atoms have been performed. Gallagher and co-workers have studied, experimentally, the interaction of microwave fields with excited alkali atoms, Gatzke *et al.* (1994), Lee *et al.* (1992), Mahon *et al.* (1991) and Gallagher *et al.* (1989). The comparison of the 10% thresholds of alkali atoms measured by Gallagher *et al.* (1989) with the hydrogen thresholds, see Fig. 1, has produced some disagreement over the underlying physical process of the ionisation dynamics. The detailed comparison given by Koch and van Leeuwen (1995, Fig. 17), shows that the simple model proposed in Gallagher *et al.* (1989), on the basis of log-log plots, is misleading.

Arndt *et al.* (1993) have published experimental results for the microwave ionisation of rubidium and the effects of additional noise was described in Blümel *et al.* (1989, 1991); some aspects of the effects of noise on hydrogen atoms is discussed by Leopold and Richards (1991a).

The unperturbed, $F = F_s = 0$, bound classical motion is elliptical with frequency and semi-major axis

$$\omega = \frac{\mu e^4}{I_0^3}, \quad a = \frac{I_0^2}{\mu e^2} \tag{4}$$

where I_0 is the principal action variable corresponding to the principal quantum number, $I_0 = n_0\hbar$. The eccentricity, ϵ, of the initial ellipse depends upon the magnitude of the angular momentum, L, $\epsilon^2 = 1 - L^2/I_0^2$. The notion of mechanical similarity, Landau and Lifshitz (1965, Sect. 10), can be used to define new units, $t = \omega\tau$ and $Z = az$ etc, so that the Hamiltonians 1 and 3 can be written in the simpler form

$$H_1 = \frac{1}{2}p^2 - \frac{1}{z} + zF_{s0} + \frac{pF_0}{\Omega_0}\lambda(t)\sin\Omega_0 t, \quad z \geq 0 \tag{5}$$

$$H_3 = \frac{1}{2}\mathbf{p}^2 - \frac{1}{r} + \mathbf{r}\cdot\mathbf{F}_{s0} + \frac{F_0\lambda(t)}{\Omega_0}\{p_x \sin\Omega_0 t + p_y \sin(\Omega_0 t + \delta)\} \tag{6}$$

where scaled frequency and field are defined to be

$$\Omega_0 = \frac{\Omega}{\omega} = \frac{\Omega I_0^3}{\mu e^4} = \frac{\Omega}{\text{GHz}}(0.00533757\, n_0)^3, \tag{7}$$

$$F_0 = \frac{F I_0^4}{\mu^2 e^6} = \frac{F}{\text{V/cm}}(0.00373535\, n_0)^4. \tag{8}$$

In these units Schrödinger's equation becomes

$$\frac{i}{n_0}\frac{\partial\psi}{\partial t} = H(\hat{\mathbf{p}}, \mathbf{r}, t)\psi, \quad \hat{\mathbf{p}} = -\frac{i}{n_0}\frac{\partial}{\partial\mathbf{r}}, \tag{9}$$

that is $\hbar \to 1/n_0$. Henceforth we usually use scaled atomic units.

These scaled units are important for three reasons. First, in the circumstances of most experiments the ionisation probabilities computed using classical dynamics depend only upon Ω_0 and F_0, and F_{s0} if $F_s \neq 0$, and not the four variables (n, Ω, F, F_s) separately; departures from this scaling therefore provide a measure of the magnitude of quantal effects, Koch and van Leeuwen(1995) and Sauer *et al.* (1992a). Second, it is the scaled frequency Ω_0 which determines the character of the motion and the way ionisation occurs; this feature will be discussed further in Sect. 2. Third, the initial principal quantum number can vary from $n_0 \simeq 20$ to $n_0 \simeq 90$ and for any given scaled frequency the ionisation probability changes from 0 to 1 as the scaled field increases from about 0.05 to at most 0.3, whilst the field strength, expressed in real units, can vary by a factor of 2000, from $10^4\text{V cm}^{-1}$to 5V cm^{-1}, a change which is unwieldy and difficult to handle by comparison.

If $F_s = 0$ the hydrogen atom is special because the unperturbed system has only one non zero frequency, ω equation (4). This is a consequence of the $O4$ symmetry of the Coulomb potential and in quantum mechanics manifests itself in the dependence of the energy levels on only the principal quantum number; this means that there is only one scaled frequency, a fact that considerably simplifies our classification scheme of the motion, Sect. 2. Another important consequence of this seems to be that for linearly polarised fields, $\delta = 0$, for scaled frequencies $\Omega_0 < 2$ and for weak fields F, which produce less than 10% ionisation, a good understanding of the ionisation dynamics of the three dimensional motion can be obtained by studying the one-dimensional system described by H_1, indeed this system, with $F_s = 0$ was first introduced by Jensen (1982, 1984) in order to help understand the experiments of

Koch on three-dimensional atoms. The 10% threshold fields, defined in the Sect. 2, for one- and three-dimensional atoms are very close for $0 < \Omega_0 \leq 1$; a comparison is given in Jensen (1992, Fig. 3, page 135). With increasing Ω_0 departures from unity of the initial ellipse eccentricity become increasingly important and this simplification is lost, see Richards (1992b); for very large values of Ω_0 and strong fields there are profound differences between one- and many-dimensional hydrogen atoms, see Richards and Szeredi (1996a, b), which is a consequence of the Coulomb singularity.

Atoms having non-Coulombic cores, but with a single excited electron, are more difficult to treat theoretically, partly because one-dimensional models are not valid and partly because the departures from the hydrogenic behaviour is limited to low angular momentum states with the consequence that classical descriptions are suspect. The problems caused by such perturbations in relation to the Quadratic Zeeman Effect are discussed by Dando *et al.* (1994). With respect to the general problem of the accuracy of classical dynamics it is worth pointing out that the Coulomb potential is special because of the correspondence identities that exist, see Norcliffe and Percival (1968a, b) and Percival (1969). The causes of these identities are probably partly responsible for the excellent agreement which we sometimes find between the experimental and classical ionisation probabilities, see for instance Fig. 1 and the more detailed comparisons given in Rath and Richards (1996). However, despite being special it seems likely that it is necessary to understand the simpler Coulomb problem as a first step in understanding the more general case.

We end this introduction with a short discussion of the difficulties of computing the solution of either Hamilton's equations or Schrödinger's equation, starting with classical dynamics.

It is a relatively straightforward matter to numerically solve Hamilton's equations once the Coulomb singularity has been taken care of: the most elegant method of removing this is by moving into extended phase space, see Synge (1960), and then making a canonical transformation to a new time variable which 'regularises' the motion near the singularity — in the unperturbed limit, $F = F_s = 0$, this is the eccentric anomaly, see for example Born (1960, Sect. 22) or Landau and Lifshitz (1965, Sect. 15). The required transformations are different for one-dimensional atoms, (Leopold and Richards, 1985), two-dimensional atoms, and three-dimensional atoms (Rath and Richards 1988 and Cornish 1984). Ionisation probabilities can be computed using the Monte Carlo method described in, for instance Leopold and Percival (1979) and Percival and Richards (1975), provided care is taken with the definition of ionisation, see Rath and Richards (1996). In principle arbitrarily accurate ionisation probabilities can be computed but, in practice, the statistical errors decrease as, $N_o^{-1/2}$, where N_o is the number of orbits used. If a total of n_i orbits of the sample ionises then the Monte Carlo estimate of the ionisation probability is $P_i = n_i/N_o$ and there is a 68% chance of the correct result lying in the range $(P_i - \sigma, P_i + \sigma)$ and a 95% chance of it lying in the range $(P_i - 2\sigma, P_i + 2\sigma)$ where $\sigma = \sqrt{P_i(1 - P_i)/N_o}$.

Thus, the computation of classical observables is feasible on relatively small computers with the main limitation being computational time, particularly for low frequency fields: the essential step of understanding the results and the dynamical behaviour is, however, much more difficult and it is necessary to use a variety of techniques such as classical perturbation theory (not to be confused with quantal perturbation theory), resonance analysis, see Sect. 3 and averaging methods, Sect. 4.

In contrast finding accurate numerical solutions of Schrödinger's equation is normally far more difficult, because of the large number of basis states needed and the necessity to include, or at least make allowance for, the continuum. The quantal three-dimensional atom has been tackled by only a few brave workers, all of whom used a wave function expanded in a subset of a complete basis. Chu and Wang (1989) computed a time-dependent Floquet approximation, with $m = 0$ and no continuum. Floquet methods have also been used by Buchleitner *et al.* (1995), coupled with a complex dilation method to include the continuum; despite the magnitude of these calculations these workers were able to compare 10% thresholds with experiment. Similar Floquet computations have been made with circularly polarised fields, Buchleitner and Delande *et al.* (1993) and Zakrzewski *et al.* (1993). Casati *et al.* (1988) have also numerically solved the time-dependent Schrödinger equation for two-dimensional atoms using a basis of unperturbed bound states. The one-dimensional system is naturally more amenable to such expansion methods and various schemes have been deployed. Casati *et al.* (1987) and Susskind and Jensen (1988), following the work of Edmonds (1973) on the quadratic Zeeman effect, used a basis of Sturmian functions as these are complete and have a discrete spectrum. Susskind and Jensen (1988) overcome the problems produced by the continuum by putting the whole system in a finite box; both of these methods are unsatisfactory if $F_s \neq 0$ and the interaction time is long. A third method was introduced by Blümel and Smilansky (1987) who project out the continuum and approximate the resulting Green's function with decay factors; the particular approximation used to implement this method seems to be valid only for low scaled frequencies. Leopold and Richards (1991b) also introduced complex parts to the energy, but based on the short-time classical or semiclassical dynamics, which is known to be accurate; the approximations used in this method are valid only for $\Omega_0 > 1$, though this is not necessarily a limitation of the method.

The fact that it is difficult to find accurate solutions of Schrödinger's equation means that it important to understand the nature of the classical dynamics and its relation with the quantum dynamics, as this can provide indications of which quantal approximations will be useful and may suggest methods of approximating Schrödinger's equation.

In the next section we provide a description of how the system behaves as the scaled frequency changes from very small, where the perturbation is almost static, to very large where single photon ionisation is possible. The purpose of this section is to show the rich variety of behaviour such systems exhibit.

In the following two sections I describe in some detail the behaviour at two frequencies where the motion, classical and quantal, is affected by two different types of resonance.

2 An Overview of the Dynamical Behaviour

The scaled frequency is the single most important parameter of a periodically driven system. For a given value of Ω_0 the value of the field strength determines whether or not the system breaks up, or whether or not the system becomes chaotic, but it is the value of Ω_0 which determines how this occurs and consequently which approximation most efficiently describes the system. In this section we describe the six different types of behaviour encountered with strongly driven hydrogen atoms. The boundaries given are specific to hydrogen, but the dynamical behaviour described can occur for any strongly driven system having one dominant scaled frequency. As with all classifications of this type the actual boundaries are not sharply defined and the system moves gradually from one type of behaviour to another showing characteristics of both at the boundaries. In this section we assume the field to be linearly polarised although much of the discussion applies to elliptically polarised fields, particularly in regime III and higher frequencies.

2.1 Regime I: Low Frequencies; Dynamical Tunnelling

For low frequencies the perturbation varies sufficiently slowly that the system adjusts to it adiabatically; the atom responds to the periodic field as if the field were a very slowly varying static field, so all efficient approximations start with the static field limit: for hydrogen atoms and for linearly polarised fields this is the separable Stark Hamiltonian,

$$H_s = \frac{\mathbf{p}^2}{2\mu} - \frac{e^2}{r} + zf. \tag{10}$$

The classical and quantal consequences of the slow variation are, however, usually different for quantum numbers currently accessible, so it is better to discuss the two dynamics separately.

In classical dynamics and in the static field limit the action variables can be taken to be (I_n, I_e, I_m) where I_e is related to the more usual action variables I_1 and I_2 obtained by separation of variables, see Born (1960) and Damburg and Kolosov 1983), by $I_e = I_2 - I_1$; a geometric interpretation of I_e is given after equation (22) below. When the field f in the Hamiltonian

H_s is allowed to vary slowly these actions are adiabatic invariants, that is they remain approximate constants of the motion; they are *not* constants of the motion, as is sometimes stated, but oscillate with the frequency of the unperturbed system and with amplitude proportional to $\dot{f}$.

If we assume that the actions are adiabatic invariants then it is a relatively trivial matter to compute the ionisation probability because there exists a critical field, $F_{\text{crit}}(I_n, I_e, I_m)$ such that if $f < F_{\text{crit}}(I_n, I_e, I_m)$ then an invariant torus with these actions exists but if $f > F_{\text{crit}}(I_n, I_e, I_m)$ it does not and the motion is unbounded. A good approximation to $F_{\text{crit}}(I_n, I_e, I_m)$ is provided by Banks and Leopold (1978).

If f increases slowly and monotonically with time then at the critical time t_I where $f(t_I) = F_{\text{crit}}(I_n, I_e, I_m)$ all orbits with the actions (I_n, I_e, I_m) ionise; thus, given any distribution of actions we can compute the adiabatic ionisation probability as a function of the field strength.

If f oscillates with a slowly increasing amplitude, as in the Hamiltonian (6), then we need to allow for the fact that changing the field direction is equivalent to changing the quantisation axis so the change in sign of f is accounted for by changing the signs of I_e and I_m, as these quantum numbers are projections of the Runge-Lenz vector, $\mathbf{A}$ equation (22), and the angular momentum vector $\mathbf{L}$, on to the field direction. With this minor adjustment we can compute the adiabatic ionisation probability.

It is very difficult to determine a priori an accurate estimate of when approximations based upon the notion of adiabatic invariance break down, and normally we are reduced to making comparisons of the adiabatic ionisation probability with exact numerical calculations.

For low frequency linearly polarised fields we find (Rath, private communication), see also Rath and Richards (1996), that the convergence to the static field limit is very slow and even for $\Omega_0 \sim 0.01$ the true ionisation probability (as computed using Monte Carlo methods) can exceed the adiabatic limit by over 15% and only for $\Omega_0 = 0.002$ do the two calculations agree to within the statistical errors. On the other hand, for circularly polarised fields, for reasons to be discussed in Sect. 4, we find good agreement for $\Omega_0 \sim 0.1$; an example is shown in Fig. 8. A detailed comparison of the ionisation process for one-dimensional systems in low frequency fields is provided by Richards (1996a).

Now consider the quantal system. The classical critical field F_{crit} also plays a role in the quantum dynamics. If f is constant and $f < F_{\text{crit}}(I_n, I_e, I_m)$ then a quasi-bound state exists, but it has a finite life-time; let such a state be $|\mathbf{n}, f\rangle \exp(iE_{\mathbf{n}}(f)t/\hbar)$. If f varies slowly we can simply replace f by $f(t)$ to obtain a time time-dependent adiabatic state

$$|\mathbf{n}, f(t)\rangle \exp\left(\frac{i}{\hbar}\int_0^t dt E_{\mathbf{n}}(f(t))\right),$$

which is an approximate solution of the time-dependent Schrödinger equation, coinciding with a Stark-state if f is constant.

Initially suppose the system is in a particular eigenstate, which, because $\lambda = 0$, coincides with an adiabatic state. As λ increases there will be transitions to other adiabatic states, but the coupling to these other states is proportional to Ω_0, see Richards (1987) and Richards *et al.* (1989a), so normally if Ω_0 is sufficiently small the system remains in an adiabatic state with a high probability.

In this quantal adiabatic limit ionisation will occur by tunnelling if $\lambda(t)F < F_{\mathrm{crit}}$ or by escape over the barrier if $\lambda(t)F > F_{\mathrm{crit}}$; the relative effectiveness of these two mechanisms depends on the principle quantum number (which determines the tunnelling rate) and the time taken for the value of $\lambda(t)F$ to reach F_{crit}; see Richards (1996a) for a discussion of the effects of the interaction time on the ionisation of ground state, one-dimensional atoms. For excited three-dimensional atoms Sauer *et al.* (1992b) have shown, by comparing experimental results with semi-classical calculations, assuming no transitions between adiabatic states, that if $n_0\Omega_0 < 1$ then the assumption of quantal adiabatic invariance leads to good agreement. Note that this criterion depends upon n_0, or in real units $\hbar$, besides Ω_0, as should be expected; the classical criterion for adiabatic invariance depends only upon Ω_0.

The condition $n_0\Omega_0 < 1$ is obtained by considering the magnitude of the coupling between the adiabatic states produced by the slow variation of the field. If the scaled frequency is too large then quantal adiabatic invariance breaks down; near this limit only a few adiabatic states are involved in the dynamics, but excitation to higher adiabatic states inevitably leads to increased ionisation probabilities as these states have both larger tunnelling rates and smaller values of F_{crit}. Thus increasing Ω_0 beyond $1/n_0$ leads to our second regime in which several, but not too many, adiabatic states are involved in the dynamics. Conversely, de-excitation to lower adiabatic states will decrease the ionisation probability.

There is, however, another method of enhancing ionisation by exciting higher adiabatic states even for very low frequencies. If $\Omega_0 \sim 1/p$, for some large integer p then there are resonance transitions between the initial adiabatic state and the adjacent excited state which enhances ionisation. These resonances exist only over a very narrow frequency band, $\Delta\Omega_0 \sim \Omega_0^{5/2}e^{-C/\Omega_0}$ for some positive constant C, Dando and Richards (1993), but their existence shows that the passage to the static field limit is not trivial, as at an infinite set of frequencies near $\Omega_0 = 0$ adiabatic invariance breaks down.

The resonances near $\Omega_0 = 1/p$, ($p \gg 1$), also affect the classical dynamics for long enough interaction times by creating a p-component island chain; however, the area of the these islands decreases exponentially with increasing p so they have no effect on the quantum dynamics for the quantum numbers accessible to current experiments.

2.2 Regime II: Low Frequencies, $n_0\Omega_0 > 1$

In this region several, but not too many, adiabatic states are needed to describe the quantal dynamics. The important point is that because relatively few states are involved the behaviour is not well approximated by classical dynamics. Moreover the resonances mentioned at the end of the previous section are broader and hence more significant.

An extensive comparison of experimental ionisation probabilities, classical Monte Carlo simulations and quantal calculations on a one-dimensional model have been published by Richards *et al.* (1989a) and show that resonances can increase the ionisation probability via excitation to higher adiabatic states or decrease it by de-excitation. An alternative theory explaining these features in terms of avoided crossings of many Floquet states, is due to Blümel and Smilansky (1987).

2.3 Regime III: Near Classical, $\Omega_0 < 1$

In this region the ionisation probability of hydrogen atoms is given accurately by classical dynamics, although there are isolated exceptions where it fails, see Fig. 1. There are no precise formulae for the boundaries of this region, but the quantum dynamics is characterised by the fact that there is no representation in which the wave function can be accurately approximated using a few states. The lower frequency boundary is clearly n_0 dependent: comparisons of classical simulations with experiments on three-dimensional hydrogen atoms, for n_0 in the range 30 to 90, suggest that for $0.1 < \Omega_0 < 1.2$ classical dynamics is accurate.

An overall picture of this agreement is obtained by comparing the classical and experimental threshold fields $F_0(P_i)$ which is the field at which the ionisation probability is P_i; in this regime the ionisation probabilities, if larger than 0.05, increase monotonically with F_0 so there is no ambiguity in the definition on $F_0(P_i)$ for $P_i > 0.05$.

In Fig. 1 we compare the experimental thresholds, the dashed lines, with those obtained from classical simulations, the solid lines, for $P_i = 0.1$ and 0.9 for a $\Omega/2\pi = 9.91\,\text{GHz}$ field, for which $\Omega_0 = (0.011465\,n_0)^3$; to aid comparison we have joined the discrete set of points at $n_0 = 32, 33, \cdots, 90$ by a continuous line.

For $0.05 \le \Omega_0 \le 0.8$ the agreement is fairly impressive except at the resonant frequencies $\Omega_0 = \frac{2}{3}, \frac{1}{2}, \frac{1}{3}, \frac{1}{4}$ etc, where there are definite suggestions of quantal effects as would be expected from the resonance analysis given in Sect. 3.

For $\Omega_0 > 0.85$ there appears to be significant disagreement, but here this particular experiment was on the edge of its working range and these results are not necessarily reliable; see Koch and van Leeuwen (1995, page 351) for a detailed discussion. Later experiments, Galvez *et al.* (1988), show much

Fig. 1. Comparison of the experimental and classical threshold fields for $P_i = 0.1$ and 0.9, as a function of the scaled frequency, Ω_0. The dashed lines represent the experimental thresholds, the solid lines the classical thresholds. In the experiments the microwave cavity frequency was 9,923 GHz and the total interaction time was about 300 field periods with a roughly 60 period switch on and off.

closer agreement for the 10% threshold up to $\Omega_0 = 1$, after which the classical thresholds are generally smaller. More recent results, Bellermann *et al.* (1996), show good agreement of the 10% and the 50% thresholds, for linearly, elliptically and circularly polarised fields for $\Omega_0 < 1.1$. At present the reasons for the differences at $\Omega_0 \sim 1$ are not known, but it is interesting to note that the field envelope of the 1988 and 1996 experiments were different and the interaction times of the latter were shorter.

In this frequency regime classical ionisation occurs via irregular orbits so the agreement seen in Fig. 1 is evidence that a strongly driven quantal systems can mimic chaotic dynamics. Further evidence of classical behaviour is obtained by comparing experimental results at different combinations of (n_0, Ω, F), remember that here $F_s = 0$, which give the same values of (Ω_0, F_0), as the classical dynamics depends only upon the scaled variables; such a comparison is given in Koch and van Leeuven (1995, Fig. 1, page 306) at $\Omega_0 = 0.44$.

It must be remembered that all the above comparisons involve an average over all initial substates of the atom which is undoubtedly advantageous for classical dynamics. It would be interesting to make the same comparison for the one-dimensional atom, with the Hamiltonian (5), in order to determine the effect of this averaging. Further, it would also be interesting, for reasons to be discussed in the section on region V, to compare the final bound state probabilities in both one- and three-dimensional cases.

2.4 Regime IV: Transition Region $1 < \Omega_0 < 2$

At the lower end of this region, $\Omega_0 \sim 1$, the classical and quantal ionisation probabilities agree reasonably well, particularly for three-dimensional atoms. The classical resonance island plays a dominant role in the quantum dynamics, though the precise connection between the classical and quantum dynamics in this region is quite subtle and will be described in more detail in Sect. 3.

At the upper end of the region, $\Omega_0 \sim 2$, the behaviour of the two dynamics is distinctly different for the quantum numbers currently accessible, that is n_0 up to 100. The reason for the differences is discussed below.

The dominant feature of the classical dynamics in this region is the resonant island around $\Omega_0 = 1$. This is a region of comparative stability in both classical and quantum dynamics and produces the maximum in $F_0(0.1)$ and $F_0(0.9)$ in Fig. 1. This island surrounds a stable periodic orbit and is bounded in phase space by an unstable manifold which affects the ionisation probabilities of the classical system at $\Omega_0 \simeq 0.8$ and the quantal system here and at $\Omega_0 \simeq 1.3$ for reasons that will be discussed in Sect. 3.

There is also a resonant island around $\Omega_0 = 2$ near which the classical dynamics is similar to that in the region around $\Omega_0 = 1$, provided the field is sufficiently weak. But it is not clear how large the quantum numbers need to be in order that this island affects the quantum dynamics as significantly as the $\Omega_0 = 1$ island.

2.5 Regime V: High Frequencies: $\Omega_0 > 2$

If $\Omega_0 > 2$ the classical and the quantal dynamics are quite different because of interference effects. Consider first the classical dynamics.

A high frequency field will cause the electron to oscillate as it moves along its unperturbed orbit. During one field oscillation the electron moves a small distance along its orbit and if the radius of curvature of the orbit is large the electron behaves as if on a rectilinear path and there is almost no net energy transfer. The dominant energy exchange between the electron and the field occurs when the orbit is least rectilinear, that is when the acceleration due to the binding potential is largest; for the Coulomb potential this occurs near the perihelion. Thus at every orbital period, as the electron passes the perihelion, it undergoes an impulsive change to its mean energy the value of which depends upon the relative phase of the field and the orbital motion, Jensen *et al.* (1988) and Richards *et al.* (1989b). It is the manner in which this sequence of impulses is combined which produces the difference between the classical and the quantal motion.

For the classical motion each impulse, if sufficiently strong, may be considered independent of the previous impulse and hence random and then the electron executes a random walk in the action space. This random walk can be approximated by the diffusion equation, Casati *et al.* (1987)), Jensen *et al.* (1989). The mathematics behind this physical description was first developed by Casati *et al.* (1984), but see also the articles by Chirikov (1991, b) and Dando and Richards (1990) for the generalisation of arbitrary one-dimensional systems.

The quantum dynamics is different only because there is a phase associated with the classical path, and the resultant interference between successive passages past the nucleus means that the events cannot always be considered independent. One consequence of these interference effects is that the selection rule induced by the periodicity of the Hamiltonian becomes important and the quantum dynamics is dominated by those states which are separated by the energy splitting $\Omega\hbar$, Jensen *et al.* (1988) and Leopold and Richards (1989, in particular Figs. 4 and 5). For instance if $\Omega_0 = 4.8328$ and $n_0 = 98$ the excited quasi-resonant states most involved in the quantum dynamics are $n = 103$, 109, 116, 125, 136, 151, 173, 208, 279, 623 although, in practice states adjacent to these are also significantly populated, see Leopold and Richards (1990). This breakdown of classical dynamics was first predicted by Casati *et al.* (1984) and was confirmed experimentally by Galvez *et al.* (1988) and Bayfield *et al.* (1989). As the principal quantum number n_0 increases – or equivalently $\hbar$ decreases, equation (9) – these interference effects become relatively less significant and the classical diffusive motion is regained. Precisely how classical dynamics is regained with decreasing $\hbar$ and fixed (Ω_0, F_0) is not clear. The localisation theory of Casati *et al.* (1984) does not seem to be consistent with numerical calculations, Leopold and Richards (1991b), and the comparisons of this theory with experiment have been severely criticised by Koch *et al.* (1990). An alternative approach of Jensen *et al.* (1989) looks promising, but by assuming the validity of the Master equation at the outset ignores the interference effects, the role of which one is trying to understand.

The addition of noise destroys these interference effects and, for Ω_0 not too large, forces the system to behave classically, Leopold and Richards (1991a) and Sirko *et al.* (1993).

One might also think that if $\Omega_0 \gg 1$ and F_0 is large the classical limit would be regained as n_0 increases. This limit, however, is rather subtle and a recent examination of one- and two-dimensional atoms, Richards and Szeredi (1995, 1996a, b), shows profound differences between these two cases. In short, for the one-dimensional case interferences dominate and make the passage to the classical limit very slow; even when a band of initial and final energies is averaged over there are noticeable differences for $n_0 = 1000$ when $F_0 = 1$ and $\Omega_0 = 15$. On the other hand the spatial average involved for two-dimensional atoms smooths these effects and for the same scaled field and frequency classical and quantum dynamics are similar for $n_0 = 80$, without an average over initial and final energies.

Finally, it is worth noting that if there are only a few quasi-resonant states then the quantum dynamics for the one-dimensional atom can be accurately approximated using only a small basis comprising the quasi-resonant states and a few adjacent states, which are essential, Leopold and Richards (1990). Such a method would probably be a good approximation for three-dimensional atoms for these frequencies.

2.6 Regime VI: The Photoelectric Effect: $\Omega_0 > n_0/2$

In this regime the photon has sufficient energy to directly ionise the atom and all vestiges of classical motion have disappeared, except possibly for very strong fields; the work of Richards and Szeredi (1996a) suggests that for sufficiently strong fields classical behaviour may predominate.

3 Resonance Island Effects

In this section we consider a one-dimensional hydrogen atom, with Hamiltonian (5), driven by a strong, near resonant frequency force, so the scaled frequency is close to unity; such a system has been studied experimentally and theoretically by Bayfield *et al.* (1996b). Since the unperturbed energy levels normally change relatively slowly with quantum number it follows from Bohr's correspondence principle,

$$\omega(I)\hbar \simeq E_{n+1} - E_n\,, \quad I = n\hbar\,,$$

that there is a band of unperturbed states for which the field frequency is close to the energy splitting, giving many strongly coupled unperturbed states. Under these circumstances a classical description is particularly helpful, even though the classical and quantal observables sometimes behave quite differently.

For the case we consider the static field is non zero so the unperturbed frequency is given by a modification of equation (4), Richards (1996b),

$$\omega_1(I) = \frac{\mu e^4}{I^3}\frac{1}{1 - 3F_{s0}(I/I_0)^4 - 8.124F_{s0}^2(I/I_0)^8} + O(F_{s0}^3), \tag{11}$$

so the scaled frequency for this system is, approximately,

$$\Omega_1 = \frac{\Omega_0}{1 - 3F_{s0} - 8.124F_{s0}^2}. \tag{12}$$

We start by showing the graphs of the classical, the dashed line, and quantal ionisation probabilities, P_i, over a range of scaled frequencies around unity (see Fig. 2). For this comparison we take $n_0 = 50$ and there is a static field present, $F_s = 16.384\mathrm{V\,cm^{-1}}$, so the scaled frequency Ω_1 is related to Ω_0, defined in equation (12) by $\Omega_1 = 1.067\Omega_0$. In addition, as P_i decreases with increasing Ω_1 and increases with increasing field strength the scaled field at each frequency is taken to be $F_0 = 0.04\Omega_0^{4/3} = 0.0367\Omega_1^{4/3}$ in order to emphasis the changes in P_i due to the dynamics rather than the increasing frequency.

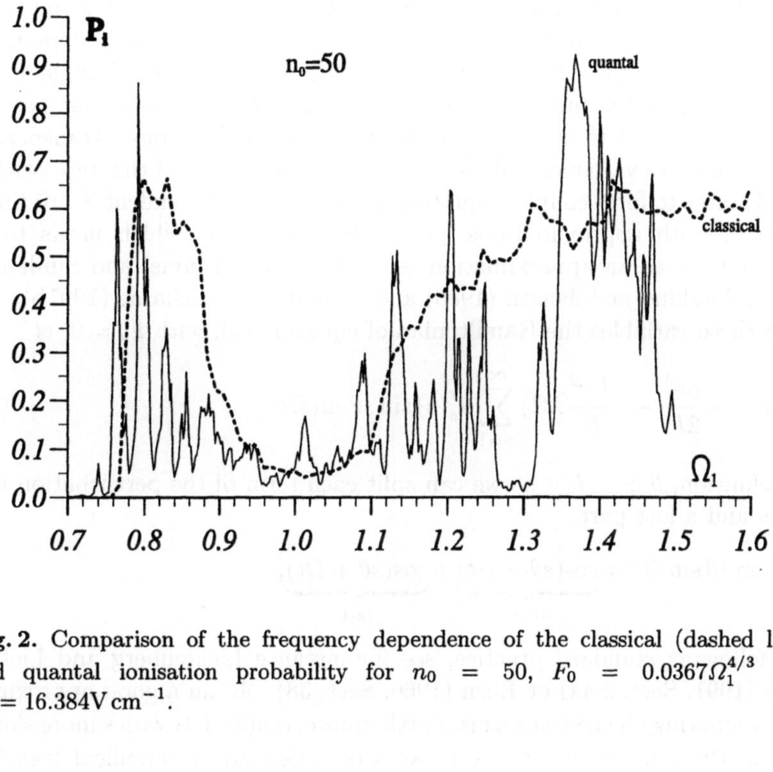

Fig. 2. Comparison of the frequency dependence of the classical (dashed line) and quantal ionisation probability for $n_0 = 50$, $F_0 = 0.0367\Omega_1^{4/3}$ and $F_s = 16.384\mathrm{V\,cm^{-1}}$.

The following points should be noted.

1) Both classical and quantal ionisation probabilities increase dramatically at $\Omega_1 \simeq \Omega_1^- = 0.742$.
2) For $\Omega_1 > \Omega_1^-$ the quantal probabilities oscillate quite dramatically. Although the variation in individual cases may appear regular the patterns change significantly with n_0 and F_{s0}.
3) The ionisation probabilities have broad local minimum at the resonance frequency $\Omega_1 = 1$.

4) As $\Omega_1 \to 1$ the quantal ionisation probabilities fluctuate less and agreement with the classical probabilities improves.
5) For $\Omega_1 \simeq 1.3$ there is a broad minimum in the quantal ionisation probability not present in the classical probability. For larger frequencies, $\Omega_1 > 1.5$, the classical probability is larger and the difference between the two increases as Ω_1 increases. This is an example of the breakdown of classical dynamics in Regime IV described in the previous section.

The slowly varying field envelope plays a crucial role in this dynamics as it can be regarded as preparing the system in a state from which ionisation occurs. Small variations in the driving frequency can change the nature of this state and/or the manner in which it interacts with other states and hence the ionisation probability. Much of this behaviour is easier to understand if we first examine the classical dynamics using the unperturbed angle-action variables, (θ, I); the full details are given in Richards (1996b), but here we set the static field to zero in order to make the analysis more transparent. The use of these variables is directly equivalent to the use of the unperturbed bound states to form coupled equations from the time-dependent Schrödinger equation; both approximations ignore the continuum, which needs to be re-introduced as an approximation, see for example Blümel and Smilansky (1987), Susskind and Jensen (1988) and Leopold and Richards (1991b).

In these variables the Hamiltonian of equation (5), with $F_s = 0$, is

$$H = -\frac{\mu e^4}{2I^2} + \frac{2Fe^2}{\Omega I}\lambda(t)\sum_{s=1}^{\infty} J_s'(s)\sin s\theta \sin \Omega t. \tag{13}$$

By definition, $\theta \simeq \omega(I_0)t$ so we can split each term of the perturbation into a slow and a fast part,

$$2\sin s\theta \sin \Omega t = \underbrace{\cos(s\theta - \Omega t)}_{\text{slow}} + \underbrace{\cos(s\theta + \Omega t)}_{\text{fast}},$$

and, following standard practice, see for instance Lichtenberg and Liebermann (1991, Sect. 2.4a) or Born (1960, Sect. 38), obtain a good approximation by ignoring all the fast terms. Furthermore, $\cos(\theta - \Omega t)$ varies more slowly than all the other terms, so we make a time-dependent canonical transformation $\phi = \theta - \Omega t$ to obtain the new Hamiltonian $K = K_p + K_1$.

$$K_p = E_0\left\{-\frac{I_0^2}{2I^2} - \Omega_0\frac{I}{I_0} + \frac{\alpha F_0 I_0}{\Omega_0 I}\lambda(t)\cos\phi\right\} \quad \text{(pendulum)}, \tag{14}$$

$$K_1 = E_0\frac{F_0 I_0}{\Omega_0 I}\lambda(t)\sum_{s=2}^{\infty} J_s'(s)\cos(s\phi + (s-1)\Omega t) \quad \text{(perturbation)} \tag{15}$$

where $\alpha = J_1'(1) \simeq 0.325$ and $E_0 = \mu e^4/I_0^2$. In these equations I_0 is the initial value of the action I and Ω_0 is the scaled frequency $\Omega_0 = \Omega/\omega(I_0)$: it is convenient to keep both I_0 and Ω_0 as independent variables and to define the action of the resonant orbit to be $I_r = I_0 \Omega_0^{-1/3}$. The numerical results obtained using this Hamiltonian are very close to those obtained using the Hamiltonian (13), but it is far more efficient to use the latter as the fast terms in H dominate the numerical integration but have negligible physical effect. As an aside we note that, for numerical work, it is generally more efficient to use the exact Hamiltonian, equation (5), in regularised coordinates than to use angle-action variables.

The pendulum Hamiltonian K_p is now to be considered the unperturbed Hamiltonian, and it is this which determines most of the features seen in Fig. 2. The dominant role of the perturbation, K_1 is to cause transitions from the pendulum states of K_p to states of larger actions from which ionisation occurs.

The motion produced by Hamiltonians of this pendulum type is described in Percival and Richards (1982, Chap. 4) and also Leopold and Richards (1994). Briefly K_p has a stable fixed point at $\phi = 0$ and $I \simeq I_r = I_0 \Omega_0^{-1/3}$, corresponding to a stable periodic orbit in the original representation. There is also an unstable fixed point at $\phi = \pi$, and $I \simeq I_r$, through which the separatrix, $\mathcal{S}$, passes, see the dashed line in Fig. 3: the separatrix divides phase space into three invariant regions with the central region containing librational motion and the outer two rotational motion. If the area contained by $\mathcal{S}$ is sufficiently large the eigenvalues of K_p reflect this difference, see Figs. 4 to 6. The number of librational states inside the separatrix when $\lambda = 1$, that is its area divided by $2\pi\hbar$, is

$$N_L = \frac{0.84 I_0}{\hbar \Omega_0} \sqrt{F_0} \quad (F_s = 0). \tag{16}$$

The resonant island, that is the interior of the separatrix, is important as its position relative to the initial state is largely responsible for the fate of that state in both classical and quantum dynamics. So we consider the evolution of the phase curves of K_p as λ changes slowly from 0 to 1. Initially $\lambda = 0$, all phase curves are rotational and there is no resonance island; we need to know which phase curves remain rotational and which are captured into the resonance island as it grows with λ

If $I_0 > I_r + N_L\hbar/2$ or $I_0 < I_r - N_L\hbar/2$ then the principle of adiabatic invariance holds and the final phase curve will also be rotational curve with action close to that of the initial action – remember that here we are ignoring the effect of K_1.

But if $I_r - N_L\hbar/2 \le I_0 \le I_r + N_L\hbar/2$ the principle of adiabatic invariance breaks down as the frequency of a phase curve becomes zero when it crosses the separatrix, that is at the instant when the motion becomes trapped in the potential well. Consider an orbit, rather than a phase curve: as λ increases the maximum of the potential $\lambda \cos\phi$ slowly increases and at some point is large enough to trap the orbit in the potential well. Only for a small time interval will this orbit not behave either like a rotational or a librational orbit in a slowly varying potential. That is, for most of the time the principle of adiabatic invariance holds; it is invalid only during the relatively short interval during which the particular orbit is trapped so, to a good approximation, the initial and final action of orbits trapped in the potential well are related by $I_f \simeq \frac{1}{2} I_i$. Consequently the separatrix of K_p is populated from the two states with actions $I_0 = I_r \pm N_L \hbar/2$.

So far we have not considered the role of the perturbation, K_1, equation (15). A casual glance at this function suggests that it will become less effective as I increases; in fact it becomes *more* effective as I increases, because it is not the magnitude of the perturbation that is most significant, but its rate of change. Below the island the unperturbed period is short so K_1 varies relatively rapidly and has little effect: above the island the unperturbed period is longer and there are resonances at $I = m^{1/3} I_r$, $m = 1, 2, \cdots$. Thus the field is more efficiently dynamically coupled to the atom above than below the island.

The separatrix of K_p is unstable so any small perturbation will destroy it to produce a stable and an unstable manifold. Thus phase curves with initial action $I_0 < I_r - N_L \hbar/2$ are relatively robust by comparison with those of initial action $I_0 \simeq I_r - N_L \hbar/2$.

This behaviour can be seen by following the evolution of a phase curve initially coinciding with the line of constant action, corresponding to the initial state, as λ slowly increases from 0 to 1. This is achieved by integrating a large number of orbits, here 1000, each with initial action $I_k(0) = I_0$ but the initial angle variable is uniformly spread over $(0, 2\pi)$, and plotting the final points $(\phi_k(T_m/2), I_k(T_m/2))$.

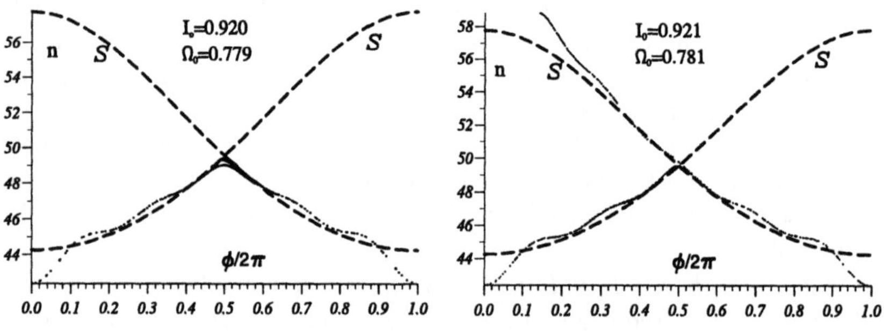

Fig. 3. Figure showing phase curves after their evolution to the peak field for the two different initial actions $I_0 = 0.920 I_r$ (LHS) and $I_0 = 0.921 I_r$ (RHS) with $I_r = 50\hbar$ ($I_r = I_0 \Omega_0^{-1/3}$); the vertical scale measures $I/\hbar$. The field strength is $F_0 \simeq 0.029$. The separatrix S of the pendulum Hamiltonian is depicted by the dashed line.

In Fig. 3 we show the result of two such calculations for $F_0 = 0.029$; for this demonstration we used the Hamiltonian defined in equations (14) and (15) but with the infinite sum truncated after five terms. On the left $I_0 = 0.920$ ($\Omega_0 = 0.779$) and the evolved phase curve coincides closely with the lower portion of $\mathcal{S}$, depicted by the dashed lines: further integration to $F = 0$ shows that all orbits collapse back to an action close to the initial action, $I_0 = 0.92$, as would be expected from the principle of adiabatic invariance. On the right $I_0 = 0.921$ ($\Omega_0 = 0.781$); again the phase curve is close to the separatrix but now some of it has wandered onto the upper half by passing through the unstable fixed point at $\phi = \pi$: in this case integration to $t = T_m$ leaves the final actions scattered over a wide range with a distinct peak near I_0 but having a long tail out to about $2I_0$.

For weaker fields the phase curve with initial action near $I_r - N_L\hbar/2$ will be wrapped round the separatrix during the slow switch-on of the field then as the field is switched off, there will be a tendency to produce a distribution of actions peaked at $I_r \pm N_L\hbar/2$. This behaviour is seen in both classical and quantal calculations and in the experiments of Bayfield (1995, Figs. 8 and 9).

Figure 3 shows that only a small change in I_0, or Ω_0, can completely change the nature of the classical motion. This behaviour is seen in the ionisation probabilities, Fig. 2 above, as a sharp increase as the frequency increases through a critical value $\Omega_1 = \Omega_1^-$ at which the equation $I_0 = I_r - N_L\hbar/2$ is satisfied; for the conditions of Fig. 2 this gives $\Omega_1^- = 0.742$, Richards(1996b).

This analysis explains the behaviour of the classical ionisation curve shown in Fig. 2, but it is not obviously related to the quantal ionisation curve. The connection is complex and can only be briefly summarised here: the full analysis is given in Richards (1996b).

In order to understand the quantum mechanics it is necessary to use a Floquet analysis on the full Hamiltonian, and the first important observation, which can be confirmed numerically, is that the exact quasi-energy levels $\mathcal{E}_k$ are similar to the energy levels of the pendulum Hamiltonian K_p, see Richards (1996b, Fig. 9). Second the field dependence of the quasi-energy levels belonging to the rotational and the librational states of K_p is different, see Fig. 4 below. Finally the perturbation K_1 mixes the states of K_p and causes avoided crossings in the Floquet states near those field strengths which satisfy $\mathcal{E}_n(\lambda F_0) - \mathcal{E}_m(\lambda F_0) = p\Omega\hbar$ for some integer p and at some time.

We need to understand the behaviour of the quasi-energy levels as a function of both the field strength and frequency and this is most conveniently achieved by plotting the locus of the complex numbers

$$z_k(F_0, \Omega_0) = (a + bF_0) \times \exp\left(i\frac{\pi}{2} - i\frac{\mathcal{E}_{n_0}(0)T}{\hbar}\right) \times \exp\left(i\frac{\mathcal{E}_k(F_r)T}{\hbar}\right) \quad (17)$$

in the complex plane, T being the field period. The last factor in this expression is the conjugate of an eigenvalue of the period operator, and lies on the unit circle. The first factor is simply a device for showing the variation with F_0; the actual values of a and b are of no consequence. The second factor just rotates the whole spectrum so that the quasi-energy associated with the initial state starts at $(0, i/2)$; this is helpful as there is a large change in the relative angles of the lines as the frequency varies and we wish to compare the results at different frequencies. In this diagram the lines are labelled by the quantum numbers of the unperturbed states. This representation of the quasi- energies is different from usual, see for instance Blümel and Smilansky (1987), Breuer *et al.* (1991) and Sirko and Koch (1995), but it exaggerates the changes of $\mathcal{E}_n$ with F_0.

As an example we show in Fig. 4 the eigenvalues of the K_p, for $n_0 = 20$, $\Omega_1 = 0.74$ and $0 \le F_0 \le 0.04$; this frequency is chosen as here the ionisation probability has a local maximum and the final population of excited states is largest for $n = 24$ and 31. These eigenvalues are qualitatively similar to those of Mathieu's equation as a careful comparison between the lines in this figure and Fig. 20.1 in Abramowitz and Stegun (1965, page 724) reveals.

The following points should be noted.

1) Each line rotates relatively little so the value of the energy changes little; for instance the initial state at $F_0 = 0$ has the value $\mathcal{E}_{20}(0) = 14.3 \times 2\pi$ but this rotates through a very small angle; the state $n = 22$ rotates through the largest angle, but even this is less than π. This behaviour is important and will be discussed next.
2) Most states are rotational and rotate anticlockwise. But the state $n = 22$ rotates clockwise and the adjacent states begin rotating anticlockwise but change direction at $F_0 = 0.005$ and $F_0 = 0.03$ respectively: the state $n = 20$ also changes direction at $F_0 = 0.035$.
   This behaviour is due to the resonance island. For very small values of the field the island is very small and hardly affects the eigenvalues; for this frequency the centre of the island is at $I_r = 22.6\hbar$ so the state $n = 22$ is 'captured' into the island almost immediately and it rotates in a different direction to the rotational states adjacent to it. When $F_0 \simeq 0.007$ the number of the states inside the island is about 2 and the state $n = 21$ changes direction. The initial state $n = 20$ changes its direction of rotation at $F_0 \simeq 0.035$, where $N_L = 4.5$ and the state $n = 23$ changes at $F_0 \simeq 0.025$ ($N_L \simeq 3.8$). Thus the shapes of the lines in this Floquet-plot reflect the classical resonant island structure.
3) The change in the argument of $z_{n_0} - z_{n_0+\Delta n}$ at $F_s = F_r = 0$ as the frequency increases from Ω_0 to $\Omega_0 + \Delta$ is about $2\pi \Delta n \Omega_0^{-2} \Delta$. Thus the order of the lines in the diagram changes quite rapidly with changing frequency and as Ω_0 increases many lines intersect the line associated with the initial state. This is the key to understanding some aspects of the behaviour seen in Fig. 2.

In Fig. 5 we show another Floquet plot for $\Omega_1 = 0.73$ and for the full Hamiltonian, but otherwise for the same parameters used in Fig. 4. In this example the dominant crossing occurs with the $n = 30$ Floquet state. Note

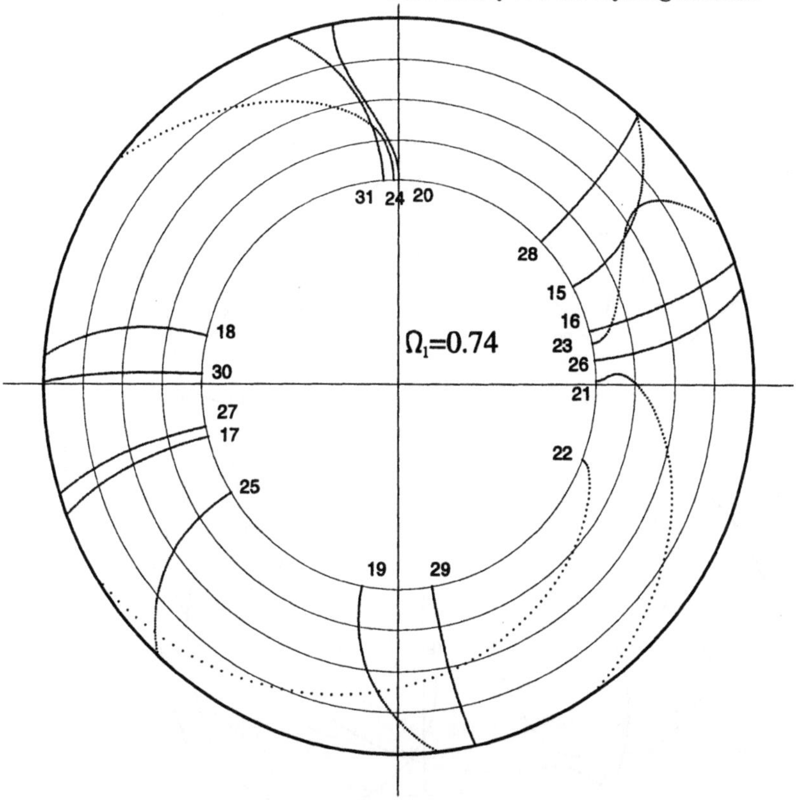

Fig. 4. Diagram showing the variation of the quasi-energy levels $\mathcal{E}_k$ of the pendulum Hamiltonian K_p, discussed in the text, through the variation of z_k, defined in equation (17), with F_0 for $F_s = 640\mathrm{V\,cm}^{-1}$, $\Omega_1 = 0.74$ and $n_0 = 20$. The inner circle is at $F_0 = 0$ and the rings are at intervals of $\Delta F_0 = 0.01$, so the outer ring is at $F_0 = 0.04$.

also that the relative ordering of the states is quite different from that shown in Fig. 4, even though the frequency is only a little different.

From the data shown in these figures, and other data not shown here, we conclude that the local maxima in the quantal ionisation probabilities seen in Fig. 2 are produced by avoided crossings of Floquet states and that these maxima are distinct for $\Omega_1 \simeq \Omega_1^-$, because the quasi-energies associated with the initial states, $\mathcal{E}_{n_0}$, do not change much as F_0 increases so the initial state is coupled by the perturbation either to only one excited state or none, depending upon the frequency. As the principal quantum number increases the density of lines in this Floquet plot increases and the number of maxima in P_i also increases.

In Fig. 4 we also notice that the quasi-energy associated with the state $n = 23$ rotates much further than the $n = 20$ state, suggesting that the

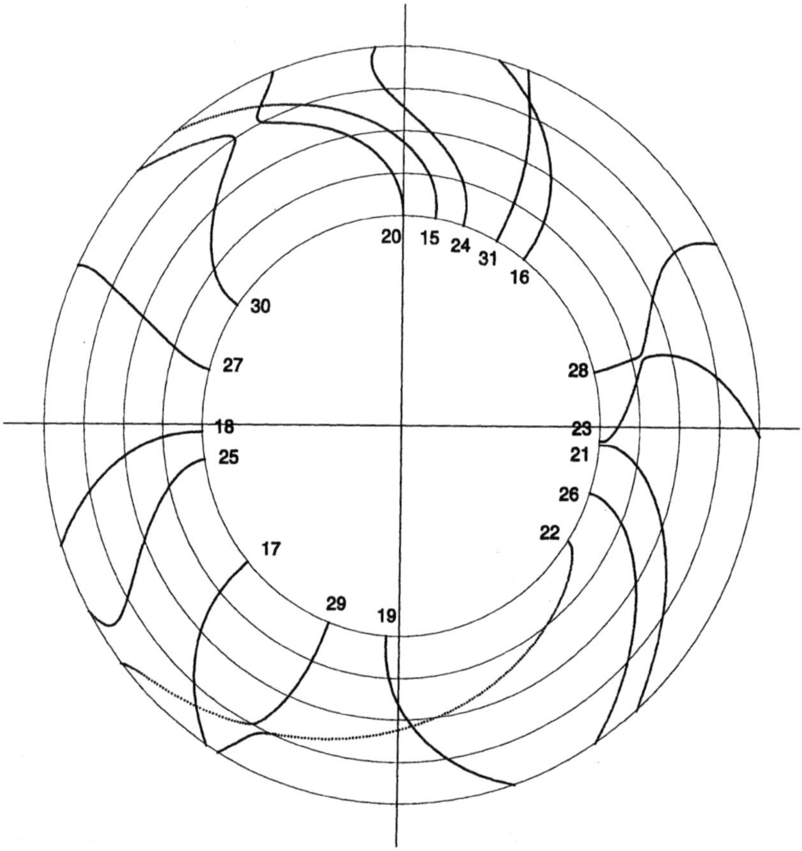

Fig. 5. Diagram showing the variation of the quasi-energy levels $\mathcal{E}_k$ of the full Hamiltonian, equation 10, through the variation of z_k, defined in equation 17, with F_0 for $F_s = 640\mathrm{V\,cm}^{-1}$, $\Omega_1 = 0.73$ and $n_0 = 20$. The inner circle is at $F_0 = 0$ and the rings are at intervals of $\Delta F_0 = 0.01$, so the outer ring is at $F_0 = 0.04$.

ionisation probability for $\Omega_1 \sim 1$ will behave differently. In Fig. 6 we show the Floquet plot at the frequency $\Omega_1 = 0.988$, ($\Omega_0 = 0.874$), for $n_0 = 50$ as this diagram shows clearly the differences in the dynamics for a state switched onto the island centre, $\Omega_1 \sim 1$, and one switched to near the separatrix, $\Omega_1 \sim \Omega_1^-$.

The important difference between this diagram and that of Figs. 4 and 5, for which $\Omega_1 \simeq \Omega_1^-$, is that now the initial Floquet state is captured into the island almost immediately and hence rotates a long way for a small increase in F_0, so crossing many excited states: thus there are now many narrow avoided crossings at relatively weak fields. For $\Omega_1 \sim \Omega_1^-$ avoided crossings are rare and each has a more distinct effect. The effect of this difference can also be

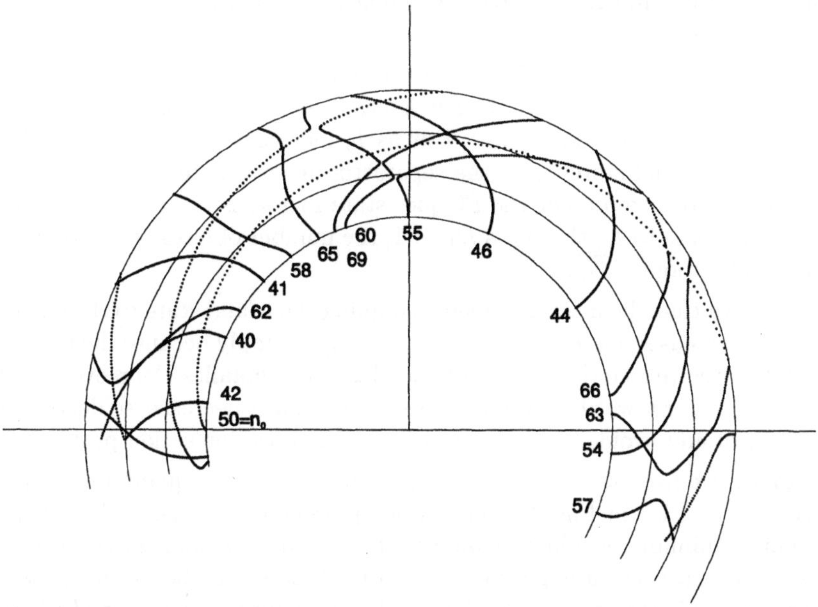

Fig. 6. A Floquet plot for $n_0 = 50$, $F_s = 29\mathrm{V\,cm}^{-1}$, $\Omega_0 = 0.8737$ ($\Omega_1 = 0.988$) using the basis [39, 69]. In this figure the initial state starts on the extreme left and rotates clockwise, though about π as F_0 increases from 0 to 0.03. As in Fig. 4 the rings are at $F_0 = 0.01$ intervals.

seen in the final state populations; for $\Omega_1 \simeq \Omega_1^-$ only a few selected excited states are populated and for $\Omega_1 \simeq 1$ many states are populated.

The difference in the behaviour of P_i for $\Omega_1 \simeq \Omega_1^-$ and $\Omega_1 \simeq 1$ is thus entirely due to the different behaviour of the Floquet states which, in turn, reflects the classical island resonance.

We have now arrived at a paradox. This analysis has shown that the quantal ionisation probability for $\Omega_1 \simeq \Omega_1^-$ is dominated by only two Floquet states, one pendulum state and one excited state. We have also shown, numerically in Fig. 2, that the classical and quantal ionisation probabilities both increase sharply at $\Omega_1 \simeq \Omega_1^-$; but when the quantum dynamics is dominated by only two states it is not normal for it to behave like the equivalent classical dynamics. In order to resolve this paradox it is necessary to consider the quantum dynamics near an avoided crossing. The details are given in Richards (1996b), where we find that the transition probability depends exponentially upon a matrix element between an excited state and a pendulum state.

Now any excited Floquet state must be strongly perturbed by the field, so will be represented by a sum of many basis states $\exp(ik\phi)/\sqrt{2\pi}$; the number of states in this sum, with a significant population, can be shown to be proportional to n_0.

Conversely if $I_0 < I_r - N_L\hbar/2$ and the initial state is switched on to a lower rotational state of K_p, then it is easily seen that this state is represented by only a few basis states. But as I_0 increases through $I_r - N_L\hbar/2$, and the librational states near the separatrix are populated, the number of states required to represent the Floquet state increases to about $3N_L$; as I_0 further increases to I_r, that is as $\Omega_1 \to 1$, the number of basis states required decreases to about $\sqrt{N_L}$.

It follows that the matrix elements coupling the island state to the excited state increases as I_0 increases through $I_r - N_L\hbar/2$, simply because the relevant quantal states reflect the nature of the classical dynamics. Moreover, as I_0 increases further towards I_r the matrix elements must decrease, so decreasing the effect of any avoided crossing and hence the total ionisation probability.

Finally, we need to explain the broad minimum in the quantal ionisation curve at $\Omega_1 \simeq 1.3$ in Fig. 2, which is not present in the classical probability. This minimum — which manifests itself as a maximum in the threshold curves measured in experiments — was observed in the experiments of Galvez *et al.* (1988) and first explained theoretically by Jensen *et al.* (1989) and Jensen (1992), who associated it with an unstable periodic orbit. It was shown by Sauer *et al.* (1992) that this local maximum in the experimental ionisation thresholds, although not present in any purely classical calculation, nevertheless scaled classically. Subsequently, Sirko *et al.* (1993) showed that state populated by starting in the two states $I_0 = I_r \pm N_L\hbar/2$, which for the parameters of that experiment corresponded to the initial quantum numbers $n_0 = 55$ and 62, have the same threshold field and behave similarly under the influence of noise.

These works show fairly conclusively that the minimum in P_i at $\Omega_1 \simeq 1.3$ is a consequence of the separatrix of K_p, equation (14), but they do not explain why the quantal state associated with the separatrix is locally more stable than adjacent states. This question is addressed by Leopold and Richards (1994) who showed that it was possible to find a representation in which a separatrix state of K_p was less strongly affected by the perturbation K_1 than adjacent states, and that this quantal stability is caused by the same dynamics that makes the separatrix unstable. The essential point is that the average motion on the separatrix is slower than on neighbouring phase curves: so an averaging approximation can be used to obtain a representation in which the dominant perturbation, due to K_1 is proportional to the mean of the momentum squared, which is smallest on the separatrix. This does not save the separatrix, which is unstable in any representation; but stability of quantal states is different and the same representation used in the quantum dynamics shows that a separatrix state is locally more stable against the influence of K_1 than adjacent states, hence the minimum in P_i at $\Omega_1 \simeq 1.3$, seen in Fig. 2.

4 Low Frequency Elliptically Polarised Fields

For our final example we consider the effect of a low frequency elliptically polarised field on an excited hydrogen atom. This example is unusual in that the resonance is between the driving frequency and one of the atomic frequencies which depends upon the field amplitude, and is zero when this is zero. This resonance is present only in three-dimensional atoms — or two-dimensional atoms confined to the plane of the field — and so is more difficult to understand. All of the following analysis is classical but almost all the effects described have been observed in experiments: for this system and for the quantum numbers involved in the experiment purely quantal calculations are beyond the capabilities of modern computers.

We now set the static field to zero, $F_s = 0$, and because the scaled frequency is small, $\Omega_0 \sim 0.1$, use the dipole Hamiltonian, equation (2), which becomes

$$H = \frac{\mathbf{P}^2}{2\mu} - \frac{e^2}{R} + F\lambda(t)\left[X \sin \Omega t + Y \cos(\Omega t - \delta_1)\right] \tag{18}$$

where $\delta_1 = \pi/2 - \delta$ is a constant which determines the polarisation of the field; for $\delta_1 = \pi/2$ the field is linearly polarised and for $\delta_1 = 0$ it is circularly polarised. Here we are interested in fields which are near the circularly polarised limit so it is better to work in a frame of reference rotating clockwise with frequency Ω, so the Hamiltonian is

$$\begin{aligned} K = \frac{\mathbf{p}^2}{2\mu} - \frac{e^2}{r} + \lambda(t)Fy + \Omega L_z + \frac{1}{2}\lambda(t)F\left\{y \sin 2\Omega t - x(1 - \cos 2\Omega t)\right\} \sin \delta_1 \\ + \frac{1}{2}\lambda(t)F\left\{x \sin 2\Omega t - y(1 + \cos 2\Omega t)\right\} \sin^2(\delta_1/2). \end{aligned} \tag{19}$$

Typically, in the cases of interest here $\delta_1 \sim 0.1$ so the last term is negligible. For low frequency fields and small values of δ_1 the dominant part of this Hamiltonian is the Stark Hamiltonian

$$K_s = \frac{\mathbf{p}^2}{2\mu} - \frac{e^2}{r} + \lambda(t)Fy, \tag{20}$$

which for weak fields has two natural frequencies

$$\omega(I) = \frac{\mu e^4}{I^3}, \quad \omega_s = \frac{3F\lambda(t)I}{2\mu e^2} = \frac{3}{2}F_0\lambda(t)\omega(I). \tag{21}$$

The larger frequency ω is that of the electron around the unperturbed Kepler ellipse, with eccentricity ϵ, and the smaller Stark frequency ω_s is half the frequency with which the Runge-Lenz vector, see Goldstein (1980)

$$\mathbf{A} = \frac{1}{\sqrt{-2\mu E}}\left(\mathbf{p} \times \mathbf{L} - \mu e^2 \frac{\mathbf{r}}{r}\right), \quad |\mathbf{A}| = \epsilon I_n, \tag{22}$$

rotates about the field direction. The action I_e, discussed in Sect. 2.1, is the length of **A** projected onto the field direction.

Initially $\omega_s = 0$ but as $\lambda(t)$ increases this frequency increases and there will be a resonance between this precession frequency and the driving frequency when $2\Omega \simeq \omega_s$. At these low frequencies ionisation occurs for $0.13 < F_0 < 0.2$, so if there is a resonance while the field is in this interval it should affect the ionisation probability.

In Fig. 7 we show graphs of the classical ionisation probability for the frequency $\Omega_0 = 0.11165$, (corresponding to $n_0 = 42$ for a 9.91 GHz field) where $\lambda(t) = \sin(\Omega_0 t/2N_f)$, with $N_f = 154$, and for various values of δ_1. These results were obtained using the Monte carlo method with 243 orbits. On the left are results for δ_1 near $\pi/2$, the linearly polarised limit, and on the right are results for small δ_1, near the circularly polarised limit. In Fig. 8 below ionisation probabilities for the circularly polarised field are shown.

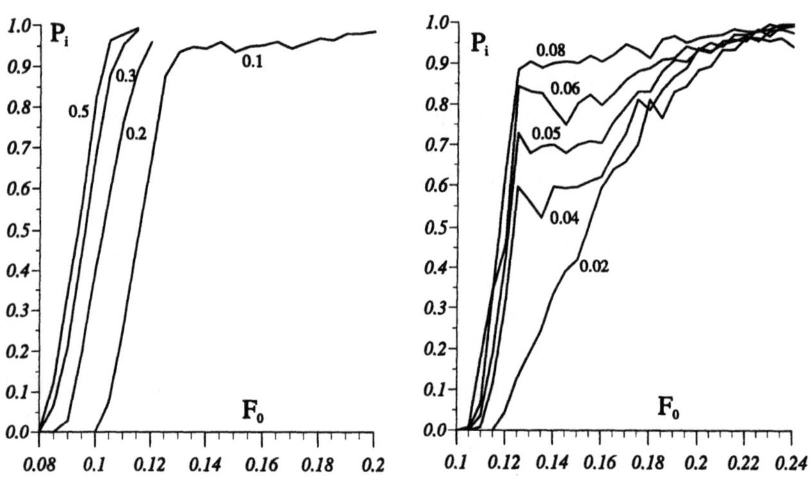

Fig. 7. Graphs of some classical ionisation probabilities for $\Omega = 0.11165$, $n_0 = 42$, obtained with a half sine field envelope having 154 field periods, for various values of δ_1, defined in equation (18); the numbers with each curve are the values of $2\delta_1/\pi$. On the LHS $\delta_1 \geq 0.1(\pi/2)$ and we see how little P_i changes if $\delta_1 > 0.2(\pi/2)$. On the RHS δ_1 is smaller. The ionisation curve for the circularly polarised field is shown in Fig. 8

The graphs show that as δ_1 decreases from $\pi/2$ the ionisation probabilities change very little and that $P_i(F_0)$ increases from 0 to 1 over the range $0.08 \leq F_0 \leq 0.115$. But at $\delta_1 = 0.05\pi$ the behaviour changes abruptly: the rapid rise as F_0 increases from 0.1 to 0.12 remains, but for $F_0 > 0.12$ the ionisation probability increases very slowly and has still not reached unity at $F_0 = 0.2$.

For higher scaled frequencies the changes in $P_i(F_0)$ with δ_1 are smoother and there are no abrupt changes as F_0 increases for fixed values of δ_1.

As δ_1 decreases further the ionisation probability changes more dramatically and for $0.08 \le 2\delta_1/\pi \le 0.04$ a shoulder develops at $F_0 \simeq 0.12$ and in some cases, for example $\delta_1 = 0.025\pi$ there is a minimum in $P_i(F_0)$ at $F_0 \simeq 0.14$; for $F_0 > 0.14$ the ionisation probability increases relatively slowly. We have checked that the minimum in $P_i(F_0)$ is real and not a consequence of statistical errors.

In Fig. 8 we extend this comparison to the circularly polarised limit and also show the ionisation probability obtained using the adiabatic method described in Sect. 2.1; even for $\Omega_0 = 0.11$ this approximation is quite good for circularly polarised fields.

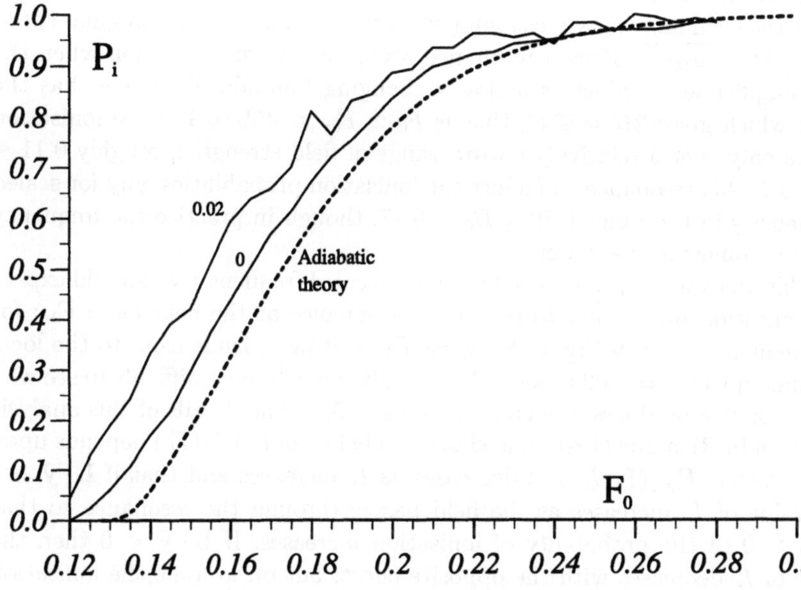

Fig. 8. Graphs of some classical ionisation probabilities for $\Omega = 0.11165$, $n_0 = 42$, obtained with a half sine field envelope having 154 field periods, for $\delta_1 = 0.02(\pi/2)$ and $\delta_1 = 0$, the circularly polarised limit. Also shown with the dashed line is the adiabatic probability discussed in Sect. 2.1.

The crude comparison between the Stark and driving frequency made above gives a clue as to why the ionisation probability has this non-monotonic behaviour but it is not entirely convincing neither does it explain the precise effect of the resonance. In order to proceed we observe that $\omega_s \ll \omega$ so that the equations of motion can be simplified by averaging over the fast Kepler motion to approximate the equations of motion in the form

$$\dot{\mathbf{X}} = \mathbf{a}(t) \times \mathbf{X}, \quad \dot{\mathbf{Y}} = \mathbf{b}(t) \times \mathbf{Y} \tag{23}$$

where

$$\mathbf{a}(t) = \left(-\frac{\delta_1\omega_s}{2}(1-\cos 2\Omega t), \omega_s\left(1+\frac{1}{2}\delta_1 \sin 2\Omega t\right), \Omega\right)$$

$$\mathbf{b}(t) = \left(\frac{\delta_1\omega_s}{2}(1-\cos 2\Omega t), -\omega_s\left(1+\frac{1}{2}\delta_1 \sin 2\Omega t\right), \Omega\right),$$

and where $\mathbf{X}$ and $\mathbf{Y}$ are related to the Runge-Lenz and angular momentum vector by

$$\mathbf{X} = \mathbf{L} - \mathbf{A}\,, \quad \mathbf{Y} = \mathbf{L} + \mathbf{A}\,.$$

In the circularly polarised limit $\delta_1 = 0$ the vectors $\mathbf{X}$ and $\mathbf{Y}$ rotate slowly about the even more slowly changing vectors $\mathbf{a}(t)$ and $\mathbf{b}(t)$ with angular speed $\overline{\omega} = \sqrt{\Omega^2 + \omega_s(t)^2}$. Thus a resonance occurs in the mean motion when the field amplitude — which is a slowly changing function of time — has the value which gives $2\Omega = \overline{\omega}(t)$, that is $F_0 \simeq F_{0r} = 2\Omega_0/\sqrt{3}$. Since ionisation occurs only over a relatively narrow range of field strengths, roughly $0.11 \leq F_0 \leq 0.2$, this resonance can affect the ionisation probabilities only for scaled frequencies in the range $0.10 \leq \Omega_0 \leq 0.17$, though in practice the frequency range is somewhat narrower.

This analysis suggests that, for a given scaled frequency, we should expect the ionisation probability to reflect the resonance at the field $F_0 \simeq F_{0r}$; for the frequency used in Fig. 7 this gives $F_{0r} = 0.13$ which is close to the local maxima in the $\delta_1 = 0.02\pi$ cases. However, it is much more difficult to see why the resonance produces a maximum at $F_0 = F_{0r}$. The details of this analysis are given in Richards (1996c); in short, the behaviour of $P_i(F_0)$ depends upon the fact that $F_{\text{crit}}(I_n, I_e, I_m)$ decreases as I_e increases and that if $\mathbf{L}\cdot\hat{\mathbf{y}} > 0$ the value of I_e increases as the field passes through the resonance so that if $F_0 > 0.13$ the probability of ionisation increases. If $\mathbf{L}\cdot\hat{\mathbf{y}} < 0$ then the value of I_e decreases with the opposite effect; but on average the ionisation probability increases. If δ_1 is too large the perturbation is larger and the ionisation probability changes from 0 to 1 over a narrow field range and $P_i(F_0) = 1$ when $F_0 = 0.13$ so the resonance has no effect.

Numerical calculations show that there are similar effects at the lower frequencies corresponding to $4\Omega = \overline{\omega}$, that is $F_0 = 2\sqrt{15}\Omega_0/3$, and $6\Omega = \overline{\omega}$, ($F_0 = 2\sqrt{35}\Omega_0/3$), but in each case δ_1 needs to be larger. At $4\Omega = \overline{\omega}$ the experimental results agree well with the classical simulations, but there are no experimental results at the lower frequency of $6\Omega = \overline{\omega}$. At present there is no theoretical explanation for these resonances.

5 Summary

In this article I have attempted to show that a relatively strong, periodic electric field acting on an excited hydrogen atom produces a rich diversity of behaviours. I have tried to show that the different effects are dependent upon the value of the scaled frequency and that these are not necessarily special to the hydrogen atom. Efficient numerical solutions of Schrödinger's equation should therefore take account of the value of the scaled frequency. Not all effects have been mentioned, for instance stabilisation and localisation, and I have concentrated on what is known rather than the many more unsolved problems.

Acknowledgements
It is a pleasure to thank Dr T Szeredi for many helpful conversations while writing this article.

References

Abramowitz M and Stegun I A 1964 *Handbook of Mathematical Functions* (Dover NY)

Arndt M, Buchleitner A, Mantegna R N and Walther H 1993 Phys. Rev. Lett **71** 3263–66

Banks D and Leopold J G 1978 J Phys **B11** 37–46

Bayfield J E and Pinnaduwage L A 1985 Phys Rev Lett **54** 313–316

Bayfield J E 1987 *Studies of the sinusoidally driven weakly bound atomic electron in the threshold region for classically stochastic behaviour* in *Quantum Measurement and chaos* pages 1–33 Eds E R Pike and S Sarkar (Plenum)

Bayfield J E and Koch P M 1974 Phys Rev Lett **33** 258–261

Bayfield J E, Casati G, Guarneri I and Sokol D W 1989 Phys Rev Lett **63** 364–7

Bayfield J E, Luie S Y, Perotti L C and Skrzypkowski M P 1995 Physica **D83** 46–54

Bayfield J E, Luie S Y, Perotti L C and Skrzypkowski M P 1996 Phys Rev **A53** R12–5

Bellermann M R W, Koch P M, Mariani and Richards D 1996 Phys Rev Lett **76** 892–5

Blümel R and Smilansky U 1987 Z Phys **D6** 83–105

Blümel R, Graham R, Sirko L, Smilansky U, Walther H and Yammada 1989 Phys Rev Lett **62** 341–4

Blümel R, Buchleitner A Graham R, Sirko L, Smilansky U and Walther H 1991 Phys Rev **A44** 4521–40

Born M 1960 *Mechanics of the Atom* (Fredrick Ungar Publishing Co.)

Breuer H P, Dietz K and Holthaus M 1991 Z fur Physik **D18** 239–48

Buchleitner A and Delande D 1993 Phys Rev lett **71** 3633–6

Buchleitner A and Delande D and Gay J-C 1995 J Opt Soc Amer **B12** 505–19

Casati G, Chirikov B V and Shepelyansky D L 1984 Phys Rev Lett **53** 2525–8

Casati G, Chirikov B V, Shepelyansky D L and Guarneri 1987 Phys Rep **154** 77–123

Casati G, Guarneri I and Shepelyansky D L 1988 J Quantum Electron **24** 1240–444

Chirikov B V 1991 *Time-dependent quantum systems* in Les Houches lectures *Chaos and Quantum Physics* 1st-31st August 1989, Eds M -J Giannoni, A Voros and J Zinn-Justin (Nort Holland)

Chu S I and Wang K 1989 Phys Rev **A39** 1800–8

Cornish F H J 1984 J Phys **A17** 323–7

Damburg R J and Kolosov V V 1983 *Theoretical studies of hydrogen atoms in electric fields* in *Rydberg states of atoms and molecules* (Cambridge University press) Eds R F Stebbings and F B Dunning.

Dando P A and Richards D 1990 J Phys **B23** 3179–204

Dando P A and Richards D 1993 J Phys **B26** 3001–16

Dando P A, Monterio T S, Jans W and Schweizer W 1994 Progr. Theor Phys, Supp. No. 116 403–6

Edmonds A R 1973 J Phys **B6** 1603–15

Gallagher T F, Mahon C R, Pillet P, Panming Fu and Newman J B, 1989 phys Rev **A39** 4545–9

Galvez E J, Sauer B E, Moorman L, Koch P M and Richards D 1988 Phys Rev Lett **61** 2011–4

Gatzke M, Broers B, Noordam L D, Watkins R B and Gallagher T F 1994 Phys Rev **A50** 2502–7

Goldstein H 1980 *Classical Mechanics* (Addison-Wesley)

Jensen R V, Leopold J G and Richards D 1988 J Phys **B21** L527–31

Jensen R V 1982 Phys Rev Lett **49** 1365–8

Jensen R V 1984 Phys Rev **A30** 386–97

Jensen R V 1992 *The Bohr Atom Revisited: a test case for Quantum chaos* pages 127–139 in *Irregular Atomic Systems and Quantum Chaos* Ed Jean-Claude Gay (Gordon and Breach)

Jensen R V, Susskind S M and Sanders M M 1989a, Phys Rev Lett **62** 1476–9

Jensen R V, Sanders M M, Saraceno M and Sundaram B 1989b Phys Rev Lett **63** 2771–5

Koch P M 1990a *Microwave Excitation and Ionization of Excited Hydrogen Atoms* in *The Ubiquity of Chaos* 75–96, Ed S Krasner (American Association for the Advancement of Science)

Koch P M 1990b *Microwave ionization of excited hydrogen Atoms* in *Chaos/Xaoc: Soviet-American Perspectives on Nonlinear Science* 441–475, Ed D K Campbell (American Institute of Physics, NY)

Koch P M 1992 Chaos **2** 131–44

Koch P M 1995 Physica **D83** 178–205

Koch P M, Moorman L and Sauer B E 1992 *Microwave Ionisation of Excited Hydrogen Atoms* pages 173–191 in *Irregular Atomic Systems and Quantum Chaos* Ed Jean-Claude Gay (Gordon and Breach)

Koch P M and van Leeuwen K A H 1995 Phys Rep **225** 289–406

Landau L D and Lifshitz E M 1965 *Mechanics* (Pergamon Press)

Lee C Y, Hettema J M, Gallagher T F and Conover C W S 1992 Phys Rev **A46** 7048–52

Leopold J G and Percival I C 1979 J Phys **B12** 709–21

Leopold J G and Richards D 1985 J Phys **B18** 3369–94

Leopold J G and Richards D 1987 J Phys **B21** 2369–82

Leopold J G and Richards D 1989 J Phys **B22** 1931–61

Leopold J G and Richards D 1990 J. Phys.**B23** 2911–27

Leopold J G and Richards D 1991a J Phys **B24** L243–8

Leopold J G and Richards D 1991b J Phys **B24** 1209–40

Leopold J G and Richards D 1993 J Phys **B26** 1519–39

Leopold J G and Richards D 1994 J Phys **B27** 2169–89

Lichtenberg A J and Lieberman M A 1991 *Regular and Chaotic Dynamics* (Springer)

Mahon C R, Dexter J L, Pillet P and Gallagher T F 1991 Phys Rev **A44** 1859–73

Norcliffe A and Percival I C 1968a J Phys **B2** 775–83

Norcliffe A and Percival I C 1968b J Phys **B2** 784–94

Percival I C 1969 *Correspondence Identities* in *Physics of the one- and two-electron atoms* (North Holland)

Percival I C and Richards D 1975 *Advances in Atomic and Molecular Physics,* **11** 1–82 (Eds D R Bates and B Bederson, Academic Press)

Percival I C and Richards D 1982 *Introduction to Dynamics* (Cambridge University Press)

Rath O and Richards D 1988 J Phys **B21** 555–71

Rath O and Richards D 1996 In preparation

Richards D 1987 J Phys **B20** 2171–92

Richards D 1989 *Microwave Ionisation of Atomic Hydrogen* in Proceedings of the International conference on *Classical Dynamics in Atomic and Molecular Physics,* Eds T Grozdanov, P Grujić and P Krstić, Brioni, Yugoslavia 1988, (World Scientific)

Richards 1992a *Classical and Quantum Dynamics of highly Excited periodically Forced Systems* in *Irregular Atomic Systems and Quantum Chaos* Ed Jean-Claude Gay (Gordon and Breach)

Richards D 1992b J Phys **B25** 1347–60

Richards D 1996a J Phys **B29** 2925–49

Richards D 1996b J Phys **B29** 5237–71

Richards D 1996c submitted to J Phys **B**

Richards D, Leopold J G, Koch P M, Galvez E J, van Leeuwen K A H, Moorman L, Sauer B E and Jensen R V 1989a J Phys **B22** 1307–33

Richards D, Leopold J G and Jensen R V 1989b J. Phys. B: At Mol Opt Phys **22** 417–33
Richards D and Szeredi T 1995 J Phys **B28** L413–21
Richards D and Szeredi T 1996a To be published in Phys Rev **A**
Richards D and Szeredi T 1996b To be published in Phys Rev **A**
Sauer B E, Bellermann M R W and Koch P M 1992a Phys Rev Lett **68** 1633–6
Sauer B E, Yoakum S, Moorman L, Koch P M, Richards D and Dando P A 1992b Phys Rev Lett **68** 468–71
Sirko L, Bellermann M R W, Haffmans A, Koch P M and Richards D 1993 Phys Rev Lett **71** 2895–8
Sirko L and Koch P M 1995 Applied Phys **B60** S195–S202
Synge J L 1960 *Principles of Classical Mechanics and Field theory* pages 1–225, Handbuch der Physik Vol III/I (Ed S Flugge, Springer-Verlag, Berlin)
Susskind S M and Jensen R V 1988 Phys Rev **A38** 711–28
Szebehely V 1967 *Theory of Orbits* (Academic Press)
Zakrzewski J, Delande D, Gay J-C and Rzążewski K 1993 Phys Rev **A47** R2468–71

Ionization of Hydrogen Atoms by Circularly Polarized Microwaves

Dominique Delande[1] and Jakub Zakrzewski[1,2]

[1] Laboratoire Kastler–Brossel, Université Pierre et Marie Curie, T12, E1, 4 place Jussieu, 75272 Paris Cedex 05, France
[2] Instytut Fizyki Mariana Smoluchowskiego, Uniwersytet Jagielloński, ul. Reymonta 4, 30-059 Kraków, Poland

Abstract. The current status of knowledge about the ionization of Hydrogen Rydberg atoms by *circularly* polarized microwaves is reviewed. A brief presentation of experimental data available in the literature is followed by a discussion of theoretical predictions based on both the classical and the quantum studies. A few misconceptions present in research papers devoted to this subject are clarified. A comparison with the linearly polarized microwave ionization problem reveals similarities and, sometimes, strong differences between the two systems. Both are of value for deepening the understanding of the nature of the ionization process as well as, more generally, of quantum systems which classically exhibit irregular or mixed regular-irregular behaviour.

1 Introduction

The ionization of highly excited hydrogen atoms by linearly polarized microwaves (LPM) has a long history starting with the pioneering experiment of Bayfield and Koch (1974). The very first understanding of the ionization process was obtained in Leopold and Percival (1979) using Monte-Carlo classical simulations, associating the ionization threshold with the onset of classical chaos in the system. Numerous studies performed since this early work treated the problem either classically or quantum mechanically, at various degrees of approximation. At the same time, new improved experiments provided a stimulus and new puzzles for the theory (for recent reviews of the theory see Casati *et al.* (1987), Casati *et al.* (1988), Chirikov (1991), Jensen *et al.* (1991), Buchleitner *et al.* (1995), experimental details may be found in Koch (1993), Koch and van Leeuven (1995), Bayfield and Sokol (1988)).

Similarly, intensive studies of the ionization by *circularly* polarized microwaves (CPM) have been triggered by the experiments carried out for alkali atoms by Fu *et al.* (1990), Gallagher (1991). Prior to 1990, circularly polarized radiation was considered mainly in the optical frequency domain with the exception of early papers considering resonant excitation, see Mostowski and Sanchez-Mondragon (1979), Meerson (1981), and the seminal work devoted to the photonic localization theory (Casati *et al.* (1988)) which, while mainly dealing with the LPM, considered also general elliptical polarization.

However, in the last five years, a number of theoretical works has appeared and it seems to be a good moment to provide a unified viewpoint on a number of phenomena observed in these studies. Especially since the authors of original papers often present quite contradictory points of view. Unfortunately, several statements in the various papers contain basic errors, mainly due to the non standard form of the Hamiltonian in the rotating frame, making the situation very confused.

The organization of the paper is as follows. In section 2, we present the basic equations, with emphasis on the transformation to the rotating frame and the resulting subtleties. A parameter of crucial importance is the scaled frequency ω_0, the ratio of the microwave frequency to the Kepler frequency of the electron on its classical orbit. In section 3, we discuss the low frequency regime ($\omega_0 \ll 1$), where the classical dynamics is mainly regular. The medium and high frequency regime ($\omega_0 > 1$) is discussed in section 4: there, the classical dynamics is mainly (but not fully) chaotic. Ionization is mainly due to chaotic diffusion, as for the linear polarization case. However, the behaviour of the system is somewhat different (as compared with linear polarization excitation). In LPM, the theory of "quantum photonic localization" developed by Casati *et al.* (1988) apparently gives accurate quantitative predictions for the ionization threshold. In contrast, in CPM, extensive classical and quantum calculations show qualitative agreement, but some quantitative disagreement with the predictions of this theory. Finally, in section 5, we discuss the importance of mixed regular-irregular dynamics in the intermediate frequency regime $\omega_0 \simeq 1$, and the opportunity that it offers to study phenomena like "chaos assisted tunneling" or "chaos assisted ionization".

2 Rotating Frame - Scaling Properties

Consider the hydrogen atom under the influence of the CPM. Choosing the coordinate frame in such a way that the polarization plane coincides with the $z = 0$ plane, the Hamiltonian of the system (in the dipole approximation and in atomic units) reads:

$$H = \frac{p_x^2 + p_y^2 + p_z^2}{2} - \frac{1}{r} + Ff(t)\big(x\cos(\omega t) + y\sin(\omega t)\big), \tag{1}$$

where $r = \sqrt{x^2+y^2+z^2}$ while F and ω are the maximal amplitude and the frequency of the microwave, respectively. A change of the sign of ω is equivalent to a change from right-hand CPM to left-hand CPM. In (1), $f(t)$ describes the microwave pulse shape, a smooth function of time, $t \in [0,T] \to [0,1]$, where T is the microwave pulse duration. Let us pass to the rotating frame defined as

$$\tilde{x} = x\cos(\omega t) + y\sin(\omega t) \tag{2}$$

$$\tilde{y} = -x\sin(\omega t) + y\cos(\omega t) \tag{3}$$

realized by a unitary operator $U = \exp(-iL_z t)$ in quantum mechanics ($L_z = xp_y - p_x y$ is the angular momentum in the polarization plane). In the new frame, the fast oscillations are removed (Bunkin and Prokhorov (1964)) and the Hamiltonian takes the form:

$$H = \frac{p_x^2 + p_y^2 + p_z^2}{2} - \frac{1}{r} + Ff(t)x - \omega L_z. \tag{4}$$

Note that in order to simplify the notation, we dropped the ˜ from the variables. In the following, we shall work consistently in the rotating frame only.

It is intuitively clear that the influence of the CPM on the motion of the electron (leading, e.g., to its ionization) will be strongest in the polarization plane. Most of the treatments present in the literature consider, therefore, a simplified two-dimensional (2D) hydrogen atom model in which the electronic motion is confined to the polarization plane. It is easy to see that classically such a model is exact for the subset of initial conditions fulfilling $z = p_z = 0$. This is no longer true in the quantum world, and the predictions obtained for the 2D and the realistic 3D atom may quantitatively differ. For simplicity, we shall restrict ourselves mostly to the 2D model.

Let us fix $f(t) = 1$ for the moment: we obtain a time-independent Hamiltonian in the rotating frame. The price to pay for such a simplification is two-fold: firstly, the Hamiltonian contains an unusual term proportional to L_z which mixes position and momentum coordinates. Hence, the usual intuition which relates potential minima to stable points and potential maxima or saddle points to unstable points is simply wrong. Much care is needed when discussing the classical dynamics. Secondly, the energy in the rotating frame differs from what it is in the original frame by ωL_z which can be large, even for vanishingly small amplitudes of the microwave. Several incorrect statements in the literature (clarified below) originate from neglecting one or other of these two important points.

The classical dynamics generated by the Hamiltonian, Eq. (4), has a simple scaling property. It is invariant under the following scaling transformation:

$$\begin{cases} \mathbf{r} \to \lambda \mathbf{r} \\ \mathbf{p} \to \lambda^{-1/2}\mathbf{p} \\ t \to \lambda^{3/2} t \\ H \to \lambda^{-1} H \\ F \to \lambda^{-2} F \\ \omega \to \lambda^{-3/2}\omega \end{cases} \tag{5}$$

for any positive parameter λ. This means that, as long as classical mechanics is studied, there is no need to use the three independent parameters H, F, ω, but that we can use scaled parameters such as the scaled field and frequency:

$$F_0 = F n_0^4 \tag{6}$$

$$\omega_0 = \omega n_0^3 \tag{7}$$

where n_0 is the initial principal quantum number. Such a scaling is well known and widely used for atoms in linearly polarized microwaves, see Koch and van Leeuven (1995). It is also valid in circular or elliptical polarization.

3 Ionization in the Low Frequency Domain, $\omega_0 < 0.5$

It is instructive to look at the studies of ionization thresholds at small frequencies ($\omega_0 < 0.5$, the upper limit being somehow arbitrary). This was the first situation studied experimentally as well as theoretically. In the corresponding situation for linearly polarized microwaves, it is nowadays well known and understood that at very low scaled frequency, the classical dynamics is mainly regular and dominated by resonances, see Koch and van Leeuven (1995). As the scaled frequency increases, a significant part of the phase space turns chaotic.

The first experiments, as mentioned in the introduction, were performed for Na atoms by Fu *et al.* (1990), Gallagher (1991). Laser prepared atoms in initial Rydberg states with low values of $|m|$ (the quantum number corresponding to L_z) were exposed to microwave pulses of frequency much lower than the Kepler frequency, $\omega \ll 1/n_0^3$, i.e. $\omega_0 \leq 0.2$. The experimentalists observed a rather sharp increase of the ionization threshold when the polarization of microwaves changed from elliptical to fully circular. For circularly polarized microwaves, the amplitude necessary to reach 50% ionization scaled with initial n_0 as $F = 1/16n_0^4$ yielding the scaled threshold amplitude $F_0 = 1/16$ to be independent of the initial n_0 and equal to the static field necessary to ionize Na atoms via the Stark effect. This is in very sharp contrast with results obtained in LPM on Na and Rb atoms, where the threshold is much lower, see Gallagher (1991), Arndt *et al.* (1991).

Hence, the natural idea is to use the static field situation to get more insight into the low frequency regime.

3.1 Static Electric Field: Integrable Stark Effect

For $\omega = 0$, both the energy and the projection of the Runge-Lenz vector $\mathbf{A}$ on the electric vector axis, A_x, are conserved. Classically the motion is restricted to tori characterized by the values of these two constants of the motion. Due to the scaling, it is sufficient to consider one energy value, e.g., $E = -1/2$. The motion at any other energy, e.g., $E_{n0} = -1/2n_0^2$ with field value F, is the same as the motion at the energy E with the scaled field $F_0 = Fn_0^4$.

The potential $V = -1/\sqrt{x^2+y^2} + F_0 x$ has a saddle point, S, located at $y = 0$, $x = -F_0^{-1/2}$ such that $V(S) = -2\sqrt{F_0}$. The presence of the saddle point allows for a simple estimate of the field required for static field ionization. The electron may ionize if its energy exceeds that of the saddle point yielding the critical field $F_c = 1/16n_0^4$ as a solution of $-1/2n_0^2 = -2\sqrt{F}$. As is well known, the approximate threshold field value is instead $F_c = 1/9n_0^4$

indicating the fallacy of the argument given above. The origin of the error is two–fold: firstly, the zero-order energy of the n_0-manifold was taken, secondly the conservation of A_x limits the phase space accessible for the electron: thus the barrier argument works only for those distorted tori which, projected onto the configuration space, approach the barrier for increasing F.

To compensate for the above shortcomings, the energy of the state becomes Stark shifted and reads (in the first order perturbation theory) $E(k) = -1/2n_0^2 + 3n_0kF/2$, where $-(n_0 - 1) \leq k \leq n_0 - 1$ is the quantum number corresponding to quantization of A_x (in parabolic coordinates which allow for the separation of the problem, k is a difference of actions in the individual coordinates, i.e. $k = n_1 - n_2$ while $n_0 = n_1 + n_2 + 1$). The n_0–manifold splits into a ladder of equidistant states. It is well known that tori supporting blue shifted states (k positive) lie far from the saddle, while extremal red-shifted states ($k \approx -n_0$ for n_0 large) approach the saddle as F increases. It is only for the latter states that the saddle energy criterion for ionization may work and indeed it yields the ionization threshold value $F_c = 1/9n_0^4$ in agreement with experiments on atoms prepared initially in a microcanonical initial mixture of states (all states of the n_0 manifold with equal weigths). A more rigorous approach requires a transition to parabolic coordinates in which the static field problem is known to be separable. Then the saddle point criterion can be used to obtain classical thresholds for all states, see Banks and Leopold (1978). The tori corresponding to other k values, although lying higher in energy, break up at higher F values, the most resistive being an extreme blue shifted state which ionizes at $F \approx 0.38/n_0^4$ (for a detailed study see e.g. Banks and Leopold (1978)).

3.2 Hydrogen Atom Versus Alkali Atoms

For alkali atoms, the potential seen by the outer electron is no longer Coulombic and the separation in parabolic coordinates is no longer valid. For non-penetrating states with large angular momentum (roughly $|m|$ larger than 2 in sodium), the electron does not go sufficiently close to the nucleus to feel the non-Coulombic character of the potential: the quantum defect is small and the situation is almost hydrogenic. On the other hand, for states with low angular momentum — like the $|m| = 0, 2$ states used in the experiments on Na atoms — the quantum defects may be large. Hence, the various parabolic series, which ionize at different thresholds in hydrogen, are mixed by the non-Coulombic potential: ionization is then dominated by the components which ionize first, namely the red-shifted hydrogenic components. Also, because of the mixing of the various series, there are only broad avoiding crossings between energy levels as the field strength is increased. Consequently, in most experimental conditions, the atomic state follows the energy levels adiabatically, keeping its energy roughly constant until it reaches the saddle point energy at $F = 1/16n_0^4$, where it ionizes.

In the presence of a circularly polarized microwave field, the Hamiltonian for the hydrogen atom in the rotating frame, Eq. (4), may be cast in the form:

$$H = \frac{(p_x + \omega y)^2}{2} + \frac{(p_y - \omega x)^2}{2} - \frac{1}{\sqrt{x^2+y^2}} - \frac{\omega^2}{2}\left(x^2+y^2\right), \tag{8}$$

which shows explicitly that canonical (p_i) and kinetic momenta (velocities) differ, i.e., $v_x = dx/dt = p_x + \omega y$. Writing the conserved energy as a function of velocities and coordinates, one is tempted to consider the remaining term as an effective potential (Farelly and Uzer (1995)):

$$E = \frac{v_x^2 + v_y^2}{2} + V_f; \qquad V_f = -\frac{1}{\sqrt{x^2+y^2}} + Fx - \frac{\omega^2}{2}\left(x^2+y^2\right). \tag{9}$$

Such an approach may be useful, but also quite misleading. For example, for $F = 0$, it is easy to see that the maximum of the radially symmetric V_f (i.e., its rim) occurs at $r_m = \omega^{-2/3}$ where its value is $V_{max} = -\frac{3}{2}\omega^{2/3}$. The potential idea intuition would then lead one to assume that trajectories at higher energies are not bound. This is true only for $\omega = 0$, while for any $\omega \neq 0$, bound trajectories of negative energy E_b in the laboratory frame have the energy (see above) $E = E_b - \omega L_z$ in the rotating frame which may be much higher than V_{max}. Consider, as before $E_b = -1/2$, then orbits with $L_z < (1 - 3\omega^{2/3})/2\omega$ stay above the rim. For small frequencies, $\omega < 0.125$ all orbits lie below the rim of the "potential". Even though the rim radius, $r_m > 4$ the orbits nevertheless extend only up to $r = 2$ for $E = -1/2$ indicating that it is not the classical turning points of the effective potential V_f that limit the motion of the electron. For higher frequencies, orbits typically lie above the rim of V_f. A similar picture holds for nonvanishing electromagnetic field and has important consequences (see section 5).

Clearly, due to the difference between kinetic and canonical momenta, the potential intuition may provide quite wrong ideas about orbit stability in the rotating frame.

The $-\omega L_z$ term in the Hamiltonian, Eq. (4), also breaks the separability in parabolic coordinates and mixes the different channels. This has important consequences for hydrogen atoms. On the other hand, in alkali atoms, as the non-Coulombic potential already breaks the separability, it can be expected that, for small frequencies, the ionization threshold will be very close to the static threshold $F = 1/16n_0^4$.

This is the argument used in Fu *et al.* (1990), who conclude, after discussion of the system in the rotating frame, that "in essence, a static field has the same effect whether the coordinate system is rotating or not" in agreement with their experimental observation on Na atoms.

This simple explanation was immediately challenged by Nauenberg (1990a) who developed quantitatively the previous reasoning. He used the unperturbed energy for the initial state motivating this choice by the following statement "an approximation which is surprizingly well satisfied by alkali Rydberg atoms in a static field". He found that the threshold is $F = 1/16n_0^4$ in the $\omega \to 0$ limit, but that a very significant decrease of the threshold takes place even for quite small ω. At the largest scaled frequency used in the experiment, $\omega_0 \simeq 0.2$, he predicts a threshold several times smaller, in contradiction with experimental results. The conclusion he obtained is that "further work is needed to understand the experimental results which cannot be explained theoretically as claimed in Fu *et al.* (1990)".

We believe that the result of Nauenberg - which underestimates the experimental result - is only a lower bound for the ionization threshold. Indeed, even when starting from a low $|m|$ state, the excitation by CPM necessarily leads to an increase of the angular momentum (by one for each photon absorbed), driving the system to a quasi-hydrogenic situation. There, the motion is mainly regular (see next section) and atoms trapped on some invariant tori may not ionize even above the saddle point energy. Hence, the effect of the non-Coulombic character might not be as dramatic as in the static field case. This is confirmed by the fact that the discrepancy between the hydrogenic prediction and the experimental result on non-hydrogenic atoms is much smaller in circular polarization than in linear polarization: in the latter case, m is a good quantum number and the outer electron continues to feel the effect of the ionic core when it gains energy. Further work should completely clarify the effect of the ionic core in circular polarization.

3.3 Hydrogen Atoms in the Low Frequency Domain

Some *qualitative* understanding of experimental results for alkali atoms may still be obtained on purely classical grounds. Griffiths and Farelly (1992) have shown that classical simulations performed for H atom ionization reproduce a strong dependence of the ionization threshold on the state of polarization. In particular, a transition from elliptical to circular polarization is accompanied by a rapid decrease of the ionization yield in agreement with the experiments of Fu *et al.* (1990), Gallagher (1991). The authors associate this behaviour with a breakdown of nearly conserved quantities as the polarization is changed from circular to elliptical.

It is quite obvious that a deeper understanding of the ionization process in CPM may be obtained by considering the H atom. Here both classical and quantum calculations are feasible, moreover, the experiments have been planned for a long time (and have already delivered the first data, see below). Thus from now on, we shall consider only the pure H atom case.

For small microwave amplitudes, one may treat the term Fx in Eq.(4) as a perturbation. A standard averaging of the perturbation (equivalent to the so called Pauli replacement, see Delande (1991)) results in the integrable Hamiltonian

$$H_{pert} = H_0 - \omega L_z - \frac{3}{2}\sqrt{-\frac{1}{2H_0}}FA_x \qquad (10)$$

as discussed by Wintgen (1991) on the quantum level and by Griffiths and Farelly (1992) classically. In complete analogy to a similar static field problem (for a review, see Delande (1991)) the Hamiltonian H_{pert} may be expressed in a separable form using two angular momenta which are linear combinations of the orbital angular momentum **L** and the Runge-Lenz vector **A** (Griffiths and Farelly (1992)).

Not suprisingly, when the microwave strength increases, most of the invariant tori described by Eq. (10) get distorted, but nevertheless survive (Griffiths and Farelly (1992), Rakovic and Chu (1995)). In sharp contrast to the linear polarization case, (Leopold and Percival (1979), Casati *et al.* (1987)), the motion of the electron in CPM for small ω_0 is almost integrable. With increasing field, the initial torus gets deformed until it breaks up at some field value leading to ionization. Such a picture suggests a strong dependence of the ionization threshold on the initial electronic state (recall the discussion of the Stark effect above). In this way, one may interpret easily the results of classical simulations of Rzążewski and Piraux (1993) who studied the dependence of the electronic motion on the frequency for initial circular states illuminated by a microwave pulse. They noticed that for low frequencies, and microwave amplitudes below the ionization threshold, the electron returned to its initial state after the interaction with a smooth pulse. They also noticed a strong asymmetry in the ionization thresholds for different orientation of the electronic motion with respect to the rotation of the microwaves. The ionization threshold for an electron rotating in the same direction as the microwave electric field was found to be roughly 3 times lower than in the counter–rotation case, i.e., for the initial energy $E_1 = -1/2$, orbits with $L_z = 1$ ionize much faster than those with $L_z = -1$.

A subsequent, also 2D, quantum calculation by Zakrzewski *et al.* (1993), confirmed the classical results of Rzążewski and Piraux (1993) providing at the same time their full interpretation. It has been shown that, for states corotating with respect to the microwave field, one may modify the saddle point criterion using an approach similar to that used for the static field — namely by taking the perturbative expression of Wintgen (1991) for the energy of the state localized in the vicinity of the saddle. In this way one obtains agreement between classical and quantum simulations as well as with the simple saddle point prediction. For states counter–rotating with respect to the microwave field, the quantum thresholds obtained were significantly lower than the classical results due to the relatively low initial $n_0 = 13$ value taken in the quantum calculations.

Almost simultaneously with the two 1993 papers referred to in the previous paragraph, came an important contribution from Kappertz and Nauenberg (1993) who provided a full "saddle" point analysis of the ionization threshold for states of arbitrary initial eccentricity (but only positive L_z). The procedure involved, in essence, an adiabatic approximation (with the averaging over the fast Kepler motion) in which the perturbative and the strong regime were matched asymptotically. Such an approach is valid, obviously, for small ω only. The authors stressed the regular character of the motion which made their analysis possible. The analytical results were compared with extensive numerical calculations and good agreement has been found for $0.02 < \omega < 0.15$. For very small frequencies, the results become sensitive to the manner in which the field is switched on (for details see Kappertz and Nauenberg (1993)), as noted also earlier by Rzążewski and Piraux (1993), Zakrzewski *et al.* (1993).

An alternative elegant approach was proposed by Rakovic and Chu (1995) who discovered that the system described by the Hamiltonian

$$H_I = \frac{p_x^2 + p_y^2 + p_z^2}{2} - \frac{1}{r} + Fx - \omega L_z + \frac{\omega^2}{18}\left(r^2 + 3z^2\right) \tag{11}$$

is integrable. The corresponding restricted 2D version is obtained by taking $z = p_z = 0$. Clearly the difference between H in Eq.(4) and H_I is proportional to ω^2 and thus small for small ω. Raković and Chu have been able to map the tori of H into those of H_I for small microwave amplitudes. Then for a chosen set of initial conditions, they adiabatically increased F following the deformations of tori of H_I until they underwent a bifurcation. The underlying idea is that the bifurcation of tori of H_I indicates a transition to unstable motion (ionization) in the close but nonintegrable system H. Numerical results have been obtained in this way for a microcanonical ensemble of initial conditions (in the non-rotating frame) in order to simulate an experiment in which the initial state is energy selected only. The ionization threshold is defined as the amplitude at which 50% ionization occurs. They find that for small ω, the threshold defined in this way behaves as $F \approx 1/6n_0^4$, in agreement with the preliminary data from the hydrogen atom experiment of Bellermann *et al.* (1994) and is very weakly dependent on the frequency. Note that the factor 1/6 is different from the corresponding 1/9 factor for a static field ionization threshold. This difference is due to the 50% definition of the ionization threshold adopted in Rakovic and Chu (1995) following the sodium experiments.

Raković and Chu emphasize that their method is an alternative to a saddle-point analysis. They compare their own results with the early, incorrect saddle point approximation of Nauenberg (1990a) claiming that the methods give different results. A more fruitful comparison could be made with Kappertz and Nauenberg (1993). While the numerical and analytical results are quite difficult to compare directly, close inspection of both papers reveals a close similarity in the threshold behaviour obtained (compare Raković and Chu's results with Fig. 7 of Kappertz and Nauenberg (1993) which shows the 50% threshold for the relevant state of maximal initial eccentricity).

To summarize this section, it seems that the ionization mechanism for H atoms in CPM is quite well understood when the frequency is small in comparison to the Kepler frequency. The system is almost integrable and different classical approaches which use the near-integrability offer reliable predictions for the ionization threshold, see Kappertz and Nauenberg (1993), Rakovic and Chu (1995). This point of view has been sharply challenged in a recent paper of Howard (1995) based on the stability of fixed points of the Hamiltonian. The classical existence and, importantly, the quantum manifestations of the existence of fixed points in the system will be discussed in section 5. To avoid repetitions, let us thus delay the discussion of Howard (1995) and its implications for the small frequency domain till then.

To close this section, we mention that, at this time, the only quantum calculation in this regime, see Zakrzewski *et al.* (1993), agrees with the corresponding classical simulations (modulo relatively small n_0 used in the calculation) in accord with the experience from LPM studies in the comparable domain of frequencies and with the predictions of the saddle-point approach.

4 Ionization in the High Frequency Domain, $\omega_0 > 0.5$

For the higher frequency domain, $\omega_0 > 0.5$, several classical studies are available (Nauenberg (1990b), Howard (1992), Gębarowski and Zakrzewski (1994), Gębarowski and Zakrzewski (1995)). They show that most invariant tori existing at low scaled frequency are destroyed - mainly through various resonances: at relatively low microwave field, the classical dynamics becomes mostly, but not fully, chaotic. The electron undergoes a chaotic diffusive motion in phase space, eventually ionizing. Most of these studies stress the diffusive character of the ionization (as in LPM) with, however, a sensitive dependence on the shape of the initial states. For high eccentricity states and large frequencies ($\omega_0 >> 1$), the appropriate Kepler map has been constructed by Casati *et al.* (1988) yielding classically diffusive motion, as well as quantum predictions for thresholds, based, in the quantum case, on the photonic localization theory (PLT).

4.1 Experimental Results

From the experimental point of view, the first results in the intermediate and high frequency domain have been obtained only very recently by Bellermann *et al.* (1996). Importantly, the authors studied hydrogen atoms and concentrated on the intermediate frequency regime $\omega_0 \simeq 1$ so that detailed comparison with the numerical quantum simulations is not easy. On the other hand, they also did some classical simulations.

The first interesting observation is that, at the lowest frequency studied ($\omega_0 = 0.5$), the experimental results agree very well with both the quantum and semiclassical calculations in Zakrzewski *et al.* (1993). In our opinion, these results show that the controversy on the low frequency regime for hydrogen atoms (see previous section) is now closed. Provided it is done carefully, the saddle point approach predicts correctly the ionization threshold, at least when the initial state is a statistical mixture of various eccentricities.

The most striking experimental observation is that, in the frequency range around the main resonance $0.6 < \omega_0 < 1.4$, the ionization thresholds for linear, elliptic and circular polarizations are very similar and display almost exactly the same dependence on ω_0. Surprisingly, in sharp contrast with the results obtained at low frequency on Na atoms, the threshold for circular polarization is slightly *lower* than in linear polarization. Classical Monte-Carlo simulations in Bellermann *et al.* (1996), using a microcanonical distribution of initial states (as is supposed to be the case in the experiment) reproduce fairly well the experimental results.

The striking similarity between LPM and CPM thresholds around the main resonance can be at least qualitatively understood using a classical analysis of the 1:1 resonance. The resulting pendulum-like Hamiltonian is indeed very similar in the two cases. Although the detailed phase space structure may be different, one has to remember that the 10% ionization threshold is dominated by states ionizing first among all the states in the n_0 hydrogenic manifold, that is by only a small part of phase space (see also next section). It happens that this gives very similar thresholds in LPM and CPM. If a careful selection of the initial state (in angular momentum) was experimentally possible, we believe that the results in LPM and CPM should be quite different.

What can be learnt from these experimental results is that, in the intermediate frequency regime, the system cannot be considered a strongly chaotic system, but that the phase space (in LPM and in CPM) is infested with residual regularities and invariant tori. Hence, the hydrogen atom in a microwave field is a prototype of mixed regular-chaotic systems rather than a fully chaotic one. This was relatively clear from classical calculations in the CPM case, but was only quite recently fully appreciated for the LPM case, see Koch and van Leeuven (1995).

4.2 Quantum Versus Photonic Localization Theory

Here, we summarize the results of our numerical *quantum* study of the CPM ionization of the hydrogen atom for a high frequency domain in the 2D model reported in Zakrzewski *et al.* (1996a) paying particular attention to a comparison with PLT (Casati *et al.* (1988)).

To summarize briefly, we consider ionization under the influence of a rectangular pulse of duration T. A fixed microwave amplitude allows us to use a diagonalization of the Floquet matrix obtained after expanding the Hamiltonian of the system in a Sturmian basis. To fully include the atomic continuum we use a complex rotation method, see e.g. Reinhardt (1982). In this way we obtain complex eigenvalues for which the real part gives the position of a Floquet eigenstate and twice the imaginary part taken with a minus sign its decay (ionization) rate Γ. Expansion of the initial state in the Floquet basis allows as to find the ionization probability in the presence of microwaves, see Buchleitner *et al.* (1995), Zakrzewski *et al.* (1996a).

The commonly accepted definition of the ionization threshold is the so called amplitude 10% threshold, $F10\%$, i.e. the value of the microwave amplitude for which 10% of atoms become ionized. Actually we use not the physical value of the microwave amplitude F (in atomic units), but rather the scaled value $F_0 = Fn_0^4$, as also used by other authors. The dependence of $F10\%$ on scaled frequency ω_0 is then used to characterize the process, see Koch (1993). Before presenting the results, let us recall the main conclusions of PLT — we shall confront its predictions with the results from our exact calculations.

The general idea of the Kepler map (Casati *et al.* (1988)) is to assume that, for large frequencies and sufficiently eccentric orbits, the energy exchange between the electron and the microwave field takes place close to the perihelion of the unperturbed elliptic trajectory. The applicability of the Kepler map for the 2D CPM ionization case is thus limited to the regime:

$$\omega_0 \gg 1 \qquad |\tilde{m}| = \frac{|m_0|}{n_0} < (2/\omega_0)^{1/3}, \tag{12}$$

where m_0, n_0 are the angular momentum quantum number and the principal quantum number respectively, describing the initial state while $\tilde{m} = m_0/n_0$ is the scaled angular momentum taking values between -1 and 1.

Then, the Kepler map describes the excitation of the atomic state when it interacts with the microwave. Classically, it has a chaotic diffusive behaviour, in agreement with classical calculations on the full system. Because of quantum interference effects, the classical diffusion is frozen in the quantum world: this is the phenomenon known as "photonic localization".

If photonic localization takes place, then the maximal excitation (up to exponential corrections) is given by the localization length (in number of photons), see Casati *et al.* (1988)

$$\mathcal{L} \approx 3.33 n_0^2 \mathcal{H}^2 F_0^2 \omega_0^{-10/3} \tag{13}$$

where

$$\mathcal{H} = 1 + \frac{\tilde{m}^2}{2} + 1.09\omega_0^{1/3}\tilde{m}, \tag{14}$$

(see Eq. (63) and Eq. (71) of Casati *et al.* (1988) with appropriate identification of symbols, in particular notice the difference of $\sqrt{2}$ in Eq. (14) due to the similar difference in the definition of the microwave amplitude).

An additional word of caution is necessary. Strictly speaking, PLT may be used for estimation of the thresholds only if the effective threshold is not high and no significant real ionization occurs (see the remarks in the Appendix of Casati *et al.* (1990)). The PLT may be modified to include an "effective ionization threshold", N, i.e. the highest principal quantum number which may be considered discrete due to experimental resolution, see Koch (1993). Typical values are $N = 90 - 160$. Since in our calculations, we may change

N (within reasonable limits) by modifying the effective size of the basis, see Zakrzewski *et al.* (1996a), we use, for the prediction of PLT, the appropriate formula given in Casati *et al.* (1990) which reads:

$$F10\% = \frac{\omega_0^{7/6}}{\sqrt{8n_0}\mathcal{H}}\sqrt{1-(\frac{n_0}{N})^2} \tag{15}$$

Having set the stage, let us present first the results for elongated states with $\tilde{m} = m_0/n_0 = 0$. Thus, $\mathcal{H} = 1$ in Eq.(14). We are left with 4 parameters: n_0, F_0, ω_0, and N. Firstly, we fix $n_0 = 48$, $N = 120$ and we study the $F10\%$ threshold as a function of ω_0. The result is shown (in the high frequency domain) in Fig. 1 as a double–logarithmic plot. Besides the existence of strong fluctuations, there is a general increasing trend with ω_0. The prediction of PLT, shown as a straight solid line, has the same trend, but obviously largely overestimates the $F10\%$ threshold. The dashed line, equal to half the PLT prediction, is in good overall agreement with the numerical results. Similar results are obtained for a different initial state $n_0 = 64$.

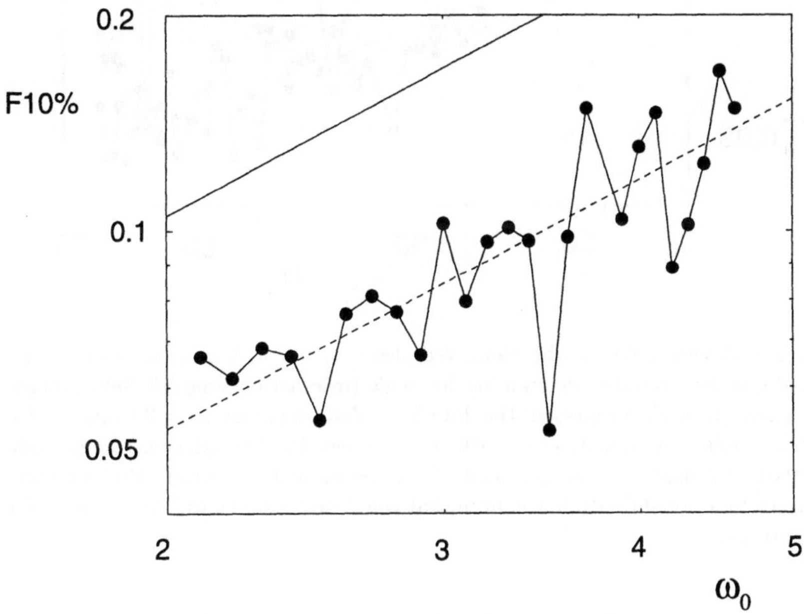

Fig. 1. $F10\%$ threshold for the elongated state $n_0 = 48, m_0 = 0$ (circles) as a function of the scaled field frequency on a log–log scale (interaction time 500 field cycles). The solid line is the PLT prediction of the photonic localization theory Casati *et al.* (1988) and the dashed line half this value. PLT predicts the correct functional dependence, but with a wrong coefficient.

The second test is to fix ω_0 and plot the $F10\%$ threshold versus n_0. For this, we take two values of $\omega_0 = 2.1$ and $\omega_0 = 2.8$. For each we change simultaneously n_0 and N to keep the ratio n_0/N fixed. The result is shown in Fig. 2 as a double–logarithmic plot. Again, the general trend is well predicted by PLT (straight solid line), but the exact value is significantly different. And again dividing the PLT prediction by 2 yields good agreement with numerical results. Clearly, PLT is in good qualitative agreement with our numerical results for elongated states but further work is needed to understand the discrepancy of a factor 1/2.

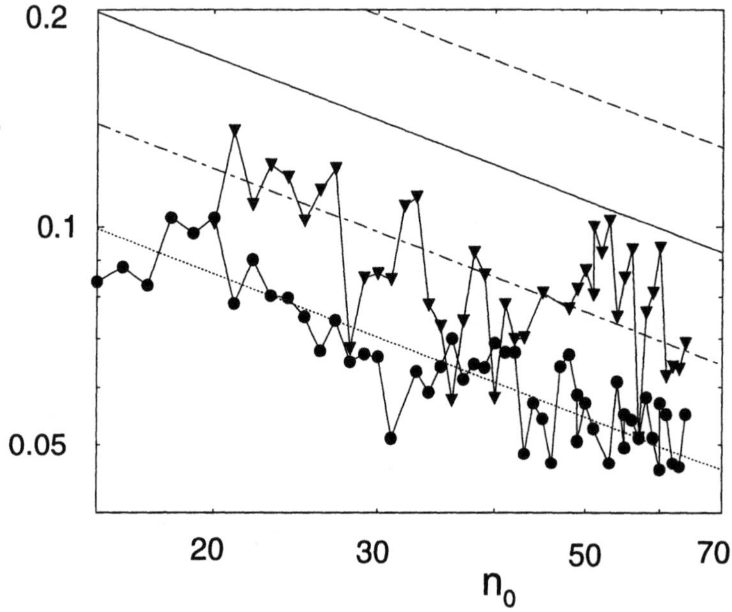

Fig. 2. $F10\%$ threshold for the elongated state $n_0, m_0 = 0$ as a function of the principal quantum number n_0 on a log–log scale (interaction time 500 field cycles). Filled circles (triangles) represent the data for scaled frequency $\omega_0 = 2.1$ ($\omega_0 = 2.8$). The PLT predictions Casati *et al.* (1988) are shown by solid (dashed) lines while the dotted (dot-dashed) lines give half of the theoretical prediction. PLT predicts the correct functional dependence, note that the "correction factor" is the same for both frequencies.

Next, notice that Eqs.(14),(15) allow us to test PLT with respect to changes in shape of initial state, i.e., changes of $\tilde{m}$. A quantitative test is presented in Fig. 3. The solid line is the prediction of PLT being also seriously off the numerical data. However, the qualitative trend is well verified (for negative and positive $\tilde{m}$). Correction by a factor 0.5, as before, yields (dashed line) a reasonable agreement with the data. A comparison with Fig. 1

shows, that at $\omega_0 = 2.1$ the fluctuations in quantum numerical data are not "favorable" and a factor of 0.5 works not too well. Therefore, we fitted the correction factor so that the numerical value and the modified PLT prediction (Casati *et al.* (1988)), exactly coincide at $\tilde{m} = 0$. Such a modification yields a dot-dashed line in excellent agreement with numerical data for $|\tilde{m}| < 0.2$. Two important factors are to be noticed: firstly, the fluctuations are much smaller than in the previous figures. An explanation for such a behaviour is given at the end of section 5. Secondly, PLT clearly breaks down around $\tilde{m} = 0.5$ whereas the prediction, Eq. (12), is that it should be valid everywhere since $\omega_0 = 2.1$. Obviously, the conditions given by Eq.(12) are not sufficiently restrictive. As a matter of fact, the Kepler map relies on the approximation that most of the energy exchange takes place in the vicinity of the perihelion. For states with significant angular momentum, this cannot be true. Hence, the applicability of the Kepler map is at best restricted to small angular momentum $\tilde{m} \ll 1$ and large frequency.

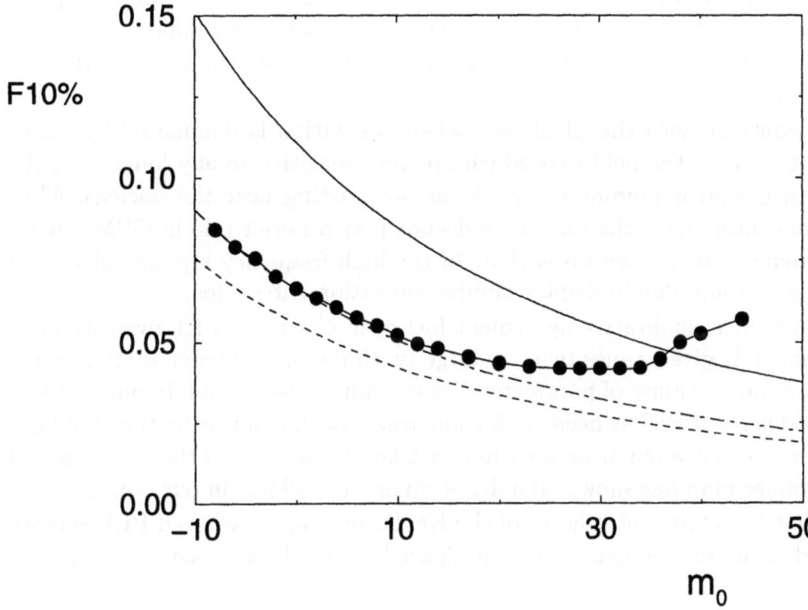

Fig. 3. The dependence of the $F10\%$ threshold for ionization on the initial angular momentum m_0 for the $n_0 = 48$ manifold at $\omega_0 = 2.1$. The PLT prediction Casati *et al.* (1988) is shown by a solid line, a dashed line gives half this value. The dot-dashed line is the result of a fit of the constant factor in the PLT prediction to numerical value at $m_0 = 0$.

The use of the Kepler map has been criticized by Nauenberg (1990b) who complained of its non-canonical character and constructed a better, canonical map. It would be most interesting, therefore, to compare quantum results with predictions of PLT based on the canonical Kepler map of Nauenberg (1990b) instead of the standard Kepler map of Casati *et al.* (1988). For high eccentricity, small microwave amplitude and high frequency, the Nauenberg Kepler map is close to the standard Kepler map and no significant difference is likely to take place. On the other hand, at medium and low eccentricities, the two maps are significantly different. The breakdown of PLT around $\tilde{m} = 0.5$ may be due either to a breakdown of the classical standard Kepler map or a failure of the localization picture itself. This interesting alternative is left for future studies since a generalization of the theory of Casati *et al.* (1988) for a map of a different functional form is unfortunately not straightforward.

The numerical results discussed in this section are extremely relevant to the interpretation of experimental results. The former show that, if a microcanonical distribution of initial states is used, the $\tilde{m} \simeq 0.5$ states will ionize first and consequently will govern the 10% ionization threshold. The numerical value obtained for the 10% threshold is in fair agreement with the experimental observation of Bellermann *et al.* (1996). However, quantum calculations and experiments have not actually been done in the same scaled frequency range. Further numerical calculations and/or experiments are needed.

In contrast with the LPM case where ionization is dominated by states elongated along the field axis which are very sensitive to any ionic core, the CPM situation is dominated by states not passing near the nucleus. Thus close encouters with the ionic core do not play a major role in CPM ionization. Hence, our prediction is that, in the high-frequency regime, alkali and hydrogen atoms should display similar ionization thresholds.

While nice qualitative agreement has been observed, PLT overestimates thresholds. It gives values twice as large (with factor of 2 being quite insensitive to detailed values of parameters) as our numerical results. In our opinion, some revision of PLT is needed. A comparison of the ionization threshold dependence on the angular momentum with the predictions of the theory based on a Kepler map has shown significant differences which, in our opinion, suggest that the regime of validity of the Kepler map approach and PLT is more limited in parameter space then anticipated originally in Casati *et al.* (1988).

5 Wavepackets Near the 1:1 Resonance - Chaos Assisted Ionization

In a widely unknown paper, Klar (1989) analyzed analytically the periodic orbits of the H atom in circularly polarized light. Being mainly interested in scattering in the presence of radiation, he predicted the existence of resonances in $e+H^-$ scattering due to trapping of resonance wavefunctions in the neighbourhood of the periodic orbits. He also pointed out that the mechanism for the localization is similar to a Wannier ridge resonance in two-electron systems thus pointing out the relation to the three-body problem.

Starting from this point, Białynicki-Birula *et al.* (1994) realized that the periodic orbits found by Klar (and having a period equal to the microwave period) are really fixed points in the frame rotating with frequency ω. Moreover, since one of the points is stable (see below), one may make a harmonic approximation around it finding the states well localized around the fixed point. In the laboratory frame, such states correspond to wavepackets orbiting around the nucleus with the microwave frequency. To test this hypothesis, they launched a Gaussian wavepacket centered at the fixed point and observed its time evolution via a numerical solution of the time-dependent Schrödinger equation. They observed that the wavepacket followed closely the fixed point for circa 20 microwave periods (limitation brought about by CPU time required) with a negligible spread.

Let us repeat the calculation of Klar (1989), Białynicki-Birula *et al.* (1994) in the rotating frame. Since 2D and 3D problems differ very little, let us present the 3D derivation. Looking at the initial Hamiltonian H, Eq.(4), one notices, that due to symmetry, one may seek the fixed point at $z = y = 0$. A condition for the fixed point, i.e., $v_i = 0$, $dp_i/dt = 0$ for $i = x, y, z$ yields immediately that $p_z = p_x = 0$, $p_{y,f} = \omega x_f$ with a subscript f indicating the fixed point. The remaining equation for dp_x/dt yields the condition

$$-F + \omega^2 x_f - \frac{|x_f|}{x_f^3} = 0 \tag{16}$$

yielding the position of fixed points as a function of F. Define the dimensionless parameter $q = (\omega^2|x_f|^3)^{-1}$ as in Białynicki-Birula *et al.* (1994). One may then easily express the fixed point position, the microwave field amplitude as well as the corresponding energy in terms of q and ω. Explicitly we may write:

$$x_{f1} = (q\omega^2)^{-1/3}; \qquad F = \frac{1-q}{q^{1/3}}\omega^{4/3}; \qquad E_f = \frac{1-4q}{2}\left(\frac{\omega}{q}\right)^{2/3} \tag{17}$$

and

$$x_{f2} = -(q\omega^2)^{-1/3}; \qquad F = \frac{q-1}{q^{1/3}}\omega^{4/3}; \qquad E_f = \frac{1-4q}{2}\left(\frac{\omega}{q}\right)^{2/3}. \tag{18}$$

For $F = 0$ we have $q = 1$ in Eqs.(17-18). For $F > 0$ (inspection of the Hamiltonian, Eq.(4) reveals that for F negative one has to interchange x_{f1} with x_{f2}) Eq.(17) is to be used for $q \leq 1$. For such a q, the corresponding fixed point is stable only for $q > 8/9$, see Klar (1989), Białynicki-Birula *et al.* (1994). The same amplitude, $F > 0$, corresponds to some $q > 1$ value in Eq.(18) yielding the position of the other fixed point (unstable for any $F > 0$). Note that the stable and the unstable fixed points are located on opposite sides of the nucleus and at almost the same distance from it (since q for realistic field values is close to unity).

The existence of the fixed point in the rotating frame is equivalent to a periodic orbit with period $T = 2\pi/\omega$ in the static frame. Thus the electronic motion is locked to the microwave frequency in a 1:1 resonance. A stable fixed point (periodic orbit) corresponds to a stable 1:1 resonance island. Note that the stable fixed point, when $F \to 0$ approaches $x_f = \omega^{-2/3}$, i.e. $x_f = 1$ for $\omega = 1$. Thus, provided the Kepler frequency and the microwave frequency match, the stable fixed point smoothly reaches the same location as that of a circular state in the microwave-free H atom (in the frame rotating with the common frequency). Note also that the corresponding energy $E_f = -3/2$ for $q = 1$ matches the energy of the circular orbit in the rotating frame.

Clearly the stability of the fixed point may strongly affect the dynamics and the ionization only for microwave frequencies quite close to the Kepler frequency (1:1 resonance). A completely different viewpoint has been presented by Howard (1995). He rederived stability conditions for the fixed points and claimed that stability of the global Hamiltonian is essential for treatment of the ionization. At the same time he strongly criticized the saddle point approach (presented above) used in the literature for low ω only. Note, however, that the condition for the fixed point to become unstable [obtained by placing $q = 8/9$ in F in Eq.(17)] which then reads $F = (\omega/3)^{4/3}/2$ (in agreement with Eq.(5.7) of Howard (1995)) leads to vanishingly small field values for very low frequencies. Consider again the scaled situation and take the electron energy (in the absence of the field) at $E_1 = -1/2$. In the rotating frame its energy is $E_r = -1/2 - L_z\omega$. For very small microwave field the maximal distance from the nucleus is about $r = 2$ while the fixed point is located at x_f given in Eq.(17). Taking the critical value $q = 8/9$, one gets for, e.g. $\omega = 0.1$: $F = 0.005364$, $x_{f1} \approx 4.8274$, and $E_f = -0.5955$. Even for the circular state, corotating with the field (which ionizes first among all states), the ionization threshold found in other classical and quantum treatments, Rzążewski and Piraux (1993), Zakrzewski *et al.* (1993), Kappertz and Nauenberg (1993), Rakovic and Chu (1995), is roughly 20 times higher. With increasing field, the unstable fixed point comes closer to the nucleus and, for the critical field value, around $F = 0.11$, the electron passes above the fixed point and ionizes. The picture nicely resembles the "just over" the barrier escape coming from the saddle-point picture.

Let us also clear another incorrect statement of Howard (1995). He claimed that his stability analysis is exact while the one presented earlier by Białynicki-Birula *et al.* (1994) was approximate. Thus his "treatment goes considerably further, in that the critical field (5.7) was derived directly from basic principles without any simplifying approximations". The discussion of fixed points presented above (following Białynicki-Birula *et al.* (1994)) is exact and leads to the same result as Howard (1995). Howard was probably misled by an additional flavor (typical for Physical Review Letters, unfortunately) of a general background discussion in Białynicki-Birula *et al.* (1994) concerning the analogy with Lagrange points in the restricted 3-body problem as well as by, may be, a bit too illustrative title of the letter referring to Trojan wavepackets.

For completeness, let us also note a recent letter by Farelly and Uzer (1995) who also rediscovered the fixed point analysis. The nice part of this letter consists of the introductory part where the effective potential is discussed both for $F = 0$ — i.e. pure, radially symmetric H atom case and for $F > 0$.

Let us come back to the wavepackets found by Białynicki-Birula *et al.* (1994). They found that Gaussian wavepackets placed in the neighbourhood of the fixed point were shape-preserving (non-dispersive) for a limited, rather short time. This interpretation was immediately challenged by Farelly *et al.* (1995) who claimed that an additional magnetic field is necessary to further stabilize the motion and make the region of linear stability in the phase space bigger. Clearly, this called for a precise and accurate determination of the wavepacket properties.

However, a different fact stimulated our work in this direction. Namely, similar wavepackets have been known to us to exist in LPM for some time, see Delande and Buchleitner (1994). They were shown to be really a *single* Floquet state (dressed state of the atom + field problem) of a full, nonlinear Hamiltonian. They were shown to follow classical dynamics for more than 10^6 field cycles (or Kepler periods of the unperturbed Coulomb motion) without visible dispersion. Using the computer code previously described in section 4.2, we have numerically calculated energies and lifetimes of "wavepacket" Floquet states, both in 2D and 3D.

To find the wave packet eigenstate among several thousands of eigenstates of H, we use classical stability analysis to write a harmonic approximation to the Hamiltonian, Eq.(4)

$$\begin{aligned} H_{\text{osc}} = E_f &+ \omega_+(a_+^\dagger a_+ + \frac{1}{2}) \\ &- \omega_-(a_-^\dagger a_- + \frac{1}{2}) + \omega_z(a_z^\dagger a_z + \frac{1}{2}) \end{aligned} \tag{19}$$

where E_f is given by Eq.(17). The $a_\pm^\dagger (a_\pm)$ are creation (annihilation) operators of the modes which entangle position and momentum variables in the polarization plane, while the harmonic binding in z direction is described by $(a_z, a_z^\dagger)$ operators. The normal modes (i.e. stability) frequencies are given by Delande *et al.* (1995), Zakrzewski *et al.* (1995)

$$\omega_\pm = \sqrt{\frac{2 - q \pm \sqrt{q(9q-8)}}{2}}\omega; \tag{20}$$

$$\omega_z = \omega\sqrt{q}. \tag{21}$$

Note that for considering the 2D quantum problem, the z-mode energy has to be suppressed (e.g. by putting $\omega_z = 0$). Let us denote by (n_+, n_-, n_z)

the wavepacket corresponding to n_i excitation of the ith mode. The "ground state" wavepacket will then be denoted by $(0,0,0)$, excited "2D" wavepackets will correspond to $(n_+, n_-, 0)$, while $n_z > 0$ will correspond to excitation in the direction perpendicular to the polarization plane. For the corresponding wavepackets obtained for the 2D quantum atom we use the notation (n_+, n_-). Inspection of the wavefunction plots (or better an analysis of the normal mode coordinates) indicates that the ω_+ mode consist mainly of "radial" excitation, while the ω_- mode is responsible for "azimuthal" excitation around the fixed point. Due to an entanglement of position and momenta in $x-y$ plane, this classification is an approximate one.

For typical plots of the wavepackets both in 3D (as isovalue plots) and in 2D, we refer the reader to the original publications (Delande *et al.* (1995), Zakrzewski *et al.* (1995)). The lifetime of the wavepacket, obtained directly from the width of the corresponding Floquet state, can easily exceed 10^6 Kepler periods, although it exhibits (see below) strong fluctuations. Clearly, recent objections (Farelly *et al.* (1995)) to wavepackets existence and non-spreading character were incorrect.

Suppose that $\omega = 1/n_0^3$. Then it can be shown that the $(0,0,0)$ wavepacket Floquet state is a superposition of atomic circular states $|n, l = n-1, m = n-1>$ centered around $n = n_0$. Similar analyses may be performed for "excited" wavepackets. Thus apart from the energy itself, one may use the overlaps with atomic states to identify the appropriate wavepacket state. This is quite useful for higher excitation of the different modes when the harmonic approximation, Eq. (19) becomes worse. While the wavepackets presented in Delande *et al.* (1995), Zakrzewski *et al.* (1995) were either of the ground state type (with all $n_i = 0$) or singly excited (one of $n_i = 1$) we present here the $(1,4)$ 2D wavepacket state for $\omega = 1/n_0^3$, $n_0 = 120$ in Fig. 4. It shows a complicated nodal pattern. Its distorted shape shows that a harmonic approximation is no longer valid. In fact, few states separate in energy the appropriate exact Floquet eigenstate from the harmonic approximation. Still the state shown lives as long as 10^6 microwave cycles and its lifetime is orders of magnitude larger than other closely lying Floquet states with comparable principal quantum number (the microwave field is quite large and the typical lifetime is of the order of few tens of microwave periods).

Another example is given in Fig. 5. The lack of simple nodal structures together with the apparent inadequacy of the harmonic approximation indicates that the state shown is localized really not on the fixed point, but rather on the *stable* periodic orbit (and tori surrounding it) emerging from the fixed point for energies higher than E_f. Such periodic orbits exist over a broad range of energies, supporting a whole ladder of states with high n_+ excitation. The state $(60,0)$ shown in Fig. 5 is an example of such states. Among the ladder of states some are quite distorted (reflecting an accidental avoided crossing occuring with other Floquet states - see below) while some

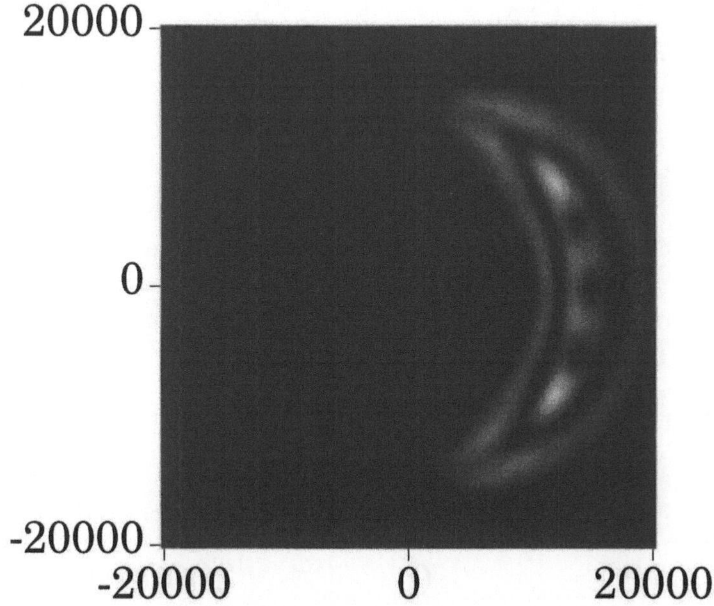

Fig. 4. Electronic density for a wavepacket eigenstate of the hydrogen atom in a circularly polarized microwave resonant with the Kepler frequency, that is $\omega = 1/n_0^3$, with principal quantum number $n_0 = 120$ and scaled microwave amplitude $F_0 = Fn_0^4 = 0.023$ (scale in Bohr radii). The wavepacket is a stationary state in the rotating frame. In the laboratory frame, it orbits around the nucleus preserving its shape, locked on the external microwave frequency. This is not the smallest (best localization) wavepacket, but an "excited" wavepacket with 4 quanta of excitation along the normal mode ω_+ and 1 quantum along the normal mode ω_-. As the normal modes entangle position and momenta coordinates, the nodal structure is not simple. The shape shows the breakdown of the harmonic approximation (see text for details).

of them remain strongly localized around the stable periodic orbit following the change of its shape with energy.

As previously discussed, there is also an unstable fixed point (in the rotating frame) on the other side of the nucleus. However, it is only quite weakly unstable and a wavepacket placed on it may keep approximately the same shape for quite a long time. Hence, one expects to find stationary states (in the rotating frame) scarred by this unstable fixed point, i.e. partially localized in its vicinity. Fig. 6 shows such a state. As expected, the electronic density is concentrated, although not fully, near the untable fixed point. Its

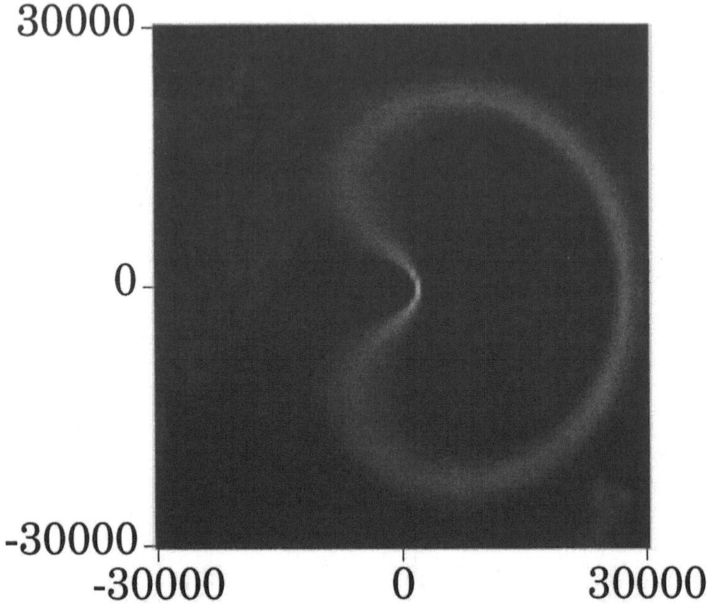

Fig. 5. Same as Fig. 4, but for a state having 60 quanta of excitation along ω_+, far beyond the domain of validity of the harmonic approximation ($n_0 = 120$, scaled frequency $\omega_0 = 1$, scaled microwave amplitude $F_0 = 0.04442$). This state is essentially localized in the vicinity of a stable periodic orbit.

lifetime versus ionization is shorter than that for the "stable" wavepacket, but is nevertheless large compared to other states.

Taking now the existence of the long–lived nonspreading quasi–classical wavepackets for granted, another important question may be asked. Is it possible to prepare them experimentally? The appropriate scheme which awaits experimental verification has been proposed in Delande *et al.* (1995). In short, it can be shown that the ground state (0,0,0) wavepacket Floquet state evolves from the circular state once the microwave field is switched on provided $\omega = 1/n_0^3$ for n_0 an integer. The state undergoes several avoided crossings during the turn on of the field. However, it can be shown, see Delande *et al.* (1995), that these avoided crossings are typically quite narrow, much smaller than the mean level spacing. Thus, one may imagine that the microwave field is turned on sufficiently slowly for an adiabatic passage to be valid in the absence of avoided crossings, while the turn on is still sufficiently fast to cross the avoided crossings diabatically. A simple Landau-Zener type estimate of the time scale on which the field should be switched on gives an estimate varying from 10 ns (for $n_0 = 60$) to 800 ns (for $n_0 = 120$), val-

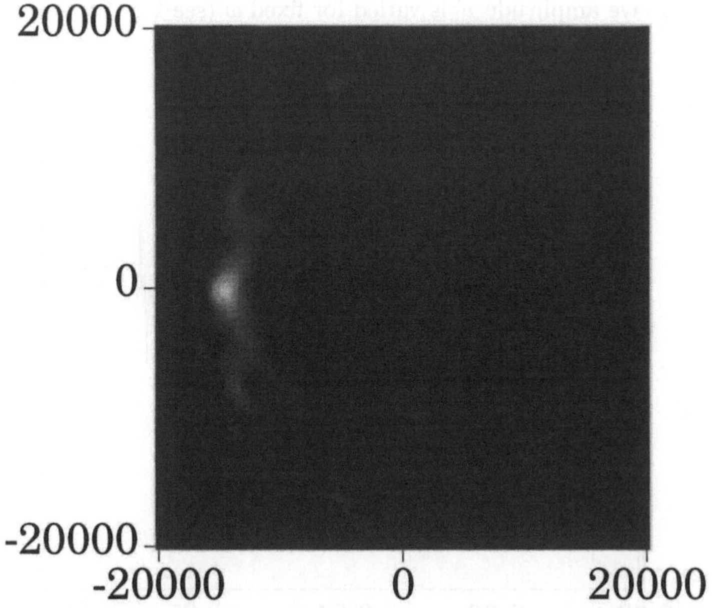

Fig. 6. Same as Fig. 4 for a wavepacket eigenstate localized on the unstable fixed point (on the other side of the nucleus). As expected, the localization on the unstable fixed point is not perfect. However, this state has a large dipole moment and a lifetime (vs. ionization) longer than neighbouring states ($n_0 = 120$, scaled frequency $\omega_0 = 1$, scaled microwave amplitude $F_0 = 0.035$).

ues which are feasible for state of the art microwave experiment. Once the wavepacket state is prepared, its detection may utilize its very long lifetime, the periodic oscillations of its large dipole moment etc. Preliminary calculations using numerical integration of the full time-dependent Schrödinger equation in the presence of a CPM with slowly increasing amplitude confirms that this method may produce almost pure wavepacket eigenstates, see Zakrzewski and Delande (1996).

Similarly, excited "wavepacket" states may be prepared, either directly, or via a subsequent excitation from the "ground state" wavepacket by, e.g. π pulses of frequency ω_+ or ω_-.

The strong localization on the fixed point (stable periodic orbit in the case of highly excited wavepackets) suggests, at first glance, that their properties, e.g. the width, are weakly and smoothly dependent on the parameters (the microwave amplitude or the frequency). This oversimplified point of view is incorrect as discussed by us recently in Zakrzewski *et al.* (1995). Small changes of ω lead to strong fluctuations of the wavepacket width which may vary over a few orders of magnitude. Similar strong fluctuations are observed when the microwave amplitude F is varied for fixed ω (see Fig. 7).

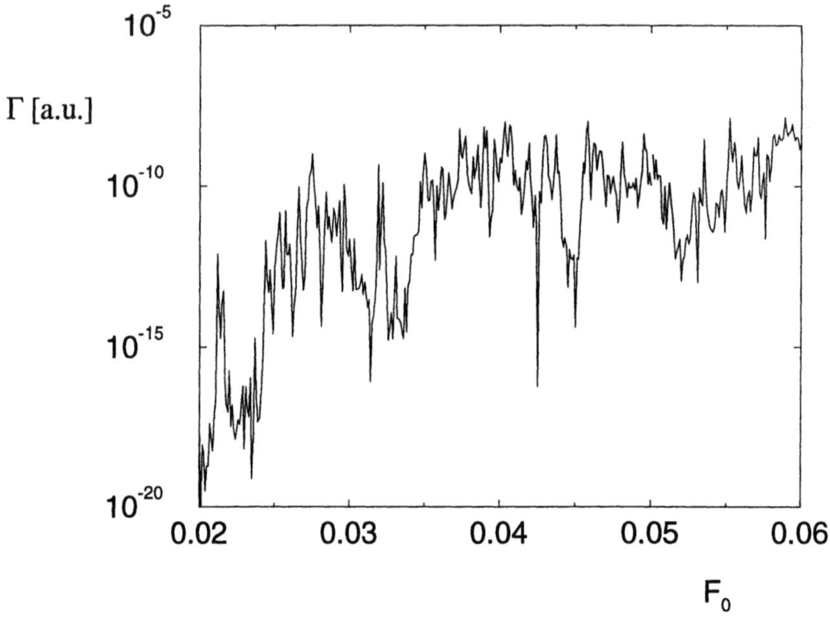

Fig. 7. Ionization rate (in atomic units) of the 2D $n_0 = 40$ wavepacket eigenstate as a function of the scaled microwave amplitude $F_0 = Fn_0^4$, for fixed microwave frequency $\omega = 1/n_0^3$. Note the general increasing trend and the huge fluctuations (several orders of magnitude) on a small scale. These "atomic conductance fluctuations" are due to small avoided crossings between the wavepacket eigenstate and neighbouring Floquet states.

The origin of these fluctuations is clear: the wavepacket Floquet state, as it evolves as a function of a parameter (microwave amplitude or frequency) changes its energy and has avoided crossings with other states lying in the chaotic part of phase space. In the vicinity of an avoided crossing, the interaction with the second state modifies strongly the width of the wavepacket state. Of course, not only the width, but also the wavefunction itself is affected - nice, strongly localized wavepackets, as depicted in Delande *et al.* (1995), Zakrzewski *et al.* (1995) are obtained only for wavepacket states well isolated from other states (i.e. not in the vicinity of some avoided crossing). However, these avoided crossings are usually extremely small as their sizes are determined by the overlap between the wavepacket state localized on a stable regular island and states lying in the chaotic sea. This overlap itself is governed by the tunneling amplitude between the regular island and the chaotic sea. Hence, the mechanism responsible for ionization of the wavepacket state is "chaos assisted ionization", that is tunneling to the chaotic sea followed by chaotic diffusion eventually leading to ionization. Hence, the situation is very similar to the "chaos assisted tunneling" phenomenon studied in Tomsovic and Ullmo (1994), Leyvraz and Ullmo (1996). The various matrix elements involved in "chaos assisted ionization" can be modeled using Random Matrix Theory. This allowed us to predict quantitatively the statistical distribution of the widths of the wavepacket states, see Zakrzewski *et al.* (1996b). This prediction is found to be in excellent agreement with the numerical results. The dramatic result, see Fig. 7, is that the widths fluctuate typically over several (3 or 4) orders of magnitude, even more than usually observed in fully chaotic systems, see Dupret *et al.* (1995).

The previous analysis provides us with clues for understanding the fluctuations in the 10% ionization threshold observed in Figs. 1 and 2, and almost absent in Fig. 3. Indeed, when, for example, the microwave amplitude changes, the Floquet states undergo plenty of avoided crossings and consequently huge changes in their widths. The total ionization yield is a sum over various individual Floquet states and consequently also displays large fluctuations. This is nothing but "atomic conductance fluctuations" (see Casati *et al.* (1990)), that is fluctuations of the transport property from an initial Rydberg state to the atomic continuum under the influence of the CPM. When the ionization yield is calculated at fixed microwave frequency and amplitude, but varying initial m_0 (angular momentum), the *same* Floquet states are involved whatever the initial state. Hence, the fluctuations vs. m_0 are much smaller, as clearly seen in Fig. 3.

6 Conclusion

In many aspects, the ionization of Rydberg states by a circularly polarized microwave field resembles the ionization by a linearly polarized microwave field. In both cases, the situation is simplified by the existence of a constant of motion. In the low-frequency regime, the classical dynamics is mainly regular and can be analyzed using static field ionization as a starting point. However, because of the transformation to the non inertial rotating frame, the analysis is tricky and has to be done very carefully. The various incorrect analyses and controversial statements in the literature can be considered, in our opinion, as definitely solved. Futhermore, for the hydrogen atom, the most recent semiclassical, quantum and experimental results agree fairly well.

On the other hand, the situation vis a vis linear and circular polarization differs greatly for non-hydrogenic atoms. In the latter case, the role of the ionic core and of collisions of the electron with it, do not seem crucial, in neither the low-frequency nor the high-frequency regime.

In the intermediate and high-frequency regime, the experimental results on hydrogen atoms in linearly or circularly polarized fields show similar behaviour. In our opinion, this is somewhat accidental and related to the absence of selection in the angular momentum of the initial state, which makes the 10% ionization threshold determined essentially by the atoms that ionize in the lowest field. The dynamics of all other states is lost in such experiments. Classical and quantum calculations show that the ionization mechanism is in both cases a diffusive chaotic motion, but the states involved are very different. In linear polarization, states elongated along the field axis are prominent and close encounters with the nucleus play a major role. In circular polarization, the more important states are the ones with intermediate angular momentum and close encounters with the nucleus are not important. This is why, in contrast with linear polarization, no important difference is expected between hydrogen and alkali atoms. A new generation of experiments, where the initial state could be precisely chosen, is required to identify unambiguously the various ionization mechanisms.

Acknowledgements

We acknowledge the important role of Andreas Buchleitner and Robert Gębarowski in obtaining some of our results reviewed here and for discussions. Laboratoire Kastler Brossel, de l'Ecole Normale Superieure et de l'Universite Pierre et Marie Curie, is unité associée 18 du CNRS. CPU time on a Cray C98 and on a Cray YMP-EL has been provided by IDRIS and by the Centre de Calcul pour la Recherche de l'Université Pierre et Marie Curie.

Support by the Polish Committee of Scientific Research under project 2P03B – 03810 as well as in the form of travel grants is acknowledged with thanks. We are grateful to French Embassy in Poland for support under billateral collaboration program No. 6209.

References

M. Arndt, A. Buchleitner, R. N. Mantegna, and H. Walther, Phys. Rev. Lett. **67**, 2435 (1991).

D. Banks and J. G. Leopold, J. Phys. B**11**, L5,37,2833 (1978).

J. E. Bayfield and P. M. Koch, Phys. Rev. Lett. **33**, 258 (1974).

J. E. Bayfield and D. W. Sokol, in Physics of Atoms and Molecules: Atomic Spectra and Collisions in External Fields, ed. K.T. Taylor, M. H. Nayfeh, and C. W. Clark, (Plenum, New York, 1988).

M. W. Bellermann, P. M. Koch, D. R. Mariani, and D. Richards, Bull. Am. Phys. Soc. **39**, 1129 (1994).

M.R.W. Bellermann, P.M. Koch D.R. Mariani and D. Richards, Phys. Rev. Lett. **76**, 892 (1996).

I. Białynicki-Birula, M. Kaliński, and J. H. Eberly, Phys. Rev. Lett. **73**, 1777 (1994).

A. Buchleitner, D. Delande, and J.–C. Gay, J. Opt. Soc. Am. B **12**, 505 (1995).

F. V. Bunkin and A. Prokhorov, Sov. Phys. JETP **19**, 739 (1964).

G. Casati, B. V. Chirikov, D. L. Shepelyansky, and I. Guarneri, Phys. Rep. **154**, 77 (1987).

G. Casati, I. Guarneri, and D. L. Shepelyansky, IEEE J. Quant. Electron. **24**, 1420 (1988).

G. Casati, I. Guarneri, and D. L. Shepelyansky, Physica A **163**, 205 (1990).

B. V. Chirikov in *Les Houches Session LII, Chaos and Quantum Physics 1989*, eds. M.-J. Giannoni, A. Voros and J. Zinn-Justin, (North-Holland, Amsterdam), (1991).

P.A. Dando, T.S. Monteiro, K.T. Taylor and D. Delande, Phys. Rev. Lett. **74**, 1099 (1995).

D. Delande, in *Chaos and quantum physics*, edited by M.-J. Giannoni, A. Voros and J. Zinn-Justin, Les Houches Summer School, Session LII (1991) (North-Holland, Amsterdam).

D. Delande and A. Buchleitner, Adv. At. Mol. Opt. Phys. **35**, 85 (1994); A. Buchleitner and D. Delande, Chaos, Solitons and Fractals **5**, 1125 (1995); Phys. Rev. Lett., **75**, 1487 (1995).

D. Delande, J. Zakrzewski, and A. Buchleitner, Europhys. Lett. **32**, 107-112 (1995).

K. Dupret, J. Zakrzewski and D. Delande, Euro. Lett. **32**, 107 (1995).

D. Farelly and T. Uzer, Phys. Rev. Lett. **74**, 1720 (1995).

D. Farelly, E. Lee, and T. Uzer, Phys. Rev. Lett. **75**, 972 (1995).

A. Franz, H. Klar, J. T. Broad, and J. S. Briggs, J. Opt. Soc. Am. B **7**, 545 (1990).

P. Fu, T. J. Scholz, J. M. Hettema, and T. F. Gallagher, Phys. Rev. Lett. **64**, 511 (1990).

T. F. Gallagher, Mod. Phys. Lett. B **5**, 259 (1991).

R. Gębarowski and J. Zakrzewski, Phys. Rev. A **50** 4408 (1994).

R. Gębarowski and J. Zakrzewski, Phys. Rev. A **51**, 1508 (1995).

J. A. Griffiths and D. Farrelly, Phys. Rev. A **45**, R2678 (1992).

J. E. Howard, Phys. Rev. A **46**, 364 (1992).

J. E. Howard, Phys. Rev. A **51**, 3934 (1995).

R. V. Jensen, S. M. Susskind, and M. M. Sanders, Phys. Rep. **201**, 1 (1991).

P. Kappertz and M. Nauenberg, Phys. Rev. A**47**, 4749 (1993).

H. Klar, Z. Phys. D**11**, 45 (1989).

P. M. Koch, in Proc. of Eighth South African Summer School in Theoretical Physics: *Chaos and Quantum Chaos*, 13-24 January 1992, Blydepoort, Eastern Transvaal, Republic of South Africa (Springer-Verlag 1993).

P. M. Koch and K. A. H. van Leeuven, Phys. Rep. **255**, 289 (1995).

J. G. Leopold and I. C. Percival, J. Phys. B**12**, 709 (1979).

F. Leyvraz and D. Ullmo, preprint (1996).

B. I. Meerson, Opt. Spectrosk. **51**, 582 (1981).

J. Mostowski and J. J. Sanchez-Mondragon, Opt. Commun. **29**, 293 (1979).

M. Nauenberg, Phys. Rev. Lett. **64**, 2731 (1990).

M. Nauenberg, Europhys. Lett. **13**, 611 (1990).

M. J. Raković and S.-I Chu, Phys. Rev. A**50**, 5077 (1994); *ibid.* **52**, 1358 (1995).

W. P. Reinhardt, Ann. Rev. Phys. Chem. **33**, 323 (1982); Y. K. Ho, Phys. Rep. **99**, 1 (1983).

K. Rzążewski and B. Piraux, Phys. Rev. A**47**, R1612 (1993).

S. Tomsovic and D. Ullmo, Phys. Rev. E **50**, 145 (1994).

D. Wintgen, Z. Phys. D **18**, 125 (1991).

J. A. Yeazell, M. Mallalieu, and C. R. Stroud, Jr., Phys. Rev. Lett. **64**, 2007 (1990); G. Alber and P. Zoller, Phys. Rep. **199**, 231 (1991).
J. Zakrzewski and D. Delande, J. Phys. B**28**, L667 (1995).
J. Zakrzewski and D. Delande, in preparation (1996).
J. Zakrzewski, D. Delande, J. C. Gay, and K. Rzążewski, Phys. Rev. A**47**, R2468 (1993).
J. Zakrzewski, D. Delande, and A. Buchleitner, Phys. Rev. Lett. **75** 4015 (1995).
J. Zakrzewski, G. Gębarowski, and D. Delande, Phys. Rev. A **54**, 691 (1996).
J. Zakrzewski, D. Delande and A. Buchleitner, to be published (1996).

Classical Orbits and Quantum Waves in Natural Atoms and in Designer Atoms

John B. Delos[1] and Charles D. Schwieters[1,2]

[1] Physics Department, The College of William and Mary, Williamsburg, VA 23187
[2] Chemistry Department, University of Pennsylvania, Philadelphia, PA 19104

Abstract. Whenever waves travel along two or more paths from a source to a detector interference effects can result. Such interference effects are seen in the absorption spectra of atoms in electric or magnetic fields. Interference is associated with waves that travel out from the atom and later return to it. Closely-related interference patterns are found in measurements of the conductance of a microscopic junction (or "designer atom"). In this latter case, interference is associated with waves that travel along various paths from the entrance to the exit lead. In this paper we discuss the close relationship between these two seemingly different phenomena.

Dedication Dieter Wintgen made many contributions to atomic theory, especially through his thoughts and his calculations that show relationships between classical and quantum descriptions of atoms. His energy, enthusiasm, and commitment can stand as a model and inspiration to every scientist. Somewhat before his untimely death, he began thinking about conduction of electrons in solids, and about the new phenomena that were beginning to appear in experiments. He again considered making use of the insights that semiclassical approximations may bring to the subject. Here we discuss the very close relationship between these phenomena. After briefly reviewing the theory of atomic absorption spectra, we describe a theory of conductance fluctuations in microjunctions. We show that both phenomena are governed by quantum interference among waves traveling along various classical paths.

1 Spectra of Atoms in Fields

Recurrence Spectroscopy – Theory The experimental and theoretical study of spectra of atoms in applied electric and magnetic fields has led to the development of "Recurrence Spectroscopy". In the first such measurement [1], hydrogen atoms initially in the $2p_z$ state were placed in a magnetic field, and the absorption spectrum to states near the ionization threshold was measured. That absorption spectrum as a function of photon energy was shown to consist of a superposition of sinusoidal oscillations. These oscillations are correlated with closed orbits of the electron as it travels on various paths that go out from the atom and later return.

We have one electron moving under the combined effects of the Coulomb field of the nucleus and the applied electric and magnetic fields (presumed to be parallel). The Hamiltonian for this system is

$$H(\mathbf{p},\mathbf{q}) = \mathbf{p}^2/2 - 1/r + B^2\rho^2/8 + Fz \ . \tag{1}$$

If the electron were a classical particle, then it would move along paths that are accelerated toward the nucleus by the $1/r$ Coulomb potential, accelerated downward along the z-axis by F, the applied electric field, and accelerated cylindrically inward by $B^2\rho^2/8$, the diamagnetic potential.

In quantum theory, the electron's state is described by a wave-function. The Green-function $G_E(\mathbf{q},\mathbf{q}')$ is the wave that arrives at $\mathbf{q}$ from a fixed-energy source at $\mathbf{q}'$. A seemingly-abstract formula in quantum mechanics relates the photoabsorption cross-section to the Green-function, and it provides a simple physical picture of the absorption process [2, 3]:

$$\mathrm{Abs}(E;B,F) = \int [D(\mathbf{q})\psi_i(\mathbf{q})]^* \, G_E(\mathbf{q},\mathbf{q}')D(\mathbf{q}')\psi_i(\mathbf{q}')\mathrm{d}\mathbf{q}\mathrm{d}\mathbf{q}' \ . \tag{2}$$

Here $\psi_i(\mathbf{q}')$ is the initial state of the electron (in our case $2p_z$), and $D(\mathbf{q}')$ is the relevant component of the dipole operator (z' for linear z-polarized light). In this formula, the function $D(\mathbf{q}')\psi_i(\mathbf{q}')$ plays the role of a "source-function." The very same function, $[D(\mathbf{q})\psi_i(\mathbf{q})]^*$ plays the role of a "detector." The Green-function propagates waves of energy E from points $\mathbf{q}'$ in the "source" to points $\mathbf{q}$ in the "detector." In this case, the "source" and "detector" are well-localized, and, most important, they are both in the same place. Hence the absorption cross-section is governed by waves that propagate out from and later return to the atom.

The physical picture is shown in Fig. 1. Light is shining on the atom in a $2p_z$ state. If the electron absorbs a photon, then it jumps into a state represented by a steady outgoing wave, which goes out from the atom in all directions (the angular distribution is governed by selection rules from the initial state). In this way, the initial-state wave-function and the dipole operator of the light constitute the "source." After the waves leave the immediate vicinity of the atom, it becomes appropriate to describe those outgoing waves using a semiclassical approximation. Instead of following the wave-fronts, we propagate the wave outward by following classical trajectories. As the trajectories move away from the atom, they are turned around by the Coulomb field and the applied fields, and some of the paths move in such a way that they return to the atom after some time. These returning trajectories carry a steady stream of returning waves, which refract and diffract around the atom to form incoming waves. These incoming waves overlap the "detector-function," and that overlap contributes to the photoabsorption cross-section. Interference among waves traveling on different closed classical orbits that go out from and return to the atom leads to the oscillations seen in the absorption cross-section.

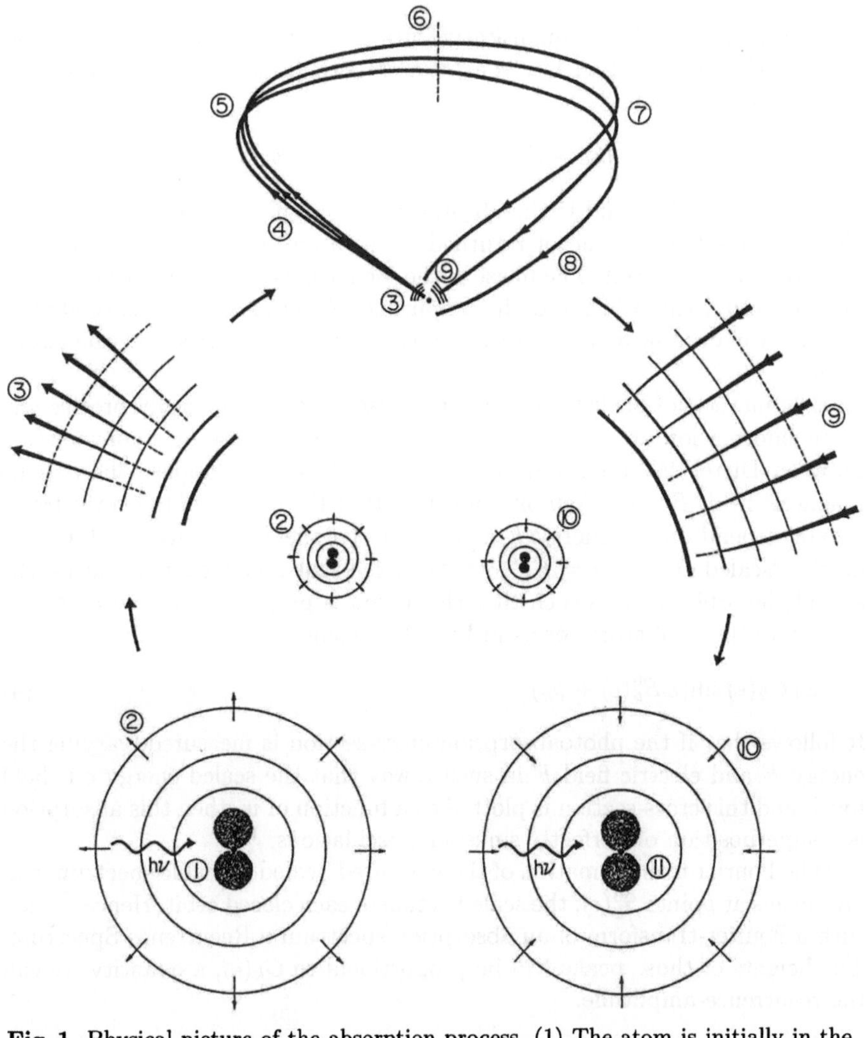

Fig. 1. Physical picture of the absorption process. (1) The atom is initially in the $2p_z$ state, with the oscillating field due to the laser present. (2) The oscillating field produces zero-energy Coulomb waves, which propagate outward in all directions. (3) For distances greater than about $50a_0$, a semiclassical approximation is appropriate, and we can propagate the wave outward by following classical trajectories. (4) A pencil of trajectories propagates outward, encounters a caustic (5), a focus (6), and another caustic (7). This group of trajectories started out in such a direction that it turned around and returned toward the atom (8). Around $50a_0$, we describe it as an incoming zero-energy Coulomb wave (9), which continues to propagate inward (10), until it overlaps with the initial $2p_z$ state (11). Interference between steadily produced outgoing and incoming waves leads to oscillations in the absorption spectrum. (The sizes of the first and last parts of the figure are about $10a_0$, the sizes of the second and fourth are about $60a_0$, and the size of the third is about $3000a_0$.)

This picture leads to a quantitative formula for the photoabsorption cross-section: the cross-section is given by a smooth background plus a sum of sinusoidal oscillations:

$$\mathrm{Abs}(E;B,F) = \mathrm{Abs}_0 + \Sigma_k C_k(E;B,F)\sin(S_k(E;B,F)+\phi_k)\ . \qquad (3)$$

The smooth background is the absorption rate that would be present if the electron were free, and never returned to the atom. Each oscillatory term is related to a closed orbit. The phase of the oscillations is related to the classical action around the orbit, and the amplitude of the oscillations is related to the electron current-density carried by returning trajectories near the closed orbit.

The method of Scaled-Variable Spectroscopy gives especially precise and clean information about closed orbits [4]. The idea arose in a conversation between Dieter Wintgen and Jörg Main in Bielefeld. Suppose there is no magnetic field, $B = 0$. Then one can show that the shapes of the trajectories do not depend on the energy E and the electric field F separately, but only on the "scaled energy" $\epsilon = EF^{-1/2}$. For fixed scaled energy, as F varies, the size of the orbit changes such that the action is proportional to $w = F^{-1/4}$. Therefore the oscillatory terms in Eq. (3) become

$$\Sigma_k C_k(\epsilon)\sin(wS_k^0(\epsilon)+\phi_k)\ . \qquad (4)$$

It follows that if the photoabsorption cross-section is measured, varying the energy E and electric field F in such a way that the scaled energy ϵ is held fixed, and this cross-section is plotted as a function of w, then this absorption is a superposition of perfectly sinusoidal oscillations. [1]

The Fourier transformation of the measured Scaled Variable spectrum will give peaks at points $S_k^0(\epsilon)$, the scaled action of each closed orbit. Hence we call such a Fourier-transform of an absorption spectrum a Recurrence Spectrum. The heights of those peaks will be proportional to $C_k(\epsilon)$, a quantity we call the recurrence-amplitude.

Recurrence Spectroscopy – Example Recurrence spectra for a Li atom in an electric field are shown in Fig. 2. Each measured peak (black) corresponds to one or more closed orbits. There is one orbit that goes up from the atom along the electric field and comes right back (the parallel orbit). When the electron returns to the vicinity of the nucleus it is turned around by the strong Coulomb field, and it retraces this orbit. The scaled action as a function of scaled energy for the parallel orbit and its repetitions are shown as the smooth curves in the action-energy plane. Toward the upper left of the figure, we see

[1] Actually, the background Abs_0 and the coefficients have a weak dependence on w. To obtain purely sinusoidal oscillations, one must subtract the background and multiply by a power of w. Details are given in Ref. [5]. There is also a scaled-variable method for magnetic fields, and for the case of both fields being present.

that the parallel orbit usually gives a relatively weak recurrence, often too weak to observe.

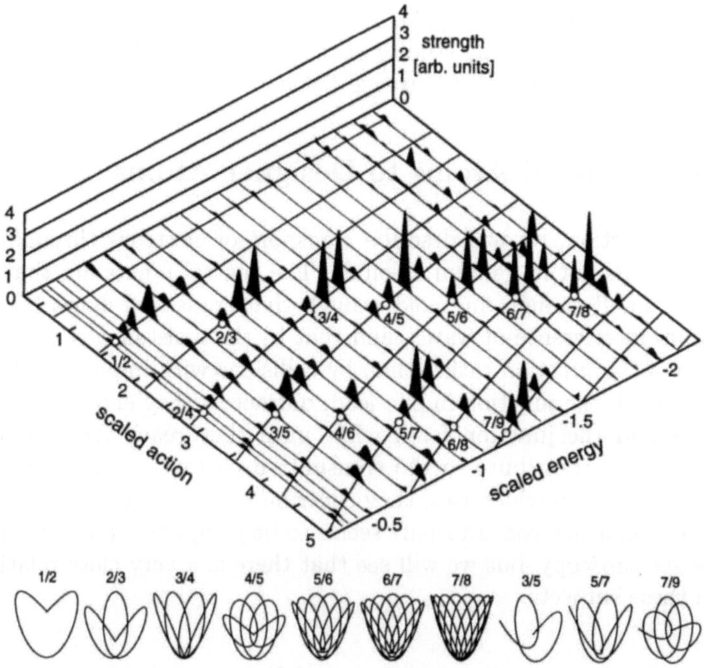

Fig. 2. The curves in the horizontal plane represent the scaled action of the parallel orbit and its repetitions as a function of scaled energy. Locations of bifurcations are marked with small open circles. New orbits created in bifurcations have almost the same action as the corresponding return of the parallel orbit. Measured recurrence strengths are shown in the vertical direction: each blackened peak is a recurrence associated with one or more returning orbits. Recurrences are especially strong at scaled energies near to (slightly lower than) bifurcations. Orbits created by bifurcation of the parallel orbit are shown along the bottom. The 2/4, 4/6, and 6/8 orbits are repetitions of the 1/2, 2/3 and 3/4 orbits, respectively, so their shapes are identical.

For $\epsilon \geq 0$ this parallel orbit is the only closed orbit, and it is unstable. For $\epsilon \leq 0$, it becomes stable, and other orbits bifurcate out of it as ϵ is lowered. Bifurcations occur when the period of stable oscillations about the parallel orbit are commensurate with its period. For example, at about $\epsilon = -0.4$ the period of oscillations transverse to the parallel orbit is twice the period of the parallel orbit, and the 1/2 orbit bifurcates from it. Close to such bifurcations, recurrences become much stronger (actually the recurrence is strongest just *after* the creation of a new orbit). The new 1/2 orbit has almost the same action as does the second return of the parallel orbit, so the measured recurrence strength is the sum of that due to the parallel orbit and that due to the new one.

Several other bifurcations (2/3, 3/4, 3/5, ...) are visible in these measurements. There are now many studies of recurrences and their bifurcations in magnetic, electric, parallel and crossed fields [6, 7, 8, 9].

2 From Natural Atoms to Designer Atoms

In the next section, we will describe transport of electrons through microscopic junctions cut in gallium arsenide. The central points are that a) the junction is small enough and cold enough that transport shows the effects of quantum interference of waves, and that b) the resistance is governed by the shape of the junction rather than by collisions with impurities. Electron waves approach the junction in one lead, diffract as they enter the junction, bounce around the junction for a while until they reach the exit lead, in which case they contribute to the transmitted current, or until they reach the entrance lead, in which case they contribute to the reflected current. At first glance, such electron transport seems to be completely different from recurrence spectroscopy, but we will see that there is a very close relationship between these subjects.

2.1 Conductance in Semiconductor Microjunctions

In recent experiments [10, 11] micron-sized two-dimensional junctions have been made such that both the quantum coherence length and the mean-free path for elastic collisions of electrons with defects are large compared to the size of the junction. The conditions are such that the electrons can be described as a two-dimensional ideal Fermi gas of noninteracting particles. In these circumstances a classical electron would bounce ballistically through the cavity; in quantum mechanics the electron wave function scatters elastically from the walls of the junction. This behavior is shown schematically in Fig. 3.

The conductance of such junctions has been measured [10, 11] and has been found to oscillate strongly as the Fermi energy or the strength of an imposed magnetic field is varied. Statistical properties of these fluctuations have been studied, and compared with predictions from random matrix theory [12, 13, 14, 15]. Also the high frequency part of the power spectrum has been studied, and its structure (power-law decay for regular systems vs. exponential decay for chaotic systems) has been compared with predictions from general semiclassical arguments. [16, 17]

Statistical properties of fluctuations may be interesting, but it is appropriate now to ask more detailed questions. Can the conductance fluctuations

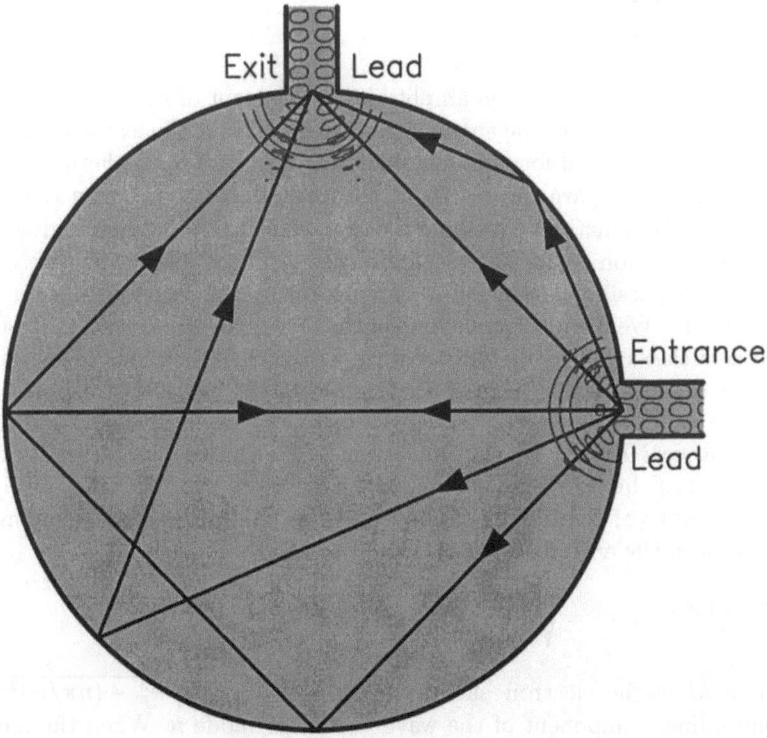

Fig. 3. Overview of the semiclassical method. A quantum wave incoming from a mode of the entrance lead diffracts as it enters the junction, and then travels along classical paths inside the junction. At the exit each trajectory contributes a term to the flux that goes out into each mode of the exit lead.

themselves be predicted? On the experimental side, will junctions be created such that the quantum fluctuations of conductance are reliably reproducible, and depend only on the lithographically observed geometry of the junction? On the theoretical side, can we develop simple formulas or algorithms that predict the fluctuations of conductance vs. Fermi energy or vs. magnetic field? In this section we describe a simple semiclassical theory that predicts some large-scale structure of conductance fluctuations. A more detailed description of the theory is given in Ref. 18.

The conductance g in a two lead junction is related to the quantum scattering matrix through the Landauer formula [19, 20, 21]. For a non-degenerate electron gas, this formula is

$$g = \frac{e^2}{\hbar} \sum_{m,n} |t_{mn}|^2 \ , \tag{5}$$

where t_{mn} is a transmission amplitude: the element of the S-matrix for scattering from mode n of the entrance lead to mode m of the exit lead. We give a general semiclassical formula for predicting t_{mn} vs. k_F, the Fermi wavenumber, and we compare the results of this formula with quantum calculations for a circular junction. We show that the semiclassical formula gives an accurate prediction of the large-scale (long wavelength) oscillations of $t_{mn}(k_F)$. It gives a possibly less accurate prediction of the large-scale structure of $|t_{mn}(k_F)|^2$. We then advance a hypothesis which, if correct, suggests that this semiclassical formula will accurately predict fluctuations of conductance in experimental realizations of quantum-coherent ballistic transport.

Semiclassical Theory We consider a junction with two straight leads of width w. Let (x, y) be coordinates along and across the entrance lead, with origin at the center of the mouth of the lead. The flux-normalized incoming wave function in the entrance lead is then

$$\psi_n^{\text{in,lead}}(x, y) = \sqrt{\frac{2M}{w\hbar k_n}} e^{ik_n x} \sin\left[n\pi(\frac{y}{w} + \frac{1}{2})\right] \ , \tag{6}$$

where M is the electron effective mass and $k_n = \sqrt{k_F^2 - (n\pi/w)^2}$ is the longitudinal component of the wave vector for mode n. When the incoming wave enters the junction it diffracts, and a circular wave radiates out from the lead into the junction. In the Kirchhoff approximation, that wave is

$$\psi_n^{\text{in,diff}}(\mathbf{q}) \approx \frac{e^{ik_F r}}{\sqrt{r}} \Theta_n(\theta) \ , \tag{7}$$

where $\Theta_n(\theta)$ is the angular distribution of these waves,

$$\Theta_n(\theta) = -n\sqrt{\pi w M/\hbar}\sqrt{\frac{k_F}{k_n}} e^{i\pi/4}(\cos(\theta) + k_n/k_F)$$
$$\times \chi_n(\frac{wk_F}{2} \sin\theta) \ , \tag{8}$$

$$\chi_n(a) = \left[(n\pi)^2 - (2a)^2\right]^{-1} \begin{cases} i\cos(a) \ , \text{n odd} \\ -\sin(a) \ , \text{n even} \end{cases} \tag{9}$$

This circular wave propagates semiclassically through the junction. Trajectories are launched from the initial circular wave, and bounce elastically from the boundary of the junction. At any arbitrary point $\mathbf{q}$ in the junction the wave function is a sum of semiclassical terms, one term for each trajectory, labeled l, that arrives at $\mathbf{q}$ from the initial circle. Each term $\psi_l^{\text{SC}}(\mathbf{q})$ is given by

$$\psi_l^{\text{SC}}(\mathbf{q}) = \psi_n^{\text{in,diff}}[(\mathbf{q}_0(\mathbf{q})]|\rho_l(\mathbf{q})|^{1/2} e^{i(k_F L_l(\mathbf{q}) - \mu_l \pi/2)} \ , \tag{10}$$

where $L_l(\mathbf{q})$ is the length of the lth trajectory, μ_l is the Maslov index associated with that trajectory, $|\rho_l(\mathbf{q})|$ is the classical density and $\mathbf{q}_0(\mathbf{q})$ is the initial point of the trajectory on the circular diffracted wave front [22].

From the wave function inside the junction, we can calculate the flux into various modes of the exit lead. This approach leads to the following result.

Semiclassical Formula Let s be a coordinate that goes around the perimeter of the junction, and straight across the leads. Let $s(\theta_0)$ be the location on the perimeter at which a trajectory starting at θ_0 lands, after some specified number of bounces. Let θ_0^l, θ_f^l be initial and final angles of the lth trajectory that goes from the center of the entrance to the center of the exit lead. We propose that each such trajectory contributes to the S-matrix a term

$$\begin{aligned} t_{mn}^{(l)}(k_F) &= -\sqrt{2\pi k_F}\frac{\hbar}{M}\mathrm{e}^{\mathrm{i}\pi/4}\Theta_m(\theta_f^l)\Theta_n(\theta_0^l) \\ &\quad\times\left|\cos(\theta_f^l)\frac{ds_l}{d\theta_0}\right|^{-1/2}\mathrm{e}^{\mathrm{i}k_F L_l-\mathrm{i}\mu_l\pi/2} \\ &= a_l\exp(\mathrm{i}k_F L_l)\ . \end{aligned} \tag{11}$$

The full derivation can be found in Ref. [18].

Path-Length Spectra: Power Spectra of Transmission and Reflection Amplitudes The power spectrum of t_{mn} is given by

$$\hat{t}_{mn}(L) = \left|\int_0^{k_F^{\max}} \mathrm{e}^{-\mathrm{i}k_F L} t_{mn}(k_F)\mathrm{d}k_F\right|^2 . \tag{12}$$

It follows from Eq. (11) that $\hat{t}_{mn}(L)$ is a set of peaks at L_l (the length of the lth trajectory from the center of the entrance to the center of the exit). The height of each peak should be proportional to the absolute square of the pre-exponential factor a_l in Eq. (11). We can therefore call this quantity the "Transmission Path-Length Spectrum".

Quantum calculations were performed for a junction consisting of a circle with radius $\sqrt{1+4/\pi}$ connected at a right angle to two leads of width $0.935\sqrt{\pi+4}$ [23]. A comparison of the quantum and semiclassical power spectra is shown in Fig. 4. The semiclassical formula well reproduces the quantum calculation of the transmission amplitude, and individual peaks can be assigned to classical paths.

The reflection amplitude r_{mn} is that element of the S-matrix corresponding to scattering from mode n of the entrance lead back into mode m of the entrance lead. Its Fourier power-spectrum constitutes the "Reflection Path-Length Spectrum". Comparison of the quantum and semiclassical $\hat{r}_{11}$ in Fig. 5A shows some agreement, but also shows that certain peaks (indicated by arrows) are missing or have the wrong magnitude.

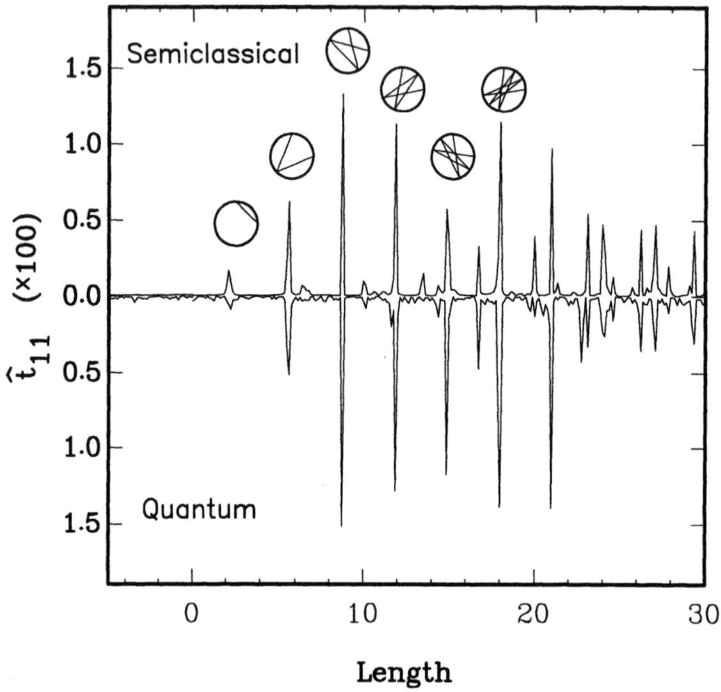

Fig. 4. Path-length spectrum for transmission, calculated by quantum and semiclassical methods. Some important peaks are labeled by their associated paths.

"Ghost Paths" In the simple semiclassical calculations (Figs. 4 and 5A), when a trajectory reaches a lead mouth it is assumed that all of the flux associated with that trajectory is absorbed: the lead casts a sharp shadow in the reflected wave. In reality, diffractive scattering blurs these shadows. A diffracted wave goes out in all directions; it only partially cancels the specularly-reflected wave, and it produces nonzero amplitude on paths that are not permitted by geometrical optics.

Therefore, diffractive scattering produces peaks associated with "ghost paths" that bounce elastically around the enclosure and diffract one or more times off a lead mouth. Diffractive reflection from an opening can be treated by Babinet's principle: the total reflected wave is the specularly reflected wave *minus* the wave produced by a small barrier coinciding with the opening.

We recalculated $r_{11}(k_F)$ including trajectories having up to 15 bounces, connected by up to two diffractive scattering events, with paths of length greater than 50 ignored. The results of a calculation including diffractive

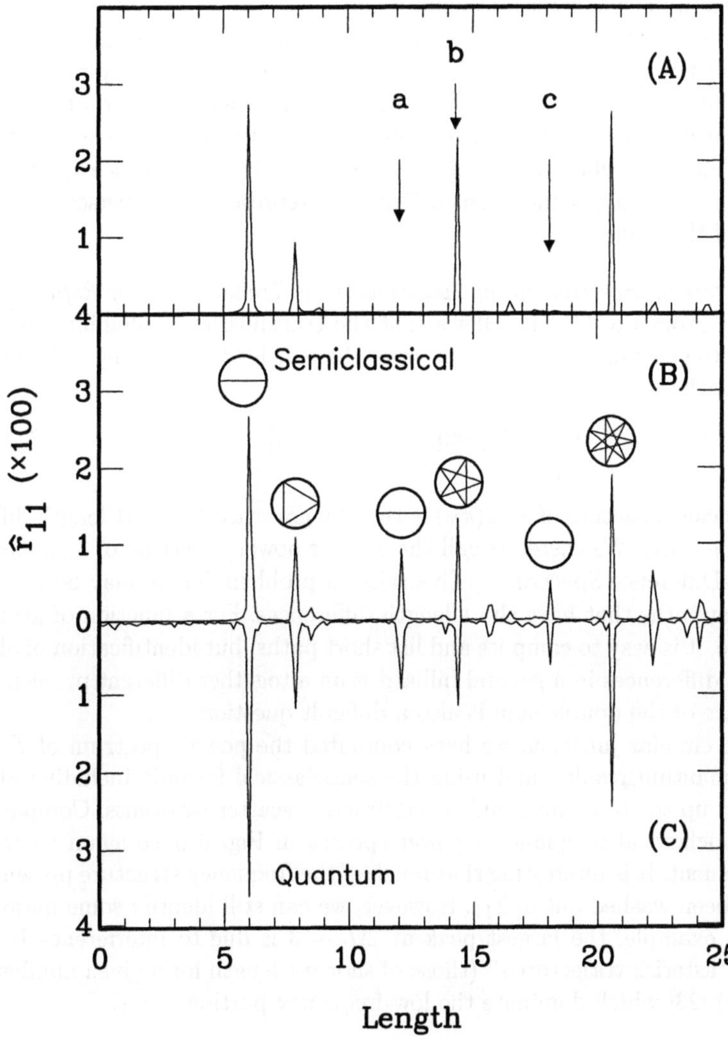

Fig. 5. Path-length spectrum for reflection, calculated by semiclassical and quantum methods. A) semiclassical without diffractive scattering from lead mouths, B) semiclassical including diffractive scattering, C) quantum. Again, the major peaks arise from identifiable paths. Peaks "a" and "c" are missing in the simple semiclassical calculation in panel A. They are "ghost paths," caused by diffractive reflection from the mouth of a lead. Waves go from the entrance lead across the junction, and are reflected back to the entrance lead; there they are partially absorbed and partially reflected with diffraction. Peak "a" represents four traverses and peak "c" six traverses of the diameter of the circle.

scattering are shown in Fig. 5B. Here there is much better agreement between the quantum and semiclassical calculations.

The lowest-frequency peak in Fig. 5A corresponds to the "straight-ahead" path, which bounces once from the far wall before going directly back out the entrance. Missing peak "a" occurs at a length exactly twice that of the straight-ahead path. This peak arises from waves which propagate across the circle and back, but which are diffractively scattered by the entrance aperture and retrace the straight-ahead path. Peak "c" represents 6 traverses of the diameter of the circle.

Length-Difference Spectra: Power Spectrum of the Transmission or Reflection Probability From Eq. (11) it follows that the transmission probability from mode n of the entrance lead to mode m of the exit lead is given by a double sum over paths

$$T_{mn}(k_F) = |t_{mn}|^2 = \sum_{l,l'} a_l^* a_{l'} \exp[ik_F(L_{l'} - L_l)] \ . \tag{13}$$

The large-scale structure of $T_{mn}(k_F)$ is thereby governed by short length differences $(L_{l'} - L_l)$. We therefore call the Fourier power-spectrum of $T_{mn}(k_F)$ a "Length-Difference Spectrum". This raises a problem. There may be pairs of very long paths that have short length differences. For a junction of arbitrary shape, it is easy to compute and list short paths, but identification of all short path differences in a general billiard is an altogether different problem. Convergence of the double sum is also a difficult question.

For the circular junction, we have computed the power spectrum of T_{11} from the quantum results, and using the semiclassical formula including all trajectories up to 15 bounces and two diffractive scattering events. Comparison of semiclassical to quantum power spectra in Fig. 6 once again shows good agreement. It is interesting that much of the frequency structure present in $\hat{t}_{11}$ has been washed out in $\hat{T}_{11}$. However, we can still identify some major peaks. For example, the largest peak at $\Delta L \sim 3$ is due to interference between the "asterisk trajectories" (those of shortest length for a given number of bounces) [23] which dominate the low-frequency portion of $\hat{t}_{11}$.

A Hypothesis About Long Paths Even in a semiconductor microstructure with no defects or impurities, thermal fluctuations disturb the electrons as they bounce around, distorting the paths and causing loss of coherence. We expect that long paths through the junction may be more disturbed by thermal fluctuations than short paths. It is credible, then, to suggest that in experimental measurements, interference fluctuations associated with short length-differences between long paths may be quenched by thermal fluctuations. If this hypothesis is correct, then for comparison with measurements, Eq. (13) should be replaced by

$$T_{mn} = T_{mn}^{\text{inc}} + \sum_{l,l'} a_l^* a_{l'} \exp[ik_F(L_l' - L_l)] f_l f_{l'} \ , \tag{14}$$

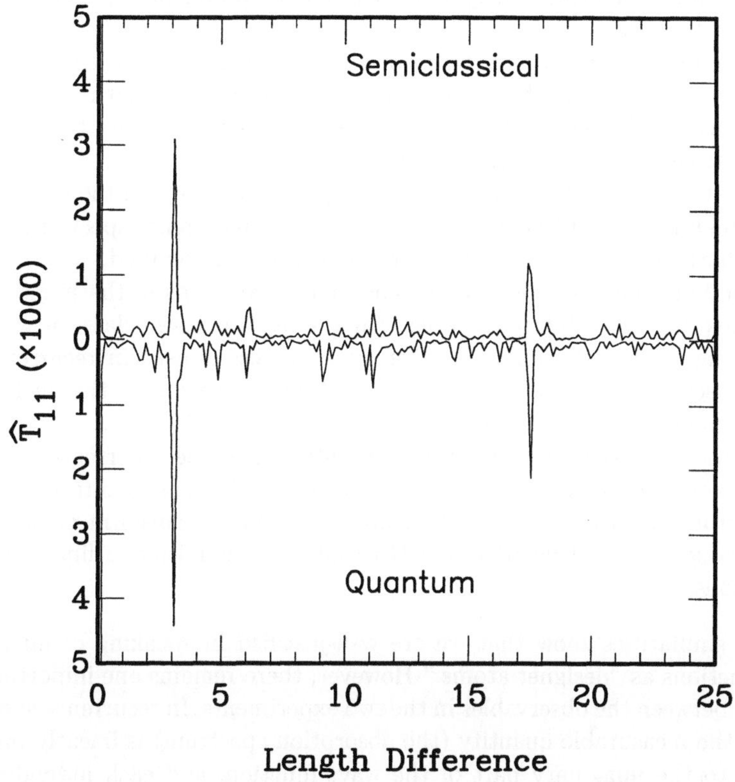

Fig. 6. Power spectra of the quantum mechanical and the semiclassical transmission coefficient modulus squared.

where T_{mn}^{inc} is an incoherent "background" contribution and f_l is a function which cuts off long paths, such as $\exp[-(L/\Lambda)^2]$. If this hypothesis is correct, it would resolve the problem of the many small ΔL contributions from long paths: neglected long paths would only contribute to the background and would not contribute to the length-difference spectra.

3 Conclusion

We have seen that there are many similarities between recurrence spectroscopy and quantum transport through microstructures. In summary, the parallels can be listed:

- An independent particle model can be used for electron transport through a junction, so even though the sample contains a huge number of electrons, we can describe transport to reasonable approximation as a "one-electron" process.

- Measurements of the conductance of microstructures show the effects of interference of waves that propagate on many different paths through the junctions. These effects are analogous to the effects of interference of waves traveling around closed orbits in atoms.
- The wavelength of the waves is small compared to the size of the junction and small compared to the relevant distances in recurrence spectroscopy, so in both cases the semiclassical approximation can be used.
- Diffraction occurs at the entrance and at the exit leads of the junction. We have to join a diffracted wave near the lead to a semiclassical wave bouncing around the junction. The same process is used in recurrence spectroscopy, where we have to join a diffracting wave near the nucleus to a semiclassical wave elsewhere.
- Both the microstructure transition amplitude and the recurrence spectrum consist of a superposition of sinusoidal oscillations, one from each path from entrance to exit. Therefore, the Fourier transform of either gives a spectrum of lengths of paths that connect initial to final wave functions.

These similarities show that we are well-justified in speaking of microscopic junctions as "designer atoms." However, there remains one important difference between the observables in the two experiments. In recurrence spectroscopy, the measurable quantity (the absorption spectrum) is linearly proportional to the imaginary part of the wave-function, and each individual path produces a sinusoidal oscillation. However in transport theory, the measurable quantity (the conductance of the junction) is proportional to the absolute square of the wave function. In this case, the transport rate also consists of a superposition of sinusoidal oscillations, but now it is one sine-wave from each *pair* of paths from entrance to exit. This is one of several effects that makes direct comparison between theory and experiment more difficult for transport than for spectroscopy.

References

1 J. Main, G. Wiebusch, A. Holle, and K. H. Welge, *Phys. Rev. Lett.* **57**, 2789 (1986).

2 M. L. Du and J. B. Delos, *Phys. Rev. A* **38**, 1896 (1988); *Phys. Rev. A* **38**, 1913 (1988).

3 E. G. Bogomol'nyi, *JETP Lett.* **47**, 526 (1988); *Sov. Phys. JETP* **69**, 275 (1989).

4 A. Holle, J. Main, G. Wiebusch, H. Rottke, and K. H. Welge, *Phys. Rev. Lett.* **61**, 161 (1988); U. Eichmann, K. Richter, D. Wintgen, and W. Sandner, *Phys. Rev. Lett.* **61**, 2438 (1988).

5 J. Main, G. Wiebusch, K. Welge, J. Shaw, and J. B. Delos, *Phys. Rev. A* **49**, 847 (1994).

6 T. van der Veldt, W. Vassen, W. Hogervorst, *Europhys. Lett.* **21**, 903 (1993).
7 M. Courtney, H. Jiao, N. Spellmeyer, D. Kleppner, J. Gao, and J. B. Delos, *Phys. Rev. Lett.* **74**, 1538 (1995); J. Gao and J. B. Delos, *Phys. Rev. A* **49**, 869 (1994).
8 J.-M. Mao, K. A. Rapelje, S. J. Blodgett-Ford, J. B. Delos, A. König, and H. Rinneberg, *Phys. Rev. A* **48**, 2117 (1993); A. D. Peters, C. Jaffé, and J. B. Delos, *Phys. Rev. Lett.* **73**, 2825 (1994); H. H. Fielding, J. Wals, W. J. van der Zande, and H. B. van Linden van den Heuvell, *Phys. Rev. A* **51**, 611 (1995); M. Courtney, *Phys. Rev. A* **51**, 4558 (1995).
9 G. Raithel, M, Fauth, and H. Walther, *Phys. Rev. A* **44**, 1898 (1991); *Phys. Rev. A* **47**, 419 (1993); G. Wiebusch, J. Main, K. Krüger, H. Rottke, A. Holle, and K. H. Welge, *Phys. Rev. Lett.* **62**, 2821 (1989); A. D. Peters and J. B. Delos, *Phys. Rev. A* **47**, 3020 (1993).
10 C.M. Marcus, A.J. Rimberg, R.M. Westervelt, P.F. Hopkins and A.C. Gossard, *Phys. Rev. Lett.* **69**, 506 (1992); C.M. Marcus, R.M. Westervelt, P.F. Hopkins and A.C. Gossard, *Chaos* **3**, 643 (1993); *Surf. Sci.* **305**, 480 (1994).
11 M.W. Keller, O. Millo, A. Mittal, D.E. Prober and R.N. Sacks, *Surf. Sci.* **305**, 501 (1994).
12 H.U. Baranger and P.A. Mello, *Phys. Rev. Lett.* **73**, 142 (1994).
13 R.A. Jalabert, J.-L. Pichard and C.W.J. Beenakker, *Europhys. Lett.* **27**, 255 (1994).
14 P.W. Brouwer and C.W.J. Beenakker,*Phys. Rev. B* **50**, 11263 (1994); *Phys. Rev. B* **51**, 7739 (1995).
15 X. Yang H. Ishio and J. Burgdörfer,*Phys. Rev. B* **52**, 8219 (1995).
16 R.V. Jensen, *Chaos* **1**, 101 (1991); W.A. Lin, J.B. Delos and R.V. Jensen, *Chaos* **3**, 655 (1993).
17 N. Argaman, *Phys. Rev. Lett.* **75**, 2750 (1995).
18 C.D. Schwieters, J.A. Alford and J.B. Delos, *Phys. Rev. B* **54**, 10652 (1996).
19 R. Landauer, *IBM J. Res. Dev.* **1**, 223 (1957); *Philos. Mag* **21**, 863 (1970).
20 M. Büttiker, *Phys. Rev. Lett.* **57**, 1761 (1986).
21 H.U. Baranger and A.D. Stone, *Phys. Rev. B* **40**, 8169 (1989).
22 J.B. Delos, *Adv. Chem. Phys.* **55**, 161 (1986).
23 H. Ishio and J. Burgdörfer, *Phys. Rev. B* **51**, 2013 (1995).

Semiclassical Approaches to Atoms in External Fields

Jörg Main

Institut für Theoretische Physik I, Ruhr-Universität Bochum, D-44780 Bochum, Germany

Abstract. Applying *closed-orbit* theory to the recurrence spectra of atoms in magnetic fields, one can interpret most, but not all, structures semiclassically in terms of hydrogenic closed classical orbits. Here we also report semiclassical approaches to phenomena in hydrogenic and nonhydrogenic atoms which can *not* be explained within conventional closed orbit theory. In particular, we analyze various types of catastrophes at bifurcations of orbits and discuss the role of *ghost orbits* living in complex phase space. Unphysical divergences of semiclassical recurrence amplitudes near bifurcations are removed by uniform semiclassical approximations which are completely determined by classical parameters of the real orbits and complex ghosts. For nonhydrogenic atoms we investigate the scattering of trajectories at the ionic core and interpret unidentified structures in nonhydrogenic spectra in terms of families of core-scattered closed orbits.

1 Introduction

Highly excited atoms in external magnetic and electric fields are nontrivial systems possessing a classically chaotic counterpart, with at least two nonseparable, strongly coupled degrees of freedom. Ever since the discovery of quasi-Landau type modulations in the spectra of barium by Garton and Tomkins (1969) atoms in magnetic fields have served as prototype systems for studying – both experimentally and theoretically – the quantum manifestations of classical chaos in real physical systems. As the corresponding classical dynamics of Rydberg atoms is chaotic, the quantitative description of their quantum features in terms of classical orbits was, and still is, a big challenge to theory as regards the development and application of semiclassical methods. An interpretation of Garton and Tomkins modulations was given by Edmonds (1970) in terms of a one-dimensional semiclassical quantization of the electronic motion perpendicular to the magnetic field axis. Since then experiments on the hydrogen atom in magnetic fields by Holle et al. (1986a) and Main et al. (1986) have revealed a variety of new types of quasi-Landau modulations which also could be qualitatively interpreted in terms of classical periodic motion. By taking advantage of a scaling property of the classical Hamiltonian the method of *scaled energy spectroscopy* (Holle et al. (1988), Main et al. (1991)) enabled a systematic search for the complete set of all types of modulations in quantum spectra related to classical periodic motion. Similar results as for the hydrogen atom were also observed in experiments on nonhydrogenic atoms in external fields, e.g. on helium by van der Veldt et al. (1993) and lithium by Courtney et al. (1995).

A decisive advance for an interpretation of structures in the photoabsorption cross section was achieved by the development of *closed orbit theory* by Du and Delos (1988) and Bogomolny (1989). The method allows, at least in low resolution, a quantitative semiclassical calculation of spectra in terms of a few parameters of closed orbits starting at the nucleus and returning back to the nucleus. A detailed comparison of experimental recurrence spectra on hydrogen with the results of closed orbit theory is presented in Main et al. (1994).

In our opinion, as a result of the intense research on atoms in external fields during the last decade, viz. experimental high resolution spectroscopy and numerical quantum calculations (Wintgen et al. (1986), Holle et al. (1986b)), and the development and application of semiclassical closed-orbit theory, the fundamental problems of this quantum system are solved to a large extent. This allows us to concentrate now on features and phenomena in spectra which can *not* be explained using conventional semiclassical theories and that have been neglected in the early studies.

Such a feature is, e.g., the observation of resonances in recurrence spectra of the hydrogen atom at positions where no closed classical orbits exist. These structures appear at energies near bifurcations of orbits, i.e. where orbits are born or vanish. Bifurcations are a typical phenomenon of classical periodic orbits in chaotic systems with nonhyperbolic Hamiltonian dynamics. A good example is the hydrogen atom in a magnetic field which undergoes a transition from regularity to chaos with increasing energy. The bifurcation scheme of this system is analysed in Mao and Delos (1992) and Sadovskií et al. (1995, 1996). Near bifurcations closed orbit theory fails, the semiclassical formulas diverge and are singular exactly at the bifurcation points. These "catastrophes" do not only occur in closed orbit theory but are well known phenomena in various fields of physics, e.g. in semiclassical scattering theory (Berry and Mount (1972), Connor (1976)), diffraction theory in optics (Berry and Upstill (1980)) or periodic orbit theory (Gutzwiller (1990), Ozorio and Hannay (1987)). Different kinds of catastrophes exist and are characterized by various forms of caustics of a bundle of lines (or more specifically in physical systems a bundle of classical trajectories or light rays). A systematic mathematical classification and analysis of structural stable caustics was achieved by the development of *catastrophe theory* (see Poston and Steward (1978)). The divergencies in semiclassical theories are removed in *uniform* semiclassical solutions (Berry and Mount (1972), Connor (1976)). Uniform approximations have been constructed e.g. for atomic and molecular scattering problems (Berry and Mount (1972), Connor (1976)), photodetachment of H^- in parallel electric and magnetic fields (Peters et al. (1994)), time dependent wave packet propagation (Beims and Alber (1993)), and for continuum

Stark spectra (Courtney et al. (1995)). In an application of periodic orbit theory to a kicked top Kuś et al. (1993) have revealed the existence and importance of prebifurcation periodic *ghost orbits*. The role of ghost orbits in the hydrogen atom and their relation to uniform semiclassical approximations has been investigated by Main and Wunner (1997).

Other features which can not be explained by hydrogenic orbits and conventional closed-orbit theory have been revealed in quantum defect R-matrix calculations (Monteiro and Wunner (1990), Dando et al. (1995)) and experimental measurements (Delande et al. (1994)) on *nonhydrogenic* Rydberg atoms in magnetic fields, i.e. changes are observed in the intensities of resonances, and, most interestingly, resonances appear at positions where no closed orbits exist in the hydrogen atom. Similar nonhydrogenic signatures have also been observed in experimental Stark spectra of lithium (Courtney et al. (1994, 1995)) and in spectra of rubidium atoms in crossed magnetic and electric fields (Raithel et al. (1994)). As general Rydberg atoms differ from hydrogen in the existence of a non-Coulombic ionic core a detailed analysis of core effects, and in particular, of core-scattering is necessary. In fact the nonhydrogenic resonance structures can be explained either by quantum mechanical scattering of waves at the ionic core (Dando et al. (1995)) or classically in terms of new types of nonhydrogenic closed orbits investigated by Hüpper et al. (1995, 1996). The decisive new start for the treatment of core effects in the framework of closed orbit theory is modelling of the ionic core by means of a suitable core potential and consideration of the classical dynamics of the highly excited electron in this model potential.

The purpose of this article is to review semiclassical approaches to atoms in external fields, and in particular the calculations and interpretations of photoabsorption spectra in terms of closed classical orbits. In Sect. 2 we discuss the classical dynamics including core-scattering in nonhydrogenic atoms and ghost orbits, i.e. the continuation of dynamics to complex phase space. In Sect. 3 we derive semiclassical closed-orbit theory and construct uniform approximations for various types of catastrophes. A discussion of results and concluding remarks follow in Secs. 4 and 5.

2 Classical Dynamics

The basic equations for the calculation of classical trajectories and a periodic orbit search for the hydrogen atom in a magnetic field are given in many papers (see e.g. Friedrich and Wintgen (1989)). Here we review briefly the main ideas only and concentrate on peculiarities of the complex continuation of phase space, the search for ghost orbits, and scattering of trajectories off the ionic core in nonhydrogenic atoms.

The nonrelativistic Hamiltonian for atoms in a magnetic field of strength B directed along the z axis reads (in atomic units, $\gamma = B/(2.35 \times 10^5\mathrm{T})$)

$$H = \frac{1}{2}\mathbf{p}^2 + V(r) + \frac{1}{2}\gamma L_z + \frac{1}{8}\gamma^2\varrho^2 = E\,. \tag{1}$$

The component of the angular momentum parallel to the field axis is conserved and we take $L_z = m\hbar = 0$ in all classical calculations. The potential $V(r)$ is the Coulomb potential in case of the hydrogen atom and is modified by a short-ranged core potential modeling the non-excited electrons of the ionic core in general Rydberg atoms. For numerical convenience in the classical calculations we assume the dependence

$$V(r) = -\frac{1}{r} - \frac{Z-1}{r}\left(1+\frac{r}{a}\right)e^{-r/a}\,, \tag{2}$$

where Z is the nuclear charge and a determines the range of the core. The potential is spherically symmetric and complies with the conditions $V(r) \to -Z/r$ and $V(r) \to -1/r$ in the limiting cases $r \to 0$ and $r \to \infty$, respectively.

A special feature of the Hamiltonian of the hydrogen atom is its scaling property with respect to the magnetic field strength. In scaled coordinates and momenta, $\tilde{\mathbf{r}} = \gamma^{2/3}\mathbf{r}$, $\tilde{\mathbf{p}} = \gamma^{-1/3}\mathbf{p}$ the classical Hamiltonian assumes the form

$$\tilde{H} = \gamma^{-2/3}H = \frac{1}{2}\tilde{\mathbf{p}}^2 - \frac{1}{\tilde{r}} + \frac{1}{8}\tilde{\varrho}^2 = \tilde{E} \tag{3}$$

The classical trajectories obtained from the scaled equations of motion do not depend on both energy and magnetic field strength but only on one parameter, the scaled energy $\tilde{E} = E\gamma^{-2/3}$. Note that the classical action scales as

$$S = 2\pi\tilde{S}\gamma^{-1/3}\,. \tag{4}$$

The scaling property is lost, in general, for nonhydrogenic atoms in magnetic fields because the size of the ionic core does not depend on the magnetic field strength. In the model potential (2), this loss manifests itself in the fact that the length parameter does not scale like $\tilde{a} = a\gamma^{2/3}$. But for a limited range of the field strengths $a\gamma^{2/3}$ may be assumed to be constant in good approximation and the same scaling laws as for hydrogen can be applied to nonhydrogenic atoms (see Hüpper et al. (1995, 1996)).

The Coulomb singularity presents an obstacle to the numerical integration of the equations of motion that follow from the Hamiltonian (3). The way out of this problem is a transformation of time $t \mapsto \tau$, with $dt = 2r d\tau$, called regularization (Stiefel and Scheifele (1971)), together with a coordinate transformation to semiparabolic coordinates $\mu = (r+z)^{1/2}$, $\nu = (r-z)^{1/2}$. The equations of motion obtained from the regularized Hamiltonian are free of singularities and allow efficient integration with the help of high-order numerical algorithms.

In a semiclassical approximation of photoabsorption spectra (see Sect. 3), closed orbits which start at and return to the nucleus are of fundamental importance. Trajectories are started at the origin $\mathbf{r} = 0$ with an angle ϑ_i

between the initial velocity and the magnetic field axis. The closed orbit search can be formulated as finding solutions (ϑ_i, T_f) so that the trajectory started with angle ϑ_i at time $t = 0$ returns back to the origin after period $t = T_f$. Numerically the problem is solved with the help of an iterative Newton algorithm and the roots (ϑ_i, T_f) are the *real* closed orbits when all parameters are defined real.

The analytic structure of the equations of motion allows a direct analytic continuation of the real phase space to the complex phase space. To search for complex closed orbits we again choose initial conditions $\mathbf{r} = 0$ but complex starting angle ϑ_i, i.e. the initial momenta become complex. With these complex initial conditions Hamilton's equations of motion can be numerically integrated from $t = 0$ to $t = T_f$. It may be noted that in general all closed orbit parameters become complex, in particular the recurrence time T_f, the classical action S, and the monodromy matrix $\mathbf{M}$, which is defined via

$$\begin{pmatrix} \delta q(T) \\ \delta p(T) \end{pmatrix} = \mathbf{M} \begin{pmatrix} \delta q(0) \\ \delta p(0) \end{pmatrix} = \begin{pmatrix} m_{11} & m_{12} \\ m_{21} & m_{22} \end{pmatrix} \begin{pmatrix} \delta q(0) \\ \delta p(0) \end{pmatrix} , \tag{5}$$

where $\delta q(t)$ and $\delta p(t)$ are small deviations in coordinate and momentum space perpendicular to the direction of the orbit (Bogomolny (1989)). An extensive search for real closed orbits of the hydrogen atom in a magnetic field and the classical bifurcation diagram is presented in Main et al. (1991) and Main et al. (1994). For classification of closed orbits we use the same nomenclature as in these papers. With increasing energy many new orbits are born in bifurcations from multiple repetitions of the fundamental orbits parallel and perpendicular to the magnetic field axis. These bifurcations occur in a systematic fashion and we call orbits bifurcated from the μ-th repetition of the parallel orbit V_μ^ν and orbits bifurcated from the μ-th repetition of the perpendicular orbit R_μ^ν with integer numbers μ and ν. Examples are the creation of the "balloon" type orbit V_1^1 at scaled energy $\tilde{E} = -0.342$ and the creation of a "pacman" shape orbit R_2^1 at $\tilde{E} = -0.316$. If the new orbits are born stable they can undergo further bifurcation and we mark orbits created in higher order bifurcations by asterisks, e.g. orbit V_2^{2*} is created in a period doubling of the balloon orbit. There also exist orbits X_μ which do not belong to the bifurcation tree of the fundamental orbits but are born "out of nowhere" in the energy-action plane mostly in saddle node bifurcations. We do not intend to review these results in more detail in this article, but rather look at peculiarities in the vicinity of bifurcations and investigate the role of complex "ghost" orbits which are important for the derivation and interpretation of uniform semiclassical approximations. We also discuss effects of the ionic core in nonhydrogenic atoms and investigate core-scattering of classical orbits.

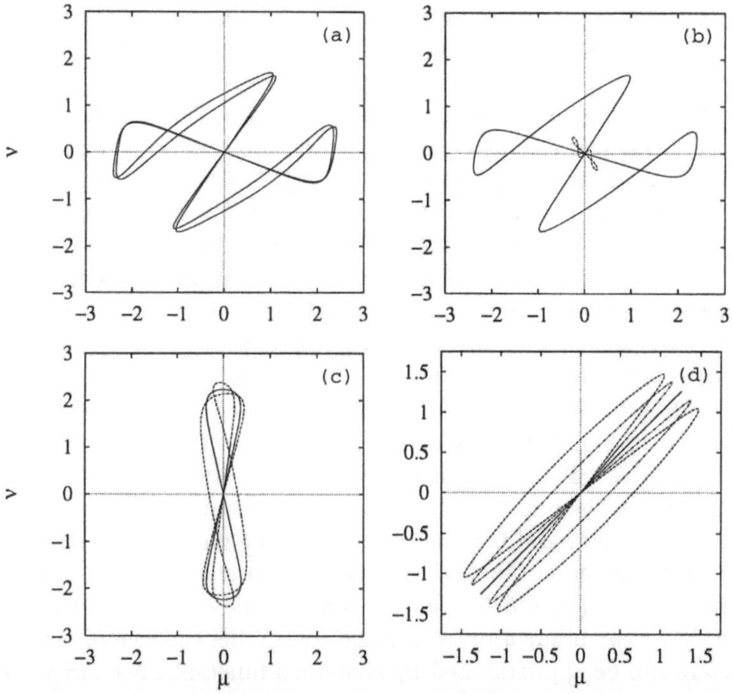

Fig. 1. Closed orbits (in semiparabolic coordinates μ, ν) at scaled energies near bifurcation points. (a) Orbits X_1^a and X_1^b at energy $\tilde{E} = -0.11$. Both orbits are born in a saddle-node bifurcation at $\tilde{E}_b = -0.11544216$. (b) Ghost orbit X_1 at energy $\tilde{E} = -0.2$. Solid line: Re μ vs. Re ν. Dashed line: Im μ vs. Im ν. (c) Balloon orbit V_1^1 at energy $\tilde{E} = -0.32$ (solid line) and orbit V_2^{2*} (dashed line) bifurcating in a period doubling at $\tilde{E}_b = -0.3420258$. (d) Perpendicular orbit R_2 (solid line), orbit R_2^1 (dashed line), and orbit R_2^{1b} (dashed dotted line) at scaled energy $\tilde{E} = -0.317$.

2.1 Bifurcations and Ghost Orbits

We look for ghost orbits in the vicinity of bifurcations of real orbits and discuss examples of the saddle node bifurcation of orbit X_1, the period doubling of V_1^1 and the bifurcation scenario of the perpendicular orbit R_1. Shapes of closed orbits near these bifurcations are shown in Fig. 1.

Saddle Node Bifurcation of Orbit X_1 The example we discuss in detail is orbit X_1 which is born in a saddle node bifurcation at scaled energy $\tilde{E}_b = -0.11544216$ and with initial and final angles $\vartheta_i = 1.2271$, $\vartheta_f = 2.4232$. At higher energies ($\tilde{E} > \tilde{E}_b$) this orbit immediately splits up into two different real closed orbits (X_1^a and X_1^b) with slightly different shapes, examples at energy $\tilde{E} = -0.11$ are shown in Fig. 1a. With decreasing energy both orbits vanish exactly at the bifurcation point and below the bifurcation energy ($\tilde{E} < \tilde{E}_b$) no real closed orbit with similar shape exists. But in the closed orbit search extended to the complex continuation of phase space the existence of prebifurcation ghost orbits could be revealed. For illustration the ghost orbit at energy $\tilde{E} = -0.2$ is presented in Fig. 1b. The real parts of semiparabolic coordinates (solid line) look similar to orbits found above the bifurcation energy (Fig. 1a) and the imaginary parts (dashed line) are usually relatively small. Note that the complex conjugate (ϑ_i^*, T_f^*) of each ghost orbit also exists, i.e. there are two closed orbits above and below the bifurcation energy which are exactly degenerate at the bifurcation point.

For the construction of uniform semiclassical approximations in Sect. 3 some closed orbit parameters, viz. the initial and final angles, the classical action, and the element m_{12} of the monodromy matrix (5) are of fundamental importance. The energy dependence of these parameters show characteristics around the bifurcation point which are related to the various types of bifurcations. In Fig. 2 we present results for the saddle node bifurcation of orbits X_1^a and X_1^b. The energy dependence of the starting angles ϑ_i is given in Fig. 2a. Above the bifurcation energy both angles are real, form a saddle at the bifurcation point and become complex for the ghost orbits at even lower energies. Fig. 2b shows the deviation of the classical action from the mean value, i.e. $\Delta\tilde{S} = \pm(\tilde{S}_2 - \tilde{S}_1)/2$ with $\tilde{S}_1$ and $\tilde{S}_2$ the classical action of orbits X_1^a and X_1^b. Near bifurcation points the energy dependence of closed orbit parameters can be approximated by analytical functions. For $\Delta\tilde{S}$ we obtain

$$\Delta\tilde{S} = \pm(\tilde{\sigma}/2\pi)\left(\tilde{E} - \tilde{E}_b\right)^{3/2} \tag{6}$$

with $\tilde{\sigma} = 4.96$ (see dashed line in Fig. 2b). The monodromy matrix elements $\tilde{m}_{12}$ of both orbits X_1^a and X_1^b vanish exactly at the bifurcation point and can be approximated at $\tilde{E} \approx \tilde{E}_b$ by

$$\tilde{m}_{12} = \pm\tilde{M}\left(\tilde{E} - \tilde{E}_b\right)^{1/2} \tag{7}$$

with $\tilde{M} = 175$ as illustrated by the solid and dashed lines in Fig. 2c. Note that the behavior of ghost orbit parameters $\Delta\tilde{S}$ and $\tilde{m}_{12}$ at energies $\tilde{E} < \tilde{E}_b$ is just the analytic continuation of Eqs. (6) and (7), i.e. the behavior of the real orbit parameters already suggests the existence of ghost orbits revealed by the more sophisticated search for ghost orbits in complex phase space.

Period Doubling of the Balloon Orbit V_1^1 The second example of a bifurcation is the period doubling of the balloon orbit V_1^1 at scaled energy $\tilde{E}_b = -0.3420258$. The balloon orbit itself is already created at lower energy $\tilde{E} = -0.3913$ in a bifurcation from the parallel orbit. A special feature of its shape is the symmetry in the initial and final angle, i.e. $\vartheta_i = \vartheta_f$. It exists below and above the period doubling energy without any spectacular change of this shape. Above the period doubling energy a new orbit V_2^{2*} which breaks this symmetry $(\vartheta_i \neq \vartheta_f)$ separates from V_1^1 and is closed roughly after two

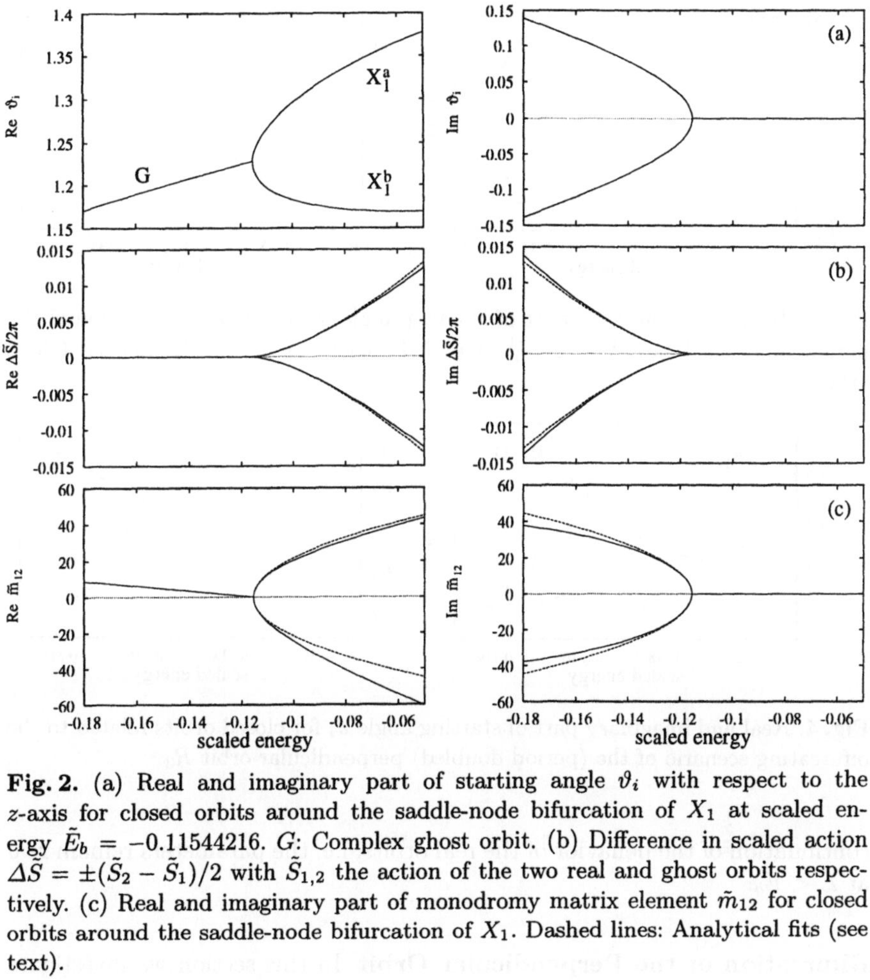

Fig. 2. (a) Real and imaginary part of starting angle ϑ_i with respect to the z-axis for closed orbits around the saddle-node bifurcation of X_1 at scaled energy $\tilde{E}_b = -0.11544216$. G: Complex ghost orbit. (b) Difference in scaled action $\Delta\tilde{S} = \pm(\tilde{S}_2 - \tilde{S}_1)/2$ with $\tilde{S}_{1,2}$ the action of the two real and ghost orbits respectively. (c) Real and imaginary part of monodromy matrix element $\tilde{m}_{12}$ for closed orbits around the saddle-node bifurcation of X_1. Dashed lines: Analytical fits (see text).

times the period of V_1^1. Shapes of the balloon orbit and its period doubling bifurcation V_2^{2*} are given in Fig. 1c. We also searched for complex orbits in the vicinity of this bifurcation and found a ghost (and its complex conjugate) at energies below bifurcation, i.e. they have complex starting angles ϑ_i (see Fig. 3). The classical action and monodromy matrix element m_{12} show a somewhat strange and unexpected behavior in the following sense. Although the ghost orbit lives in complex phase space these parameters remain exactly real even far away from the bifurcation point. The difference $\Delta\tilde{S}$ in action between orbits V_2^{2*} and V_2^2 is approximately a parabola and the monodromy matrix elements $\tilde{m}_{12}$ of orbits V_2^2 and V_2^{2*} are approximately linear functions in energy with $\tilde{m}_{12}(\tilde{E}_b) = 0$. Note that as for the saddle node bifurcation the energy dependence of ghost orbit parameters $\Delta\tilde{S}$ and $\tilde{m}_{12}$ is just the analytic

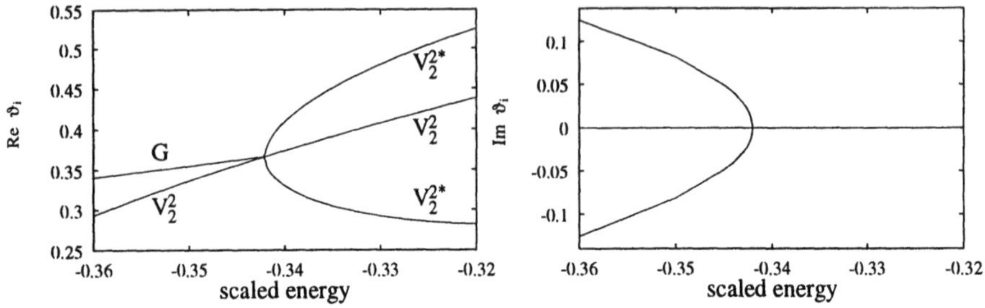

Fig. 3. Real and imaginary part of starting angle ϑ_i for closed orbits around the period doubling bifurcation of V_1^1 at scaled energy $\tilde{E}_b = -0.3420258$. G: Ghost orbit.

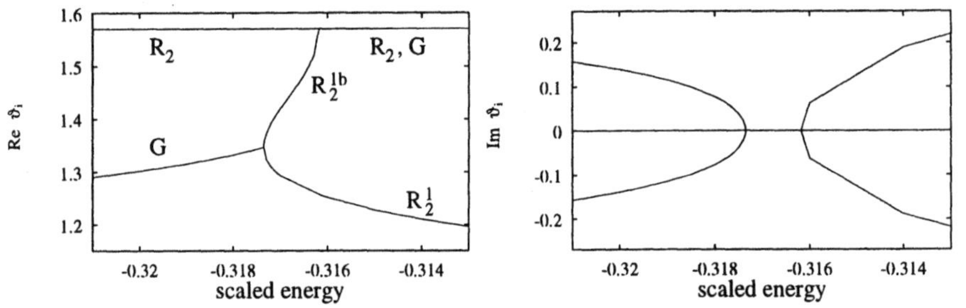

Fig. 4. Real and imaginary part of starting angle ϑ_i for closed orbits related to the bifurcating scenario of the (period doubled) perpendicular orbit R_2.

continuation of the behavior of the real orbits, i.e. the parameters remain real at $\tilde{E} < \tilde{E}_b$.

Bifurcation of the Perpendicular Orbit In this section we investigate real and ghost orbits related to the period doubling of the perpendicular orbit R_1. This third example of closed orbit bifurcations is more complicated because various orbits with similar periods undergo two different elementary types of bifurcations at nearly the same energy. The shapes of the real orbits at scaled energy $\tilde{E} = -0.317$ are drawn in Fig. 1d, R_1 (solid line $\mu = \nu$) is the orbit perpendicular to the magnetic field axis, and the dashed and dashed dotted lines represent orbits named R_2^1 and R_2^{1b} (or "pacmen" in Mao and Delos (1992)).

The structure of bifurcations and the appearance of ghost orbits can be seen clearly in the energy dependence of the starting angles ϑ_i in Fig. 4. Orbits R_2^1 and R_2^{1b} are born in a saddle node bifurcation at $\tilde{E}_b^{(1)} = -0.31735345$, $\vartheta_i = 1.3465$. Below the bifurcation energy we find (in analogy to the saddle node bifurcation discussed in Sect. 2.1) an associated ghost orbit and its complex conjugate. Orbit R_2^{1b} is real only in a very short energy interval ($\Delta\tilde{E} \approx 0.001$) and is then involved into the next bifurcation at $\tilde{E}_b^{(2)} = -0.31618537$, $\vartheta_i =$

$\pi/2$. This is the period doubling bifurcation of the perpendicular orbit, R_1 which exists at all energies ($\vartheta_i = \pi/2$ in Fig. 4). The period doubling is similar to the bifurcation of V_1^1 but with a reversed energy dependence. The real orbit R_2^{1b} separates from R_1 at energies *below* the bifurcation point, i.e. a real orbit vanishes with increasing energy. Consequently associated ghost orbits are expected at energies *above* the bifurcation, i.e. $\tilde{E} > \tilde{E}_b^{(2)}$, and indeed such "postbifurcation" ghosts have been found. Its complex starting angles are also shown in Fig. 4.

Because the energy separation between the saddle node bifurcation and period doubling is extremely small the classical action and monodromy matrix elements can not be fitted separately by the corresponding equations of the previous sections but have to be approximated globaly by somewhat more complicated formulae. But still the whole bifurcation scenario can be well described by only very few parameters. A more detailed analysis and analytic fits can be found in Main and Wunner (1997).

2.2 Classical Core Scattering in Nonhydrogenic Rydberg Atoms

Exact quantum mechanical calculations and experimental investigations have revealed resonances in recurrence spectra of general Rydberg atoms at positions where no closed orbits exist in the hydrogen atom. The fundamental question arise whether these structures are pure quantum effects 'beyond periodic orbits' as stated by Dando et al. (1995) or can also be explained and interpreted semiclassically in terms of new types of classical orbits. To answer this question we have performed extensive classical calculations on nonhydrogenic atoms with an ionic core modelled by the potential $V(r)$ (Eq. 2) and for various values of the nuclear charge Z and core size a (see Hüpper et al. (1995, 1996)). Indeed, scattering by the core potential leads to the creation of a huge number of nonhydrogenic closed orbits. The basic mechanism for the creation of the orbits is the following. A trajectory starting in almost the same initial direction as a hydrogenic closed orbit leaves the nucleus radially and evolves hydrogen-like outside the region of the ionic core. As the orbit returns close to the nucleus it is scattered by the core potential. The electron is deflected and, in contrast to the hydrogen case, the direction of the outgoing electron does not agree with the incoming direction but may now almost agree with the initial direction of another hydrogenic closed orbit. Outside the ionic core the electron again evolves hydrogen-like and may then be scattered a second time, third time, etc. until finally it exactly returns to the nucleus where the orbit is closed. Thus the nonhydrogenic closed orbits appear to be composed of two or more hydrogenic orbits. Some examples of core-scattered closed orbits at scaled energy $\tilde{E} = -0.3$ are presented in Fig. 5. The nuclear charge Z and length parameter a in the model potential (2) have been adjusted to quantum defect $\mu_1 = 0.5$ of the nonhydrogenic atom. The core-scattered orbits seem to be composed of hydrogenic orbits interlinked by the scattering at the core and can be grouped into families $h_1 \oplus h_2 \oplus h_3 \oplus \cdots$

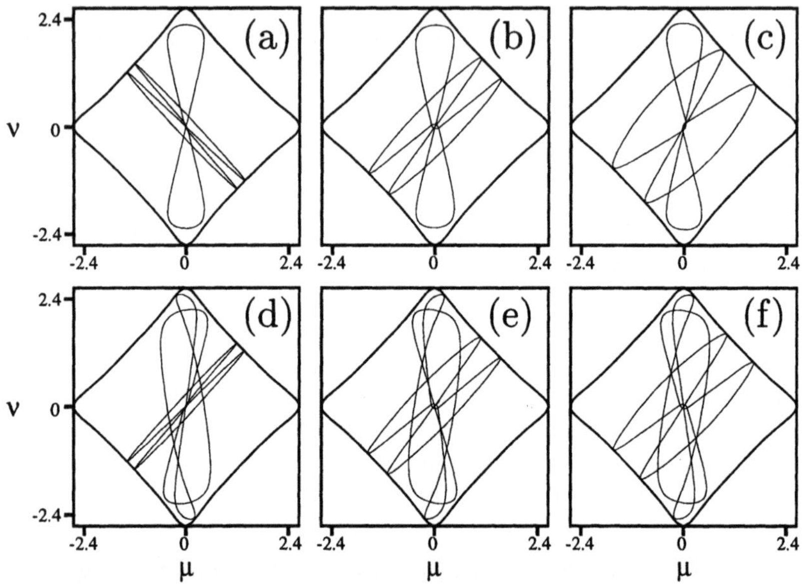

Fig. 5. Core-scattered orbits at scaled energy $\tilde{E} = -0.3$; quantum defect $\mu_1 = 0.5$. (a) - (c) Orbits of family $R_2^1 \oplus V_2^1$; (d) - (f) Orbits of family $R_2^1 \oplus V_2^{2*}$.

where h_i denotes the name of a primitive hydrogenic orbit and $\oplus$ indicates the scattering at the core potential. With this notation the orbits in Fig. 5 (a) - (c) belong to the family $R_2^1 \oplus V_2^1$ and Fig. 5 (d) - (f) present members of family $R_2^1 \oplus V_2^{2*}$.

3 Closed-Orbit Theory and Uniform Semiclassical Approximations

The link between classical trajectories and photoabsorption spectra is made by semiclassical *closed-orbit* theory. The theory was originally developed independently by Du and Delos (1988) and Bogomolny (1989) for the hydrogen atom in a magnetic field and has become a powerful tool for the interpretation of experimental spectra and exact quantum calculations as well. The agreement between quantum and semiclassical results serves as a measure for our understanding of atoms in external fields and deviations are usually an indication of new physical effects which are not yet adequately considered in the semiclassical theory. Examples are the observation of unidentified peaks in the recurrence spectra of hydrogenic and nonhydrogenic atoms which are related to complex "ghost" orbits or core-scattered orbits. Indeed, these effects can be described and understood within a generalized closed-orbit theory. Here, we can not present a complete derivation of closed-orbit theory but briefly review the basic ideas. Our main interest is to construct uniform semiclassical approximations which are free of singularities at bifurcation points of closed orbits where standard closed-orbit theory breaks down. The uniform solutions are directly related to various types of *catastrophes* (Poston and Steward (1978)) formed by the bundle of returning trajectories. We study some examples and demonstrate the role of ghost orbits in quantum photoabsorption spectra. We also include core-scattering effects in the semiclassical theory.

3.1 Physical Picture

The rationale of the semiclassical description of photoabsorption by atoms in strong magnetic fields is the following: An electron in a low-lying initial state $|\psi_i\rangle$ is excited to a Rydberg state, or a continuum state above the ionization threshold. One finds a distance r_0 from the nucleus where the semiclassical description of the wavefunctions becomes a good approximation, but the Lorentz forces are still negligibly small compared to the Coulomb attraction forces. Beyond r_0, the outgoing Coulomb wave describing the electron in the final state is propagated along classical trajectories. These trajectories obey the complete classical dynamics of the Hamiltonian (1), i.e. including effects of the magnetic field which become important at large distances from the nucleus. The combined action of the magnetic field forces and the Coulomb forces may cause trajectories to return to the nucleus and here core-scattering becomes important for nonhydrogenic atoms and is considered via the potential (2) of the ionic core in the classical equations of motion. A finally returning trajectory may have been scattered several times at the ionic core. The returning waves interfere with the initial state in the dipole matrix element, and this interference gives rise to characteristic modulations in the photoabsorption cross section. Consequently, the semiclassical expression for the oscillator strength is found to be composed of two parts, a smoothly varying continuous background f^0 related to the initially excited *outgoing* Coulomb wave, and an oscillatory part from the contributions of all *returning* waves $\psi_m^{\mathrm{ret},k}$ related to closed orbits k (m designates the conserved magnetic quantum number):

$$f = f^0 + \sum_{\mathrm{cl.o.}\ k} f_k^{\mathrm{osc}} \tag{8}$$

with

$$f_k^{\mathrm{osc}} = -\frac{2}{\pi}(E - E_i)\,\mathrm{Im}\langle\psi_i|D|\psi_m^{\mathrm{ret},k}\rangle\ . \tag{9}$$

Every closed orbit (returning exactly to the nucleus at an angle ϑ_f with respect to the z axis) will be surrounded by an ensemble of similar, almost-closed orbits and they give a contribution to the semiclassical wavefunction $\psi_m^{\mathrm{ret},k}(r,\vartheta)$ depending basically on the classical action and Jacobi matrix of the bundle of orbits. Inserting the semiclassical returning waves into (9) the fluctuating part of the oscillator strength is obtained as

$$f_k^{\rm osc} = \tag{10}$$
$$2(E-E_i)\sqrt{\sin\vartheta_i\sin\vartheta_f}\mathcal{Y}_m^\delta(\vartheta_i)\tilde{\mathcal{Y}}_m^\delta(\vartheta_f)\,{\rm Im}\left\{\mathcal{A}\exp\left(i\left[S_m^k-\frac{\pi}{2}\mu^k+\frac{\pi}{4}\right]\right)\right\},$$

where S_m^k is the classical action of the exactly closed orbit k, μ^k is the Maslov index, and the angular functions $\mathcal{Y}_m^\delta(\vartheta)$ and $\tilde{\mathcal{Y}}_m^\delta(\vartheta)$ depend on the initial state ψ_i, dipole operator D, and quantum defects δ_l of the ionic core and are defined in appendix A. The complex amplitude $\mathcal{A}$ is defined by

$$\begin{aligned} I(r) &\equiv \int_0^\pi \frac{\exp\left(i\left[\Delta S^{{\rm in},k}(r,\vartheta)-\frac{\pi}{2}\Delta\mu^{{\rm in},k}\right]\right)}{|\det[\partial(\mu,\nu)/\partial(\tau,\vartheta_i)]^{{\rm in},k}|^{1/2}}\,d\vartheta \\ &= \mathcal{A}\,\frac{\exp\left(-i\left[\sqrt{8r}-\frac{3}{4}\pi\right]\right)}{4\pi\,(2r)^{1/4}} \end{aligned} \tag{11}$$

and can be determined by evaluating $I(r)$ at some value of r outside the ionic core region where the semiclassical approximation is valid but magnetic field effects are still negligible. Eq. (11) is the matching condition of the left hand side semiclassical wavefunction to the right hand side asymptotic expression of a quantum mechanical incoming Coulomb wave and is the basis for the subsequent uniform approximations of the amplitude $\mathcal{A}$. The decisive quantities entering into $\mathcal{A}$ are the differences in classical action $\Delta S^{\rm in}$, and Maslov index, $\Delta\mu^{\rm in}$ between an incoming trajectory and the exactly closed orbit and the Jacobi matrix $\partial(\mu,\nu)/\partial(\tau,\vartheta_i)$ (in semiparabolic coordinates μ,ν) in the neighborhood of a returning (exactly closed) orbit. In the case of bifurcations, i.e. catastrophes, the behavior of these quantities differs significantly from that in the standard situation of isolated returning orbits, as will become evident in the following sections.

3.2 The Standard Situation: Isolated Returning Orbits

We start our discussion with the standard situation of nonbifurcating isolated returning orbits. We pick some closed orbit k, and will drop the index k for simplicity in what follows. In semiparabolic coordinates the returning (incoming) trajectories are straight lines in the vicinity of the origin, inclined at an angle $\vartheta_f/2$ with respect to the μ-axis. For the incoming bundle of trajectories we find

$$\Delta S^{\rm in}(r,\vartheta) = -\sqrt{8r}\cos\frac{\vartheta-\vartheta_f}{2} \overset{\vartheta\approx\vartheta_f}{=} -\left[\sqrt{8r}-\sqrt{\frac{r}{8}}\,(\vartheta-\vartheta_f)^2\right] \tag{12}$$

$$\det\left(\frac{\partial(\mu,\nu)}{\partial(\tau,\vartheta_i)}\right)^{\rm in} = 2\,m_{12} \tag{13}$$

$$\Delta\mu^{\rm in} = -1\,, \tag{14}$$

with m_{12} an element of the monodromy matrix (5). Inserting Eqs. (12) to (14) into Eq. (11) we obtain (with the integration range of the stationary phase integral formally extended to $\pm\infty$):

$$I(r) = |2m_{12}|^{-1/2} \int_{-\infty}^{+\infty} \exp\left(-i\left[\sqrt{8r} - \sqrt{\frac{r}{8}}(\vartheta - \vartheta_f)^2 - \frac{\pi}{2}\right]\right) d\vartheta$$

$$= \sqrt{\pi}2^{1/4}r^{-1/4}\frac{1}{\sqrt{|m_{12}|}} \exp\left(-i\left[\sqrt{8r} - \frac{3}{4}\pi\right]\right) \tag{15}$$

$$\Rightarrow \mathcal{A} = 2(2\pi)^{3/2}\frac{1}{\sqrt{|m_{12}|}} . \tag{16}$$

The contribution of a nonbifurcating isolated returning orbit to the oscillator strength then reads (with the scaled monodromy matrix element $\tilde{m}_{12} = \gamma^{1/3}\, m_{12}$)

$$f^{\mathrm{osc}} = 2(E - E_i)\sqrt{\sin\vartheta_i \sin\vartheta_f}\mathcal{Y}_m^\delta(\vartheta_i)\tilde{\mathcal{Y}}_m^\delta(\vartheta_f)$$

$$\times\, 2(2\pi)^{3/2}\gamma^{1/6}\frac{1}{\sqrt{|\tilde{m}_{12}|}} \sin\left(S_m^0 - \frac{\pi}{2}\mu^0 + \frac{\pi}{4}\right) . \tag{17}$$

This is the result of conventional closed-orbit theory. Obviously the oscillator strength diverges, i.e. conventional closed orbit theory fails, at bifurcations of orbits, where $\tilde{m}_{12} = 0$. The reason for the occurrence of the divergence is that, in such a case, the returning trajectories are not parallel straight lines (in semiparabolic coordinates) but form a more complicated *catastrophe* structure. We shall investigate the fold, cusp, and butterfly catastrophes.

3.3 The Fold Catastrophe

An example for the occurrence of a fold catastophe in the hydrogen atom in a magnetic field is the creation of the closed orbit X_1 through a saddle-node bifurcation at scaled energy $\tilde{E}_b = -0.11544216$. In semiparabolic coordinates the central returning orbit is a straight line, inclined at an angle $\vartheta_f/2$ with respect to the μ-axis. It is most convenient to introduce rotated semiparabolic coordinates

$$\xi = \mu\cos(\vartheta_f/2) + \nu\sin(\vartheta_f/2) = \sqrt{2r}\cos((\vartheta - \vartheta_f)/2)$$

$$\eta = \nu\cos(\vartheta_f/2) - \mu\sin(\vartheta_f/2) = \sqrt{2r}\sin((\vartheta - \vartheta_f)/2) , \tag{18}$$

where the ξ and η axes are now parallel and perpendicular to the returning orbit. The family of nearby returning orbits form a fold in rotated semiparabolic coordinates as illustrated in Fig. 6a. The fold moves with energy. Below the bifurcation energy the nucleus is located outside the bundle of trajectories, i.e. there is no closed orbit exactly returning to the nucleus. At energies above the bifurcation point we find the nucleus inside the bundle of trajectories and there are two exactly closed orbits. For the classical action and Jacobi determinant we find in this situation

$$\Delta S^{\mathrm{in}}(r,\vartheta) \overset{\vartheta\approx\vartheta_f}{=} -\left[\sqrt{8r} - \eta_0(\vartheta - \vartheta_f) - \frac{c}{12}(\vartheta - \vartheta_f)^3\right] \tag{19}$$

$$\det\left(\frac{\partial(\mu,\nu)}{\partial(\tau,\vartheta_i)}\right)^{\mathrm{in}} \overset{r\gg 0,\ \vartheta\approx\vartheta_f}{=} a\sqrt{8r} , \tag{20}$$

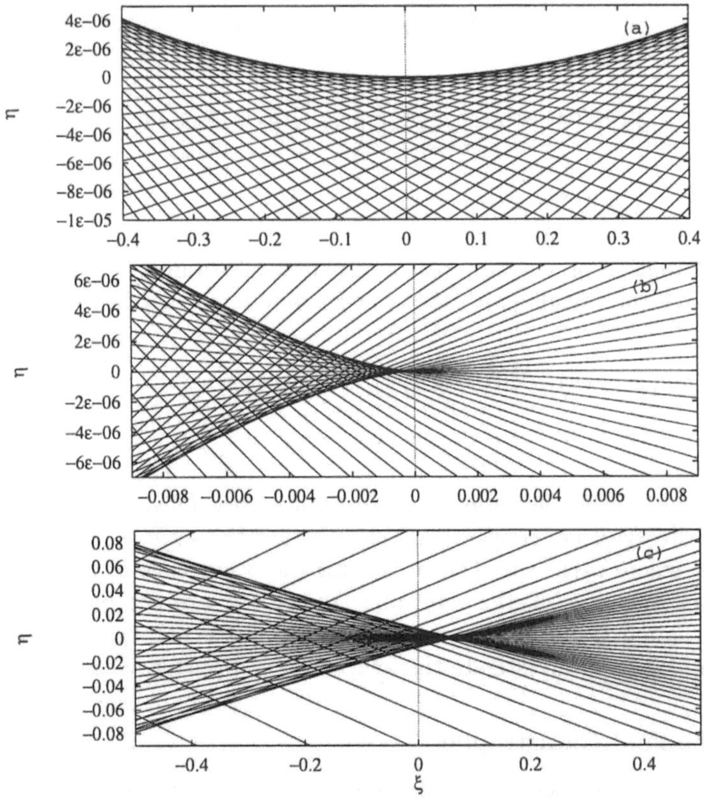

Fig. 6. (a) Fold catastrophe of returning orbits (in rotated semiparabolic coordinates ξ, η) related to the saddle-node bifurcation of orbit X_1 at bifurcation energy $\tilde{E}_b = -0.11544216$. (b) Cusp catastrophe related to the period doubling bifurcation of the balloon orbit V_1^1 at bifurcation energy $\tilde{E}_b = -0.3420258$. (c) Butterfly catastrophe (bifurcation of the perpendicular orbit) at $\tilde{E}_b = -0.31618537$.

i.e. in the direction of the returning orbit the Jacobi determinant exhibits the behavior of a focus of orbits at the origin. Inserting Eqs. (19) and (20) into (11) and extending formally again the integration range of the stationary phase integral to $\pm\infty$, we can express $I(r)$ in terms of the Airy function (Abramowitz and Stegun (1964)) and finally obtain the amplitude

$$\mathcal{A} = 2^{5/2}\pi^2 |a|^{-1/2} \exp\left(-i\frac{\pi}{4}\right) (4/c)^{1/3} \operatorname{Ai}\left((4/c)^{1/3}\eta_0\right) . \tag{21}$$

The contribution of a fold catastrophe to the closed orbit sum (8) now reads

$$f^{\mathrm{osc}} = 2(E - E_i)\sqrt{\sin\vartheta_i \sin\vartheta_f}\,\mathcal{Y}_m^\delta(\vartheta_i)\tilde{\mathcal{Y}}_m^\delta(\vartheta_f) \\ \times 2^{5/2}|a|^{-1/2}(4/c)^{1/3} \operatorname{Ai}\left((4/c)^{1/3}\eta_0\right) \sin\left(S_m^0 - \frac{\pi}{2}\mu^0\right) . \tag{22}$$

In Eq. (22) the parameters a, c, and η_0 are still undetermined but it can be shown that they are directly related to closed classical trajectories. Applying the asymptotic formula for the Airy function in the limit $c\eta_0 \ll 0$ (i.e. at energies far above the bifurcation energy) Eq. (22) takes the form of the standard solution of closed-orbit theory (17) and the contributions of two closed orbits can be identified. The closed orbit parameters of these two orbits must be consistent with Eqs. (6) and (7) of the classical analysis and the parameters a, c, and η_0 can be determined by comparison. The final result for the contribution of a fold catastrophe reads

$$f^{\mathrm{osc}} = 2(E - E_i)\sqrt{\sin\vartheta_i \sin\vartheta_f}\,\mathcal{Y}_m^\delta(\vartheta_i)\tilde{\mathcal{Y}}_m^\delta(\vartheta_f)\; 2^{7/2}\pi^2\gamma^{1/9}\,(3\tilde{\sigma}/2)^{1/6} \times |\tilde{M}|^{-1/2}\,\mathrm{Ai}\left((3\tilde{\sigma}/2)^{2/3}\gamma^{-2/9}(\tilde{E}_b - \tilde{E})\right)\sin\left(S_m^0 - \frac{\pi}{2}\mu^0\right) . \tag{23}$$

It is also illustrative to investigate the asymptotic behavior of Eq. (23) in the limit $c\eta_0 \gg 0$, i.e. at energies $\tilde{E}$ below the bifurcation energy $\tilde{E}_b$, where no orbit returns to the origin as this will reveal the complex "ghost" orbits discussed in Sect. 2.1. Applying the asymptotic form of the Airy function at $\tilde{E} \ll \tilde{E}_b$ Eq. (23) reads

$$f^{\mathrm{osc}} = 2(E - E_i)\sqrt{\sin\vartheta_i \sin\vartheta_f}\,\mathcal{Y}_m^\delta(\vartheta_i)\tilde{\mathcal{Y}}_m^\delta(\vartheta_f)\; 2(2\pi)^{3/2}\gamma^{1/6}|\tilde{M}|^{-1/2} \times (\tilde{E}_b - \tilde{E})^{-1/4}\,\mathrm{Im}\left\{\exp\left(i\left[S_m^0 + i\tilde{\sigma}\gamma^{-1/3}(\tilde{E}_b - \tilde{E})^{3/2} - \frac{\pi}{2}\mu^0\right]\right)\right\} . \tag{24}$$

Formally this equation coincides with (17), but for a closed orbit with complex action and imaginary monodromy matrix element:

$$S = S_m^0 + i\gamma^{-1/3}\tilde{\sigma}\left(\tilde{E}_b - \tilde{E}\right)^{3/2} \tag{25}$$

$$m_{12} = i\gamma^{-1/3}\tilde{M}\left(\tilde{E}_b - \tilde{E}\right)^{1/2} . \tag{26}$$

This is precisely the behavior found in the investigation of ghost orbits at energies below a saddle-node bifurcation (see Sect. 2.1). Obvioulsy, as one moves away from the bifurcation point to smaller energies, the positive imaginary part of S grows and thus the contribution of the ghost orbit to the semiclassical oscillator strength is damped exponentially. In the classical calculations, the complex conjugate ghost with negative imaginary part of the action also exists. In the above semiclassical formulae this complex conjugate ghost would produce an unphysical exponential increase of the amplitude at energies below the bifurcation point. Thus a by-product of the derivation of uniform semiclassical formulae is that ghost orbits of this type have no physical meaning. In other words, they must not be included in the standard formula (17) since they do not appear in the asymptotic expansion of the uniform approximation of the oscillator strength. The uniform solution (23) for the fold catastrophe has the advantage of connecting both asymptotic situations, the contribution of the physical ghost orbit at energies below the saddle-node bifurcation and the two real orbits at energies above the bifurcation, without exhibiting any singularity at the bifurcation point.

3.4 The Cusp Catastrophe

The second fundamental catastrophe is a cusp as it occurs for example at the period doubling bifurcation of the balloon orbit V_1^1 at scaled energy $\tilde{E}_b = -0.3420258$. The bunch of orbits forming the cusp is shown in Fig. 6b. The position of the nucleus moves along the ξ-axis with varying energy. At energies $\tilde{E} < \tilde{E}_b$ there is only one exactly closed orbit whereas above the bifurcation point three orbits are returning exactly to the nucleus. For the action and Jacobi determinant of the incoming trajectories we find (at $r \gg 0$, $\vartheta \approx \vartheta_f$)

$$\Delta S^{\rm in}(r,\vartheta) = \pm\sqrt{8r} + \frac{1}{4}\xi_0(\vartheta-\vartheta_f)^2 - \frac{1}{16}c(\vartheta-\vartheta_f)^4 \tag{27}$$

$$\det\left(\frac{\partial(\mu,\nu)}{\partial(\tau,\vartheta_i)}\right)^{\rm in} = a\sqrt{8r}\,. \tag{28}$$

Inserting these equations into (11) and following a similar procedure as described in the previous section for the fold catastrophe, we obtain the contribution of a cusp catastrophe to the closed-orbit sum (8)

$$\begin{aligned} f^{\rm osc} &= 2(E-E_i)\sqrt{\sin\vartheta_i\sin\vartheta_f}\,\mathcal{Y}_m^\delta(\vartheta_i)\tilde{\mathcal{Y}}_m^\delta(\vartheta_f)\;8\pi\gamma^{1/12}\tilde{\sigma}^{1/4}|\tilde{M}|^{-1/2} \\ &\times \mathrm{Im}\left\{\exp\left(i\left[S_m^0-\frac{\pi}{2}\mu^0\right]\right)\Phi\left(2\tilde{\sigma}^{1/2}\gamma^{-1/6}\left(\tilde{E}_b-\tilde{E}\right)\right)\right\}\,, \end{aligned} \tag{29}$$

where

$$\begin{aligned} \Phi(\pm x) &\equiv \int_{-\infty}^{+\infty}\exp\left(-i\left[\pm x t^2+t^4\right]\right)dt \\ &= \frac{\pi}{2}\sqrt{\frac{x}{2}}\exp\left[i\frac{x^2}{8}\right]\left\{\exp\left[-i\frac{\pi}{8}\right]\mathrm{J}_{-\frac{1}{4}}\left(\frac{x^2}{8}\right)\mp\exp\left[i\frac{\pi}{8}\right]\mathrm{J}_{\frac{1}{4}}\left(\frac{x^2}{8}\right)\right\} \end{aligned} \tag{30}$$

is a special case of Pearcey's integral (Pearcey (1946)) which can be solved analytically in terms of Bessel functions. Note that the parameters $\tilde{E}_b$, $\tilde{\sigma}$, and $\tilde{M}$ in (29) are determined directly from classical calculations (Main and Wunner (1997)). The uniform approximation (29) includes the contributions of one real closed orbit which exists both below and above the bifurcation point and two additional closed orbit contributions at energies $\tilde{E} > \tilde{E}_b$ without exhibiting a singularity at the bifurcation point. In the investigation of closed orbits around the period doubling bifurcation (Sect. 2.1), in addition to the real orbits two ghost orbits have been found at energies below the bifurcation energy with the somewhat strange property that the classical action S and monodromy matrix element m_{12} both remain real although the coordinates and momenta in phase space are complex. By analyzing the asymptotic behavior of the uniform solution (29) it has been shown (Main and Wunner (1997)) that, in contrast to the case of the fold catastrophe, these ghost orbits do not contribute to the semiclassical oscillator strength, i.e. they are both without physical meaning in the photoabsorption process.

3.5 The Butterfly Catastrophe

As an example of even more complicated catastrophes we mention in this section the bifurcation of R_2, i.e. the second return of the perpendicular orbit around scaled energy $\tilde{E} = -0.316$. A detailed classical analysis (see Sect. 2.1) exhibits a sequence of two different elementary types of bifurcations at nearly the same point. At scaled energy $\tilde{E}_b^{(1)} = -0.31735345$ two closed orbits namely R_2^1 and a second orbit with similar shape which we call R_2^{1b} are created by a saddle-node bifurcation related to a fold catastrophe. At only very slightly higher energy $\tilde{E}_b^{(2)} = -0.31618537$ the orbit R_2^{1b} already vanishes in a period doubling bifurcation with the second return of the perpendicular orbit R_2, i.e. in a catastrophe where neighboring returning orbits form a cusp. The energy spacing between both bifurcations is so small that they usually can not be treated as isolated and neither of the uniform formulas (23) for the fold or (29) for the cusp can be applied in a semiclassical calculation of oscillator strengths. The correct uniform approximation must describe both bifurcations simultaneously, i.e. a more complicated type of catastrophe, here namely the butterfly catastrophe (see Fig. 6c), must be investigated. The complete derivation of the uniform semiclassical formulae is presented in Main and Wunner (1997). Here we just give the final result for the contribution of a butterfly catastrophe to the total oscillator strength,

$$f^{\rm osc} = 2(E-E_i)\sqrt{\sin\vartheta_i \sin\vartheta_f}\,\mathcal{Y}_m(\vartheta_i)\mathcal{Y}_m(\vartheta_f)\,\pi\gamma^{1/18}3^{1/6}k^{1/3}(32\tilde{\sigma}/\tilde{M})^{1/2}\times$$
$$\mathrm{Im}\left\{\exp\left(i\left[S_m^0-\frac{\pi}{2}\mu^0\right]\right)\Psi\left(-3^{1/3}k^{2/3}\gamma^{-2/9}\tilde{\sigma}(\tilde{E}-\tilde{E}_b^{(2)}),-(9k)^{1/3}\gamma^{-1/9}\right)\right\} \tag{31}$$

where the phase integral

$$\Psi(x,y) \equiv \int_{-\infty}^{+\infty} \exp\left(-i\left[xt^2+yt^4+t^6\right]\right)dt \tag{32}$$

is an analytic function in both variables x and y. The parameters k, $\tilde{\sigma}$, and $\tilde{M}$ are determined classically from the behavior of the action S and monodromy matrix element m_{12} of closed orbits in the vicinity of the bifurcation. The uniform solution (32) includes the contribution of one real (viz. the perpendicular) orbit at energies $\tilde{E} < \tilde{E}_b^{(1)}$ and three real orbits (R_2 and orbit R_2^1 traversed in both directions) at energies $\tilde{E} > \tilde{E}_b^{(2)}$. In the classical analysis complex ghost orbits have been discovered both below and above the bifurcation energy. By studying the asymptotic behavior of the uniform approximation (32) it has been shown (Main and Wunner (1997)) that the ghost orbits at energies above the bifurcation have no physical meaning but one ghost below the bifurcation point contributes to the semiclassical oscillator strength. This orbit is called a "hidden ghost" because its contribution is exponentially damped with decreasing energy and usually only gives a small correction to the much stronger peak of the real orbit which appears at nearly the same position in the recurrence spectrum.

4 Results and Discussion

4.1 Uniform Semiclassical Approximations

We now discuss the effects of bifurcations and ghost orbits to photoabsorption spectra of atoms in external fields and compare the uniform semiclassical results derived in the previous chapter to solutions of standard closed orbit theory. Semiclassically each closed orbit generates a modulation in the photoabsorption cross section. Here we are not interested in the energy and field dependence of these modulations which is directly related to the classical action S of closed orbits but in the behavior of its *amplitudes* especially in the vicinity of bifurcations. In the following we also drop the prefactor $2(E-E_i)\,(\sin\vartheta_i \sin\vartheta_f)^{1/2}\,\mathcal{Y}_m^\delta(\vartheta_i)\tilde{\mathcal{Y}}_m^\delta(\vartheta_f)$ in formulae for the oscillatory part $f^{\rm osc}$ of the photoabsorption cross section. This prefactor depends on the initial state and excitation process (see Appendix A) and is for low lying initial states a slowly varying function of the initial and final angles of closed orbits. The amplitude $A(\tilde{E},\gamma)$ can now be defined as

$$A(\tilde{E},\gamma) = \left|\sum_k \mathcal{A}_k(\tilde{E},\gamma)\exp\left(i\left[S_m^k(\tilde{E},\gamma) - \frac{\pi}{2}\mu^k + \frac{\pi}{4}\right]\right)\right| . \tag{33}$$

By taking the absolute value instead of the imaginary part in (33) we suppress for a graphical presentation of results the high oscillatory modulations caused by the energy and field dependence of the classical action $S_m^k(\tilde{E},\gamma)$. Nevertheless the exponential function in (33) is important when applying (33) to non uniform standard solutions of closed orbit theory with coefficients $\mathcal{A}_k$ given by Eq. (16), i.e. (in scaled variables)

$$\mathcal{A}_k = 2(2\pi)^{3/2}\gamma^{1/6}\left(\tilde{m}_{12,k}(\tilde{E})\right)^{-1/2} . \tag{34}$$

In this case we add in (33) the contributions of all real closed orbits and ghosts k which are included in the corresponding uniform approximation. Interference effects between these orbits are now considered by the exponential function. Furthermore the classical action of ghost orbits may be complex which results, for a positive imaginary part of the action, in an exponential damping of ghost orbit contributions. Note that the energy dependence of amplitudes (33) can be measured experimentally with *scaled energy spectroscopy* (Holle et al. (1988), Main et al. (1994), Courtney et al. (1995)) by following the peak heights of resonances in the $(\tilde{E},\tilde{S})$ diagram of Fourier transform action spectra.

The Fold Catastrophe is related to a saddle node bifurcation. The amplitudes $A(\tilde{E},\gamma)$ around the bifurcation of orbit X_1 are presented in Fig. 7a at constant magnetic field strength $\gamma = 4\cdot 10^{-6}$. The dashed line shows the

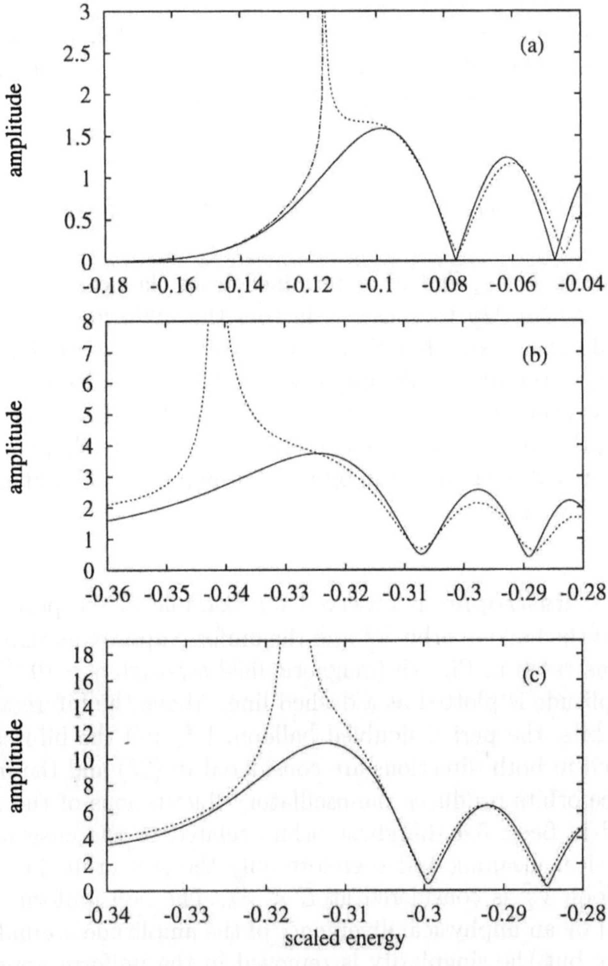

Fig. 7. Semiclassical amplitudes. (a) X_1 (fold catastrophe), magnetic field strength $\gamma = 4 \cdot 10^{-6}$ atomic units; (b) V_2^2 (cusp catastrophe), $\gamma = 10^{-7}$; (c) R_2 (butterfly catastrophe), $\gamma = 10^{-7}$. Dashed lines: Amplitudes in the standard model. Dashed dotted line (in (a)): Ghost orbit contribution. Solid lines: Uniform approximations.

superposition of the two real orbits X_1^a and X_1^b at energies $\tilde{E} > \tilde{E}_b$. The oscillatory structure of the amplitude is a result of strong interference between these two orbits. At the bifurcation energy the amplitude diverges. Note that the amplitude at energies below the bifurcation point is zero in the standard formulation of closed orbit theory, i.e. when only real orbits are considered. The dashed dotted line is the extension when ghost orbit contributions are also included. We only consider the ghost with positive imaginary part of complex action because only this ghost has a physical meaning (see discussion in Sect. 3.3). The amplitude decreases exponentially in $-(\tilde{E}_b - \tilde{E})^{3/2}$

with decreasing energy but also exhibits an unphysical divergence at the bifurcation point. The solid line in Fig. 7a is the uniform approximation for the amplitude which we obtain from Eq. (23) as

$$\begin{aligned} & A(\tilde{E},\gamma) \\ & = 2^{7/2}\pi^2\gamma^{1/9}\,(3\tilde{\sigma}/2)^{1/6}|\tilde{M}|^{-1/2}\,\left|\mathrm{Ai}\left((3\tilde{\sigma}/2)^{2/3}\,\gamma^{-2/9}(\tilde{E}_b-\tilde{E})\right)\right| \end{aligned} \tag{35}$$

with parameters $\tilde{E}_b$, $\tilde{\sigma}$, and $\tilde{M}$ determined from classical trajectory calculations in Sect. 2.1. Up to constant factors the uniform approximation to the amplitude is an Airy function of the energy difference $\tilde{E}_b - \tilde{E}$. Both the oscillatory structure of the amplitude at $\tilde{E} > \tilde{E}_b$ and the exponentially damped ghost orbit tail are well reproduced, but in addition the singularity at the bifurcation point is now completely removed. Note that the maximum of amplitude is not located at the bifurcation point $\tilde{E} = \tilde{E}_b$ but is shifted to somewhat higher energy.

The Cusp Catastrophe is observed for example in the period doubling bifurcation of the balloon orbit V_1^1 and the uniform approximation to the amplitude is illustrated in Fig. 7b (magnetic field strength $\gamma = 10^{-7}$). The non uniform amplitude is plotted as a dashed line. Above the bifurcation energy three real orbits, the period doubled balloon, V_2^2, and the bifurcated orbit, V_2^{2*}, traversed in both directions are considered in (33) and the interference of these three orbits produces the oscillatory fluctuations of the amplitude. As discussed in Sect. 3.4 the ghost orbits related to the cusp catastrophe have no physical meaning and therefore only the real orbit, i.e. the period doubled balloon V_2^2 is considered at $\tilde{E} < \tilde{E}_b$. The non uniform solution is characterized by an unphysical divergence of the amplitude around the bifurcation energy but the singularity is removed in the uniform approximation (solid line in Fig. 7b)

$$A(\tilde{E},\gamma) = 8\pi\gamma^{1/12}\tilde{\sigma}^{1/4}|\tilde{M}|^{-1/2}\,\left|\Phi\left(2\tilde{\sigma}^{1/2}\gamma^{-1/6}\left(\tilde{E}_b-\tilde{E}\right)\right)\right| \tag{36}$$

obtained from Eq. (29) and with parameters $\tilde{E}_b$, $\tilde{\sigma}$, and $\tilde{M}$ completely determined by classical calculations. The energy dependence is given by a special case of Pearcey's integral, $\Phi(x)$ (see Eq. 30). In contrast to the fold catastrophe (Fig. 7a) there is a real instead of a ghost orbit below the bifurcation point and therefore the amplitude decreases $\sim (\tilde{E}_b - \tilde{E})^{-1/2}$ rather than exponentially with decreasing energy. Also the modulations of amplitudes at $\tilde{E} > \tilde{E}_b$ are less pronounced, particularly there are no energies with vanishing amplitude, i.e. complete destructive interference between all closed orbit contributions. Oscillations in periodic orbit contributions related to a cusp catastrophe of returning trajectories have been experimentally verified in Stark spectra of lithium atoms by Courtney et al. (1995).

The Butterfly Catastrophe is related to a more complicated bifurcation scenario as found e.g. around the period doubling of the perpendicular orbit (see Sect. 2.1) and the uniform approximation to the amplitudes is obtained from (32). Results for $\gamma = 10^{-7}$ are presented in Fig. 7c. The solid line is the uniform approximation and for comparison the dashed line shows the non uniform solution of standard closed orbit theory. The modulations of amplitudes at energies above the bifurcation point are caused by the interference of three closed orbits, namely the period doubled perpendicular orbit, R_2, and the "pacman" orbit, R_2^1, traversed in both directions. Ghost orbits exist at these energies but have no physical meaning (see Sect. 3.5). Below the bifurcation point there is one real orbit, R_2, and in addition a ghost orbit (and its time reversal) contributing to the semiclassical photoabsorption spectrum. As in case of a fold catastrophe the amplitude of the ghost orbit is exponentially damped with decreasing energy but now the ghost is not isolated and its contribution is surpassed by that of the perpendicular orbit, i.e. the ghost is "hidden" behind the real orbit.

4.2 Core-Scattering in Nonhydrogenic Atoms

Nonhydrogenic resonance structures in recurrence spectra have been discovered, e.g., in R-matrix quantum calculations of a model atom by Monteiro and Wunner (1990) and in experiments on helium by Delande et al. (1994). In this experiment the helium atom was investigated with high resolution constant scaled-energy spectroscopy. Metastable $1s2s\,^3S_1$ He atoms were excited with CW laser light π-polarized with respect to the magnetic field axis to final states with magnetic quantum number $m = 0$. The scaled spectra are Fourier transformed to action spectra where resonance positions are related to the action of closed classical orbits. The experimental recurrence spectrum at scaled energy $\tilde{E} = -0.7$ and results of an exact quantum mechanical R-matrix calculation are shown in Fig. 8b and c adopted from Delande et al. (1994). Most peaks can be identified by hydrogenic closed orbits V_μ^ν and R_μ^ν bifurcating from the parallel and perpendicular orbit respectively. But some nonhydrogenic resonances structures are detected at large actions and marked by arrows in Fig. 8b and c. Detailed comparisons between between experimental data, exact quantum calculations and semiclassical closed-orbit theory for nonhydrogenic atoms in magnetic fields can be found in Hüpper et al. (1995, 1996). Here, we compare the recurrence spectra of helium with rigorous semiclassical results and explain the nonhydrogenic structures in terms of families of core-scattered closed orbits. In the calculations we chose nuclear charge $Z = 2$ and length parameter $a = 0.33$ in the potential (2) to model the helium atom with quantum defect $\mu_1 = 0.0684$ (all quantum defects μ_l for higher angular momenta are negligibly small). The semiclassical recurrence spectrum is presented in Fig. 8a. The agreement between the semiclassical results, the experimental spectrum, and the R-matrix quantum calculations

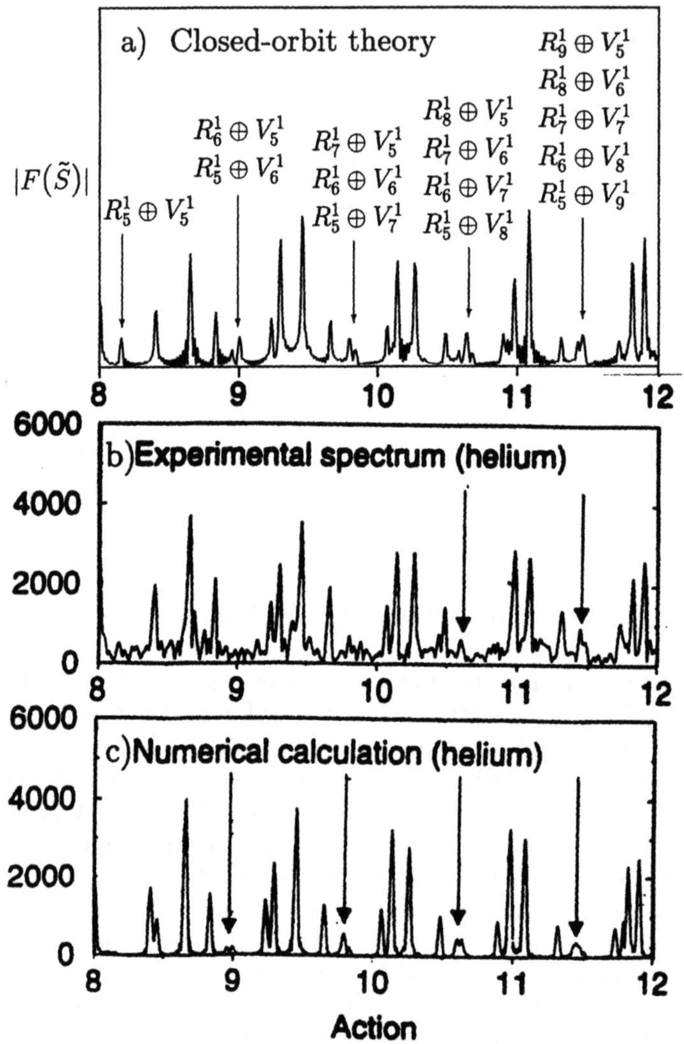

Fig. 8. Recurrence spectra of the helium atom (final states $m^\pi = 0^-$) at scaled energy $\tilde{E} = -0.7$. (a) Semiclassical calculation; (b) Experimental spectrum; (c) R-matrix quantum calculation. Nonhydrogenic resonances are marked by arrows and labelled with families of single scattered closed orbits.

is strikingly good even for the very details in which the helium atom deviates from hydrogen. Experimentally two new resonances appear at scaled action $\tilde{S} = 10.65$ and $\tilde{S} = 11.45$ which can now be identified with families of core-scattered orbits. There are several possibilities to combine two or more orbits to approximately achieve these actions. In Fig. 8, we have indicated all types of double scattered orbits that we have actually found. There is even a much greater number of multiple core-scattered orbits, e.g. the orbit

$R_7^1 \oplus V_7^1$ has almost the same action as $R_1 \oplus R_6^1 \oplus V_7^1$. Any combination of primitive hydrogenic orbits R_μ^ν and V_μ^ν existing at scaled energy $\tilde{E} = -0.7$ seems to be possible. However, every additional core-scattering leads to an enormous loss of recurrence amplitude and orbits scattered more than three times can be neglected in the periodic orbit sum of the semiclassical spectrum. In the quantum mechanically calculated spectrum of helium (Fig. 8c) there appear two further resonances at $\tilde{S} = 9.0$ and $\tilde{S} = 9.8$ dominantly built up from double-scattered orbits of families $R_\mu^1 \oplus V_\nu^1$ with an index sum $\mu + \nu = 11$ and $\mu + \nu = 12$ respectively. We also find an additional resonance at $\tilde{S} = 8.15$, which is very weak in the quantum spectrum. The corresponding core-scattered orbit consists of two hydrogeneous orbits R_5^1 and V_5^1 that do not exist at scaled action $\tilde{E} = -0.7$, but are born at slightly higher scaled energy.

5 Conclusion

Highly excited atoms in external fields are fundamental quantum systems which possess in many cases a classically chaotic analogue. The analysis of their quantum spectra is of fundamental importance for a deeper understanding of phenomena related to the so-called *quantum chaos*. The development and application of *closed-orbit* theory was the decisive step for a semiclassical calculation and interpretation of photoabsorption spectra. In its original version formulated by Du and Delos (1988) and Bogomolny (1989) closed-orbit theory was able to reproduce most structures in hydrogenic and even nonhydrogenic atoms but suffered from singularities at energies where orbits undergo bifurcations and also could not explain structures related to effects of a nonhydrogenic core. The theory has now been considerably improved in two directions. First, the modelling of an ionic core allows us to study classical core-scattering of trajectories returning close to the nucleus and deviations between hydrogenic and nonhydrogenic recurrence spectra can be interpreted in terms of core-scattered closed orbits. Second, unphysical singularities at bifurcation energies of closed orbits are removed in uniform semiclassical approximations and the analysis of these uniform solutions also reveals the role and physical importance of "ghost" orbits in complex phase space.

Acknowledgments: I thank F. Haake and G. Wunner for stimulating discussions, B. Hüpper for calculations on nonhydrogenic atoms, and M. Kaluža for a careful reading of the manuscript. The work was supported by Deutsche Forschungsgemeinschaft (SFB 237). I am grateful to Alexander von Humboldt-Stiftung for a Feodor-Lynen scholarship, and to Prof. Howard Taylor for his kind hospitality at the University of Southern California in Los Angeles.

A Angular Function $\mathcal{Y}_m^\delta(\vartheta)$

The angular functions $\mathcal{Y}_m^\delta(\vartheta)$ and $\tilde{\mathcal{Y}}_m^\delta(\vartheta)$ depend on the initial state ψ_i, dipole operator D, and phase shifts δ_l of the ionic core. They are linear superpositions of spherical harmonics

$$\mathcal{Y}_m^\delta(\vartheta) = \sum_{l=|m|}^{\infty} (-1)^l e^{i\delta_l} \mathcal{B}_{lm} Y_{lm}(\vartheta, 0) \tag{37}$$

$$\tilde{\mathcal{Y}}_m^\delta(\vartheta) = \sum_{l=|m|}^{\infty} (-1)^l e^{i\delta_l} \mathcal{B}_{lm}^* Y_{lm}(\vartheta, 0) , \tag{38}$$

with coefficients $\mathcal{B}_{lm}$ defined by the overlap integrals

$$\mathcal{B}_{lm} = \int d^3x \, (D\psi_i)(\mathbf{x}) \sqrt{2/r} \, \mathrm{J}_{2l+1}\left(\sqrt{8r}\right) Y_{lm}^*(\vartheta, \varphi) \tag{39}$$

where $\mathrm{J}_\nu(x)$ are Bessel functions. For more details see Gao et al. (1992) and Hüpper et al. (1995, 1996).

References

Abramowitz M. and Stegun I. A. (1964): *Handbook of Mathematical Functions*, Natl. Bureau of Standards

Beims M. W. and Alber G. (1993): Phys. Rev. A **48**, 3123

Berry M. V. and Mount K. E. (1972): Rep. Prog. Phys. **35**, 315–397

Berry M. V. and Upstill C. (1980) in *Progress in Optics*, Vol. 18, ed. E. Wolf, North-Holland, p. 257–346

Bogomolny E. B. (1989): Sov. Phys. JETP **69**, 275

Connor J. N. L. (1976): Mol. Phys. **31**, 33

Courtney M., Jiao H., Spellmeyer N., and Kleppner D. (1994): Phys. Rev. Lett. **73**, 1340; Courtney M., Spellmeyer N., Jiao H., and Kleppner D. (1995): Phys. Rev. A **51**, 3604

Courtney M., Jiao Hong, Spellmeyer N., Kleppner D., Gao J., and Delos J. B. (1995): Phys. Rev. Lett. **74**, 1538

Delande D., Taylor K. T., Halley M. H., van der Veldt T., Wassen W., Hogervorst W. (1994): J. Phys. B **27**, 2771

Dando P. A., Monteiro T. S., Delande D., and Taylor K. T. (1995): Phys. Rev. Lett. **74**, 1099

Du M. L. and Delos J. B. (1988): Phys. Rev. A **38**, 1896 and 1913

Edmonds A. R. (1970): J. Phys. (Paris) Colloq. C4 **31**, 71

Friedrich H. and Wintgen D. (1989): Phys. Rep. **183**, 37

Gao J., Delos J. B., and Baruch M. (1992): Phys. Rev. A **46**, 1449; Gao J. and Delos J. B. (1992): *ibid.* **46**, 1455

Garton W. R. S. and Tomkins F. S. (1969): Astrophys. J. **158**, 839

Gutzwiller M. (1990): *Chaos in Classical and Quantum Mechanics* (Springer, New York)

Holle A., Wiebusch G., Main J., Hager B., Rottke H., and Welge K. H. (1986a): Phys. Rev. Lett. **56**, 2594

Holle A., Wiebusch G., Main J., Welge K. H., Zeller G., Ertl T., and Ruder H. (1986b): Z. Phys. D **5**, 279

Holle A., Main J., Wiebusch G., Rottke H., and Welge K. H. (1988): Phys. Rev. Lett. **61**, 161

Hüpper B., Main J., and Wunner G. (1995): Phys. Rev. Lett. **74**, 2650 and (1996): Phys. Rev. A **93**, 744

Kuś M., Haake F., and Delande D. (1993): Phys. Rev. Lett. **71**, 2167

Main J., Wiebusch G., Holle A., and Welge K. H. (1986): Phys. Rev. Lett. **57**, 2789

Main J., Wiebusch G., and Welge K. H. (1991): Comm. At. Mol. Phys. **25**, 233–251

Main J., Wiebusch G., Welge K. H., Shaw J., and Delos J. B. (1994): Phys. Rev. A **49**, 847

Main J. and Wunner G. (1997): Phys. Rev. A **55**, 1743

Mao J. M. and Delos J. B. (1992): Phys. Rev. A **45**, 1746

Monteiro T. S. and Wunner G. (1990): Phys. Rev. Lett. **65**, 1100

Ozorio de Almeida A. M. and Hannay J. H. (1987): J. Phys. A **20**, 5873

Pearcey T. (1946): Phil. Mag. **37**, 311

Peters A. D., Jaffé C., and Delos J. B. (1994): Phys. Rev. Lett. **73**, 2825

Poston T. and Steward I. N. (1978): *Catastrophe theory and its applications* (Pitman, London)

Raithel G., Held H., Marmet L., and Walther H. (1994): J. Phys. B **27**, 2849

Sadovskií D. A., Show J. A., and Delos J. B. (1995): Phys. Rev. Lett. **75**, 2120; Sadovskií D. A. and Delos J. B. (1996): preprint (submitted to Phys. Rev. E)

Stiefel E. L. and Scheifele G. (1971): *Linear and Regular Celestial Mechanics* (Springer)

van der Veldt T., Vassen W., and Hogervorst W. (1993): Europhys. Lett. 21, 903

Wintgen D., Holle A., Wiebusch G., Main J., Friedrich H., and Welge K. H. (1986): J. Phys. B **19**, L557

Two-Electron Atoms: From Resonances to Fragmentation

Jan-Michael Rost[1] and Gregor Tanner[2]

[1] Fakultät für Physik der Universität Freiburg, Hermann Herder Str. 3, D–79104 Freiburg, Germany
[2] Niels Bohr Institute, Blegdamsvej 17, DK-2100 Copenhagen Ø, Denmark

Extra Phasen? - ach was, die mendeln sich im semiklassischen Limes sowieso raus!
Dieter Wintgen

Abstract. The semiclassical description of two-electron dynamics is reviewed and recent developments are highlighted. We discuss the semiclassical quantization of intermittency, which leads to a modified summation of periodic orbits in the Gutzwiller trace formula for Rydberg series of resonances. Furthermore, we present a semiclassical S-matrix formulation of Coulomb scattering, which can describe two-electron escape. Thereby, we cover an energy range of the two-electron system from the ground state across the complete fragmentation threshold into the two-electron continuum. We will illustrate the results for the helium atom, electron-hydrogen and electron-helium scattering.

1 Introduction: Semiclassical Description of Atoms

Quantum mechanics started in the beginning of the century with a successful semiclassical description of an integrable system, the Bohr–Sommerfeld quantization of the hydrogen atom. However, the failure of the pioneers of quantum mechanics to obtain a stable ground state for the simplest two-electron system, the helium atom (Born 1925), prompted the development of the full quantum theory by Schrödinger, Heisenberg and others.

Over the last decade there has been a revival of semiclassical concepts, in particular, to understand highly excited correlated motion of electrons in atoms (see the contributions in this book). Surprisingly, it turned out that even the ground state of helium can be obtained with modern semiclassical methods to an accuracy which is comparable to simple variational techniques (Ezra *et al.* 1991). In the case of Coulomb interactions it is no longer out of reach that semiclassical theory can predict the lifetime of resonances (Tanner *et al.* 1996). Correspondingly, semiclassical scattering theory can treat electron-electron interactions non–perturbatively (Rost 1995). Thereby, quantitative and qualitative results are obtained which can deepen our understanding of collision dynamics whose description so far has been based mainly on numerous perturbation series in terms of interaction potentials.

The aim of this contribution is to highlight two of the present activities regarding a semiclassical understanding of two–electron atoms. They are complementary to each other and complementary to a semiclassical description in terms of short periodic orbits (PO) which has been used commonly in the context of Gutzwiller's trace formula so far (e.g. Wintgen *et al.* 1992). The shortest periodic orbits dominate the Fourier transform of a quantum spectrum which is most impressively demonstrated by the so called scaled energy spectroscopy (Burgdörfer *et al.* this volume, Friedrich and Wintgen 1989). The existence of Rydberg series in Coulomb problems is, however, linked to the long range interaction between the particles and thus to orbits converging towards the classical ionization[1] threshold.

In a recently developed alternative approach the summation over PO's in the trace formula has been guided by the quasiseparable limit of two-electron dynamics, when one electron is far from the nucleus compared to the other. Quantum mechanically, this limit corresponds to long living resonances organized in Rydberg series. The inner electron is excited to an (approximately fixed) principal quantum number N while the excitation of the outer electron of quantum number n gives rise to the series with $n \to \infty$. Summing over subseries of PO's, which scan over the almost regular classical dynamics in the quasiseparable limit, leads to an EBK-like quantization of individual series of Rydberg resonances.

The second method to be discussed simulates scattering experiments semiclassically (Rost 1994, 1995, Rost and Heller 1994). We define the semiclassical scattering amplitude constructed from classical orbits which start asymptotically at infinite distance from the target and end again far away from the collisional center after the collision. Hence, this approach does not make use of periodic orbits and can also be applied for positive energies of the two-electron system where no periodic orbits exist. For negative energies, in the regime of resonances, the S-matrix formulation can describe collective phenomena such as simultaneous excitation of many overlapping resonances where each individual resonance is no longer resolved in the cross section.

We will see that these two semiclassical perspectives supplement each other and reveal complementary features of two-electron dynamics. For the sake of clarity we will discuss specific two-electron systems, namely the paradigmatic helium atom and electron-helium as well as electron-hydrogen scattering.

[1] To avoid confusion we will use *ionization* in this paper for two-body break-up, i.e. when the two-electron system breaks up into a one-electron atom or ion and a free electron. *Fragmentation* will be used for three-body break-up into an ion and two free electrons.

2 Resonances from Periodic Orbit Quantization

The semiclassical trace formula derived by Gutzwiller (see Gutzwiller (1990)) and Balian and Bloch (Balian and Bloch 1972) in the early 70'ties relates the classical periodic orbits of generic, non–integrable Hamiltonian systems to the corresponding quantum spectra. It took a decade and more until these ideas spread out to become established as a fundamental concept to understand the connection between quantum mechanics and the underlying classical dynamics. Atomic physics served as a catalyst for this development. Especially the work on hydrogen in a constant magnetic field (Friedrich and Wintgen 1989) and later on the helium atom (Wintgen *et al.* 1992) elucidated the power of periodic orbit theory both in calculating and interpreting quantum spectra.

After a short introduction into the semiclassical theory we will analyze the rich structure of the helium resonance spectrum in terms of the classical dynamics of the underlying 3–body Coulomb problem. Classical considerations together with semiclassical periodic orbit expressions allow for a computation of series of individual resonances in helium.

2.1 The Periodic Orbit Trace Formula: a Short Review

A stationary quantum system can be described by the Green's function (in energy representation), which is a solution of the inhomogeneous Schrödinger equation

$$\left(E - \hat{H}(q', \frac{\hbar}{i}\frac{\partial}{\partial q'})\right) G(q', q; E) = \delta(q' - q). \qquad (1)$$

The Green's function exhibits poles at the eigenenergies or resonances of the system with residue determined by the corresponding eigenfunctions at q and q'. Semiclassically, the Green's function $G(q', q; E)$ can be written as sum over all classical paths from $q \to q'$ on the energy manifold, i.e.

$$G(q', q; E) = \frac{2\pi}{(2\pi i\hbar)^{\frac{f+1}{2}}} \sum_{q \to q'} \sqrt{\left|\frac{D}{\dot{q}'\dot{q}}\right|} \exp(\frac{i}{\hbar}S(q', q; E) - i\nu\frac{\pi}{2}). \qquad (2)$$

Here, f denotes the number of degrees of freedom and $S(q', q; E) = \int_q^{q'} p\,dq$ is the action along the classical path. The amplitude contains the velocities $\dot{q}$, $\dot{q}'$ at the end and starting points together with the determinant

$$D(q', q; E) = \det\frac{\partial^2 S}{\partial\tilde{q}\partial\tilde{q}'}, \qquad (3)$$

where $\tilde{q}$ denotes a coordinate system perpendicular to the classical orbit. The integer ν counts the caustics along the classical path and corresponds to the number of singularities in the amplitude. The semiclassical approximation works well as long as pure quantum effects such as tunnelling or diffraction can be neglected.

The information about the quantum spectrum is contained in the trace of the Green's function

$$g(E) = \mathrm{Tr}G(E) = \int dq\, G(q,q;E) \sim \sum_n \frac{1}{E - E_n}. \tag{4}$$

Inserting the semiclassical expression (2) into (4), only such trajectories contribute which close in coordinate space. The phase $S(q,q;E)$ itself is stationary along periodic orbits of the classical system. The periodic orbit contributions depend sensitively on local properties of the dynamics of the underlying classical system. Closed expressions for the trace as sum over all classical periodic orbits can be given only for very restricted classes of dynamical systems. These are on the one hand integrable problems, where periodic orbits occur as continuous families on tori with rational winding numbers. Here, the integration in (4) along the families can be done analytically (Berry and Tabor 1976) and the resulting periodic orbit formula coincides essentially with the semiclassical Einstein–Brillouin– Keller (EBK) – quantization, the multidimensional generalization of WKB (Gutzwiller 1990, Berry and Tabor 1976). Extensions of the EBK–formalism to locally almost integrable structures in phase space, e.g. stable islands, are discussed in Bohigas *et al.* (1993) and Wintgen and Richter (1994). (See also Ozorio de Almeida and Hanney (1987) for corrections due to the non–integrability of the dynamics).

The Gutzwiller trace formula is valid if a stationary phase approximation of the integral in (4) can be applied to all periodic orbits. This implies, that all periodic orbits are unstable and isolated in the sense that the integration around the saddle of the action function $S(q,q;E)$ in (4) can be extended to infinity. This condition, called *hard chaos* (Gutzwiller 1990), is thus a highly local property of the dynamics. Note, that this is a much stronger restriction than e.g. the K-property (Schuster 1989) usually used to characterize chaos (Tanner 1996). Provided the hard chaos condition holds, the trace of the Green's function can be written as

$$g(E) \sim -\frac{i}{\hbar}\sum_{po} T_{po} \sum_{r=1}^{\infty} \frac{1}{\sqrt{|\mathbf{M}_{po}^r - \mathbf{1}|}} \exp(ir\frac{S_{po}(E)}{\hbar} - ir\nu_{po}\frac{\pi}{2}). \tag{5}$$

The first sum is taken over single repetitions of all periodic orbits of the system. The action S is taken along the orbit, the Monodromy or stability matrix $\mathbf{M}$ describes the linearized dynamics of the phase space in the neighborhood of the orbit after one repetition. The winding number ν counts twice the turns of the stable or unstable eigenvectors of $\mathbf{M}$ around the periodic trajectories.

The poles in the Green's function are collectively formed by all trajectories (or periodic orbits) joining the same area in phase space. Disconnected or loosely connected parts in the dynamics give uncorrelated contributions to the semiclassical expressions and lead to non – or weakly – interacting subspectra.

Semiclassical methods are thus a very sensitive tool to study structures in quantum spectra. High dimensional, non–separable quantum systems may be decomposed by classical consideration into problems of lower dimension which may be treated by semiclassical methods themselves. We will demonstrate this in the next sections for a system of fundamental interest, the helium atom.

2.2 The Structure of the S–Helium Spectrum

The helium atom as a 3–body Coulomb problem shows an amazingly rich structure both in its classical dynamics and in its quantum features. The electron–electron interaction plays a crucial role on all energy scales in the spectrum. Good quantum numbers arise from the conservation of the total angular momentum $\mathbf{L}$, and the discrete symmetries $\mathbf{r}_1 \leftrightarrow \mathbf{r}_2$ (e^- exchange) and $(\mathbf{r}_1, \mathbf{r}_2) \leftrightarrow (-\mathbf{r}_1, -\mathbf{r}_2)$ (parity), where $\mathbf{r}_1, \mathbf{r}_2$ denotes the position vector of the electron 1, 2, respectively. Helium spectra can thus be characterized by quantum numbers ${}^{2s+1}L^{\pi}$, where s denotes the total spin, (which is linked to e^- exchange due to the anti–symmetry of the complete wave function) and π = even/odd denotes parity. We will restrict ourselves in the following to non–relativistic helium in the infinite nucleus–mass approximation. We will concentrate on the ${}^1S^e$–helium spectrum (i.e. $\mathbf{L} = 0$ and wave functions symmetric with respect to $\mathbf{r}_1 \leftrightarrow \mathbf{r}_2$ exchange). An overview over the enormous literature on classical and quantum aspects of helium in general can be found in Rost and Briggs (1991), Wintgen *et al.* (1992, 1994) (and references therein).

Quantum Results. The spectrum of ${}^1S^e$ – helium (Fig. 1) consists of families of Rydberg series, each of them converging towards an ionization threshold, where the He$^+$ ion with nuclear charge $Z=2$ is left in some excited state with $E_N = -Z^2/(2N^2)$ (in atomic units). Only the series below the $N = 1$ threshold consists of bound states, all other states are resonances. The number of series with a common threshold increases as N. To label the states various approximate quantum numbers have been proposed based on group theoretical arguments (Herrick 1983) or on the existence of adiabatic invariant coordinates (Rost and Briggs 1991). These schemes have been successful in classifying resonances (also for $L \neq 0$) as obtained from full quantum calculations (e.g. Ho 1986, Bürgers *et al.* 1995) as well as from experiments (e.g. Domke *et al.* 1995). Three approximate quantum numbers (N, k, n) describe the ${}^1S^e$-states of helium. Here, $n \geq N$ counts the successive states in each Rydberg series. The index $k = -N+1, -N+3, \ldots, N-3, N-1$ denotes a Stark–type quantum number, which can be justified by considering the limit $n \to \infty$, where the inner electron remains quantized in the (approximately) constant field of the outer electron (Rost and Briggs 1991). The expectation value for the inter–electronic angle θ between the two electron position vectors $\mathbf{r}_1$ and $\mathbf{r}_2$ is thus expected to follow the quantization rule

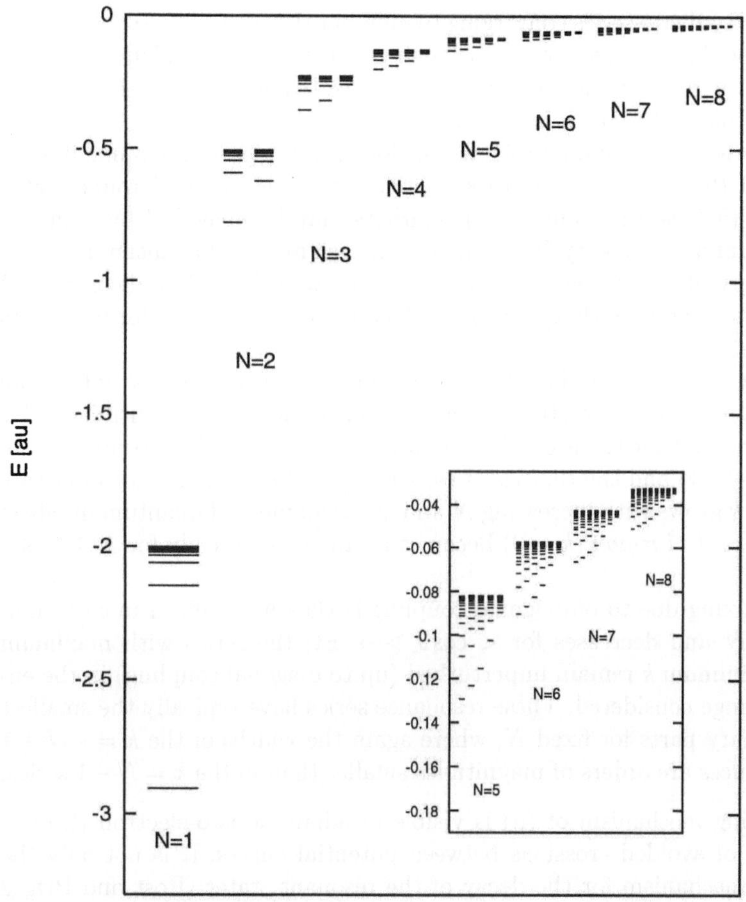

Fig. 1. The $^1S^e$ – spectrum of helium with data from Bürgers (1995). The quantum number k decreases from left to right in each manifold N.

$$< \cos\theta > \underset{n\to\infty}{\longrightarrow} -\frac{k}{N}. \tag{6}$$

The quantum energies decrease in each N – multiplet for fixed n and increasing k which reflects the decrease in the $e^- - e^-$ repulsion when going from $< \theta > \approx 0^o$ for minimum k to $< \theta > \approx 180^o$ for maximum k, (see Fig. 1).

Summarizing the results of recent quantum calculations for doubly-excited states in $^1S^e$–helium up to $N = 10$ one obtains the following qualitative picture for the coupling mechanism between Rydberg–series for different (N, k) values (Bürgers *et al.* 1995):

(i) Approximate quantum numbers can be uniquely assigned for $N \leq 3$ where Rydberg–series converging to different thresholds N do not overlap.
(ii) The coupling between states belonging to different multiplets N is maximum for series along the diagonal $\Delta = N - k$ in an (N, k) lattice as it is schematically shown in Fig. 2.
The series with maximum k have the lowest energies within a manifold N so that the (N, k) = (4,3) series is first perturbed by the "ground state" of the (5,4) series. Individual perturbers may be identified by standard quantum defect theory for small N. The assignment of quantum numbers becomes difficult when more than one series with $N' > N$ and constant Δ starts to overlap with a series N, (which happens at $N \geq 8$ for maximum k).
(iii) Offdiagonal perturbations become important, when series N with small absolute values of k start to overlap with the $N - 1$ multiplet. This happens first for the ground state of the (7,0) series which couples both, to the (6,–1) and the (6,+1) series, see Fig. 2. The coupling for moderate values k grows with increasing N and an assignment of quantum numbers in the *central region* ($k \approx 0$) becomes meaningless already for $N \geq 9$, see Fig. 2.
(iv) The mixing due to offdiagonal coupling is thus nonuniform in each multiplet N and decreases for $< \cos\theta > \rightarrow \pm 1$; the series with maximum and minimum k remain unperturbed (up to diagonal coupling) in the energy range considered. These resonance series have typically the smallest imaginary parts for fixed N, where again the widths of the $k = -N + 1$ resonances are orders of magnitudes smaller than in the $k = N - 1$ series.

The coupling mechanism of (ii) is visible in adiabatic two-electron theories as a series of avoided crossings between potential curves. It is not only the dominant mechanism for the decay of the resonant states (Rost and Briggs 1991), it also supplies the main route to fragmentation at low energies (Rost *et al.* 1989). In section 4.2, we will show that the energy dependence $\Gamma_N(E)$ of widths of resonances for states with $N = n$, the so called "saddle resonances", follows the same power law $|E|^\alpha$ as the fragmentation cross section for $|E| \rightarrow 0$ (i.e. $N \rightarrow \infty$).

Classical Interpretation. The behavior described above can qualitatively be understood by classical considerations. The dynamics of classical helium for angular momentum $L = 0$ is restricted to a plane and suitable coordinates are the distances (r_1, r_2) of the two electrons from the nucleus together with the angle θ spanned by the position vectors of the electrons (with the nucleus at the origin). The classical 3–body Coulomb problem is unbound even for total energy $E < 0$ in the sense that one electron can always escape to infinity by leaving the other electron in a Kepler ellipse around the nucleus. The collinear configurations $\theta = 0^o$ and $\theta = 180^o$ with $\dot{\theta} = 0$ are 4–dimensional invariant subspaces of the full 6–dimensional phase space. More importantly,

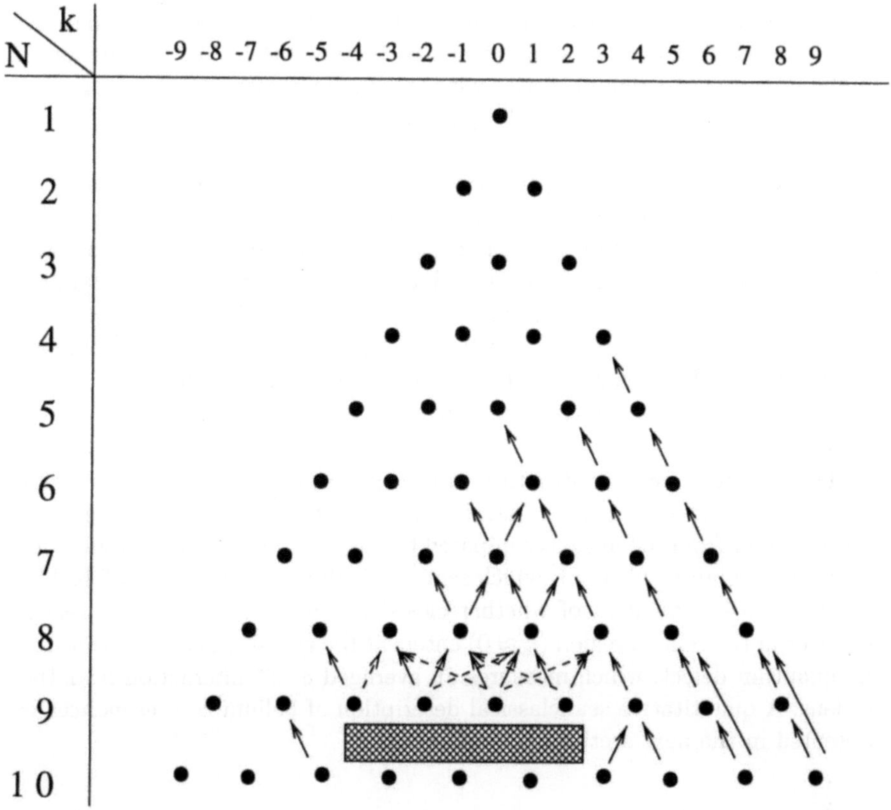

Fig. 2. Schematic representation of the interaction in the ${}^1S^e$ – spectrum of helium; the dots stand for Rydberg–series with approximate quantum numbers (N, k), the arrows indicate a coupling between different series.

however, both configurations are (mostly) stable under small perturbations in the variable θ (Richter et al. 1993). Trajectories starting slightly off the collinear configuration show vibration in the θ coordinate. The dynamics in the neighborhood of the collinear subspaces is thus essentially decoupled from the rest of the full phase space. The coupling between all coordinates in the equations of motion increases when θ approaches 90° and the dynamics becomes ergodic and chaotic on the full 5–dimensional energy manifold. A closer inspection of the collinear configurations itself shows, that the $\theta = 0^o$ phase space with both electrons on the same side of the nucleus is dominated by one stable island, whereas the dynamics in the $\theta = 180^o$ configuration is completely chaotic (Richter *et al.* 1993).

The interactions in the quantum spectrum can now be interpreted with the structure of the classical dynamics. The Rydberg series with maximum and minimum k, i.e. with maximum and minimum expectation values $< \theta >$

are associated with the corresponding collinear configurations. A coupling along the off–diagonals is suppressed due to the stability of the dynamics in the θ degree of freedom. Series with maximum k interact only along the diagonal, i.e. with states coinciding approximately in $< \cos\theta > \sim -\frac{N-1}{N} \sim -1$. The different k series start to mix in the region $k \approx 0$ or $< \theta > \approx 90^o$, where all degrees of freedom are strongly coupled. States with minimum k or $< \theta > \approx 0^o$ have the smallest widths due to the fact, that the dynamics of the corresponding collinear configuration is stable forming a closed torus in the full phase space. Ionization is thus possible only through dynamical tunnelling or by Arnold diffusion. In the collinear configuration $\theta = 180^o$, classical ionization occurs dominantly along the axis, preserving the collinear configuration. The ionization probability is again suppressed compared to off–collinear dynamics where classical ionization happens into a whole interval θ.

The dominant regular structure in the spectrum, the Rydberg series for $n \to \infty$, originates from the separable limit of the Hamiltonian, when one electron is far from the nucleus compared to the other. The classical dynamics is almost regular in this limit, which semiclassically gives rise to an EBK–like quantization independent of whether closed tori exist or not. The classical behavior in the central region ($k \approx 0$) enters at the thresholds N only through the quantum defect, which measures an averaged e^-e^- interaction near the nucleus. A quantitative semiclassical description of helium S – resonances is presented in the next section.

2.3 Periodic Orbit Quantization Including Intermittency

Our analysis in the preceding section suggests a connection between the invariant collinear dynamics and the Rydberg series with minimum and maximum k quantum numbers in each multiplet N. Indeed, such a connection has been confirmed by a semiclassical quantization of the two collinear configurations including the dynamics in the coordinate θ in linear approximation. The collinear subspaces are reduced by one degree of freedom compared to the dynamics in the full S–helium system which simplifies semiclassical calculations considerably. The stable configuration with both electrons on the same side of the nucleus could be treated by EBK–quantization (Wintgen and Richter 1994) in good agreement with the full quantum results for series with $k = -N + 1$. The collinear dynamics with electrons on different sides of the nucleus is fully chaotic. Semiclassical eigenvalues corresponding to resonances with large k can be obtained by applying periodic orbit theory together with suitable regularization techniques (e.g. Ezra *et al.* 1991, Wintgen *et al.* 1992). The s-wave helium is another reduction of the two-electron dynamics to two degrees of freedom. The classical and semiclassical aspects of this model have been investigated in detail by Friedrich and coworkers (Handke *et al.* 1993, Draeger *et al.* 1994, Ihra et al. 1995).

The motion in the full 6–dimensional phase space has to be taken into account for a semiclassical treatment of the central region in Fig. 2. The total phase space is largely unexplored (except from the dynamics on the so called Wannier saddle [2] $r_1 = r_2$; $p_{r_1} = p_{r_2}$, which also forms an invariant subspace of the full dynamics). A complete quantitative semiclassical description is still beyond what can be done by semiclassical techniques. Recently, Simonovic and Gruijc (1995) studied the near collinear dynamics beyond the linear approximation, which leads to a slight improvement in a simplified single periodic orbit quantization.

In this section, we will review the semiclassical methods to quantize chaotic collinear helium emphasizing the influence of the dynamics in the nearly regular limit, when one electron is far from the nucleus compared to the other. Hence, from now on, we will restrict ourselves to the collinear dynamics $\theta = \pi$, $\dot{\theta} = 0$ described by the classical Hamiltonian

$$H = \frac{p_1^2}{2} + \frac{p_2^2}{2} - \frac{Z}{r_1} - \frac{Z}{r_2} + \frac{1}{r_1 + r_2} = E, \tag{7}$$

with $Z = 2$ for helium. The dynamics is invariant under the exchange $r_1 \leftrightarrow r_2$. A direct quantization of the symmetric or anti–symmetric subspectra with respect to e^- – exchange can be performed by restricting the dynamics to the fundamental domain $r_1 \geq r_2$. Symmetric or anti–symmetric wavefunctions obey then Neumann or Dirichlet boundary conditions on the axis $r_1 = r_2$, which can be treated as a hard wall for the classical dynamics.

The Hamiltonian (7) exhibits hard chaos in the sense described in section 2.1, i.e. all periodic orbits are unstable and isolated. Furthermore, a symbolic description exists which assigns a binary sequence to each trajectory based on the rules:

0 if the trajectory does not hit the axis $r_1 = r_2$ between two consecutive events $r_2 = 0$,
1 otherwise.

Perturbations perpendicular to the collinear configuration correspond to a stable bending mode in the coordinate θ. The motion in θ is described in linear approximation by the Jacobian (or Monodromy – matrix) of the full flow in phase space integrated along a trajectory in the collinear plane, (see Richter *et al.* (1993) for details). Collinear helium is thus an ideal model to apply Gutzwiller's trace formula (5).

---

[2] The dynamics on the Wannier saddle shows mixed phase space structure, which is mostly unstable with respect to perturbations perpendicular to the saddle; the so called Langmuir–orbit is, however, stable with respect to all phase space coordinates (Richter *et al.* 1993). The relatively small phase space volume associated with this island suggests localized states not before $N \approx 500$. However, the influence of the Langmuir orbit has been detected already in resonant states with $N \approx 10$ (Müller and Burgdörfer 1993). See also Burgdörfer *et al.* 1997 — the following chapter of this volume.

For actual calculations, it is more convenient to switch to the spectral determinant, which itself can be written in terms of the trace, i.e.

$$D(E) = \exp \int_0^E dE' \operatorname{Tr} G(E') \sim \prod_n (E - E_n). \tag{8}$$

By expanding the denominator in (5) and summing over the repetitions of periodic orbits, one obtains a semiclassical representation of the spectral determinant as product over all periodic orbits (Voros 1988). For collinear helium, it takes on the form

$$D(E) \sim \prod_{m=0}^{\infty} \prod_p \left(1 - t_p^{(m)}\right), \tag{9}$$

where the single periodic orbit weights are given as (Wintgen *et al.* 1992)

$$t_p = \frac{1}{\sqrt{\Lambda_p}} \exp\left(2\pi \mathrm{i}(z\tilde{S}_p - \frac{n_p}{2} - (m + \frac{1}{2})\sigma_p)\right). \tag{10}$$

The classical dynamics is scale invariant under a change of the energy E; the energy dependence enters in (10) only through the action, which is written here as $S_p(E) = 2\pi\, z(E)\, \tilde{S}_p$ with $z = Z/\hbar\sqrt{-2E}$, and $\tilde{S}_p$ is the action at energy $E = -1$. Λ_p and $\exp(\pi i \sigma_p)$ denote the eigenvalues of the Monodromy matrix for the dynamics *in* and *perpendicular* to the collinear plane. The integer number n_p is the symbol length of the periodic orbit and corresponds to the winding number in the unstable degree of freedom. Equivalently, σ_p is called the winding number in the stable degree of freedom. The index m originates formally from the expansion of $det(\mathbf{1} - \mathbf{M})$ in (5) and can be associated with excitations in the bending mode θ. It is thus connected to the quantum number k, i.e. one obtains $k = N - 1 - 2m$ for $0 \le m \ll N$.

The product (9) as well as the sum (5) are absolutely convergent only for Im $E > h_c > 0$. Suitable representations with larger analyticity domain can be given for uniformly hyperbolic systems, where

$$\lambda_p = \frac{\log \Lambda_p}{T_p} \underset{T_p \to \infty}{\longrightarrow} \lambda \tag{11}$$

for long periodic orbits and λ corresponds to the Lyapunov exponent of the full dynamics. Multiplying out the periodic orbit product (9) and reordering the terms with respect to increasing total action (Berry and Keating 1990, Sieber and Steiner 1991) or, equivalently, total symbol length (Cvitanović 1988, Tanner *et al.* 1991) leads to huge cancellations in the sum due to the selfsimilarity of the classical dynamics. This ensures the convergence of the resulting sum in the energy region of interest and reads, for a system with a binary symbolic code,

$$\begin{aligned} D(E) = 1 - t_0 - t_1 - [t_{01} - t_0 t_1] \\ - [(t_{001} - t_0 t_{01}) + (t_{011} - t_{01} t_1)] - \dots . \end{aligned} \tag{12}$$

The leading uncompensated terms, the *fundamental* terms (Artuso et al. 1990), are followed by a sequence of corrections or *shadowed* contributions which drop exponentially for large total symbol length n. They contain information about the fine structure of the dynamics which becomes import with increasing energy, i.e. with decreasing wavelengths in the wavefunctions. A semiclassical quantization with the help of (12) has been performed for the collinear helium and yields the lowest states in Rydberg series with maximum quantum number $k = N - 1$ down to the ground state of helium ($N = 1$) (Ezra *et al.* 1991, Draeger *et al.* 1994).

A closer analysis of the classical dynamics reveals, however, that the condition (11) is not strictly satisfied for helium due to the separable limit of the Hamiltonian (7) for $r_1 \to \infty$, i.e.

$$H \underset{r_1 \to \infty}{\longrightarrow} \frac{p_1^2}{2} + \frac{p_2^2}{2} - \frac{Z-1}{r_1} - \frac{Z}{r_2}. \tag{13}$$

Although the motion in the collinear subspace exhibits hard chaos nearly regular dynamics takes place in the phase space region where one electron is far from the nucleus compared to the other. As a consequence, the local Lyapunov exponent (11) approaches zero for periodic orbits running further and further out in the regular region. A generic trajectory alternates between chaotic behavior in the region $r_1 \approx r_2$ and slow laminar motion out in the channel $r_1 \gg r_2$ until the classical electron finally escapes to infinity with nonzero kinetic energy. Such a behavior is called intermittency. The dynamics can be visualized in a Poincaré section, see Fig. 3, where we choose $r_2 = 0$ as the surface of section. The phase space has indeed a torus–structure for large r_1. The tori do not close, however, for small r_1 but end in a chaotic sea near the nucleus.

The expansion (12) of the spectral determinant emphasizing the shortest orbits in the system indeed misses the main feature in the spectrum, the existence of Rydberg series. The Rydberg series as a highly ordered structure in the spectrum are linked to the classical dynamics in the regular channels. The chaoticity of the central region plays a crucial role only to describe the mixing and interference between overlapping series.

The regular behavior for $r_1 \to \infty$ enters the symbolic dynamics through the symbol '0'. The symbol '1' is always related to motion in the near core region, whereas a '0' can occur for arbitrary r_1. A single '0' in the symbol code contributes differently to the action or stability exponent of a periodic orbit depending on how far away the outer electron is from the nucleus. The building blocks of the classical dynamics are thus not the symbols '0' and '1' alone, but the number of '0's occurring successively in a row after hitting the $r_1 = r_2$ axis. A new expansion scheme of the spectral determinant was proposed recently (Tanner and Wintgen 1995, Tanner *et al.* 1996), which incorporates intermittency directly. The binary alphabet is changed to an infinite letter alphabet according to

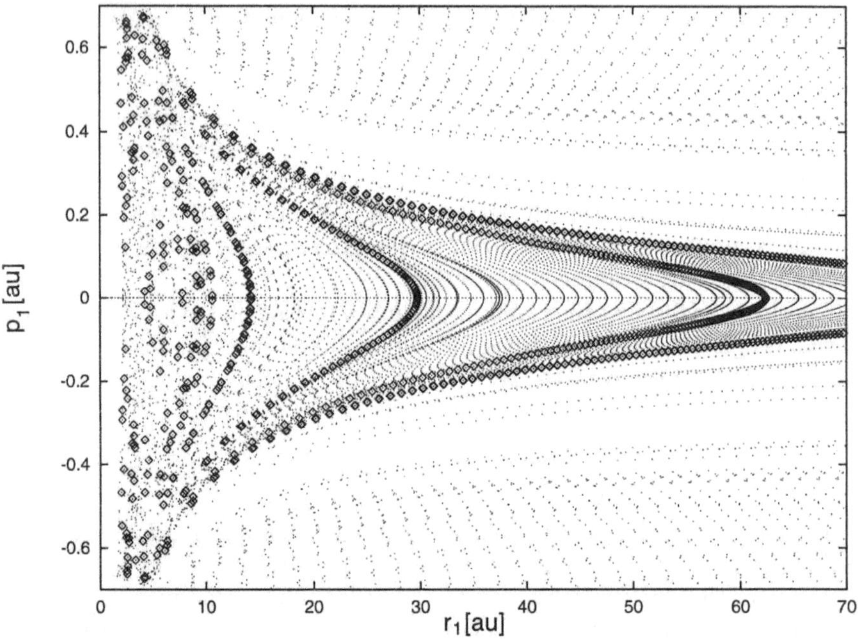

Fig. 3. Poincaré surface of section $r_2 = 0$ for the chaotic collinear configuration; the points surrounded by squares correspond to one specific trajectory interchanging several times between the chaotic region near the nucleus and the regular channel for large r_1 before escaping finally to infinity.

$$10^{n-1} \to n; \quad 10^{n-1}10^{m-1} \to nm \quad \ldots,$$

where '0^n' stands for n '0's occurring successively in a row. The expansion is now written in terms of infinite sums summing over the regular tails explicitly, i.e.

$$\begin{aligned} D(E) = 1 - \sum_{n=1}^{\infty} t_n - \sum_{m=1}^{\infty}\sum_{n=1}^{\infty} \frac{1}{2}(t_{mn} - t_m t_n) \\ - \sum_{k=1}^{\infty}\sum_{m=1}^{\infty}\sum_{n=1}^{\infty} (\frac{1}{3} t_{kmn} - \frac{1}{2} t_{km} t_n + \frac{1}{6} t_k t_m t_n) \\ - \sum_{l=1}^{\infty}\sum_{k=1}^{\infty}\sum_{m=1}^{\infty}\sum_{n=1}^{\infty} \ldots . \end{aligned} \tag{14}$$

The fundamental term is build up by the periodic orbit family with code '10^{n-1}'. The various quantities entering the periodic orbit weights (10) can be written as smooth functions of n, i.e.

$$\tilde{S}(n) = n\left[1 + (\frac{Z}{Z-1}n)^{-\frac{2}{3}}\right]^{\frac{3}{2}} + s_0 + s_{2/3}n^{-2/3} + R_s \tag{15}$$

$$\sigma(n) = n + \sigma_0 + \sigma_{2/3} n^{-2/3} + R_\sigma$$
$$\log \Lambda(n) = \frac{5}{3} \log n + l_0 + l_{2/3} n^{-2/3} + R_l \,,$$

where R denotes a remainder decreasing faster than $n^{-2/3}$. The leading terms are obtained explicitly from the asymptotic behavior $n \to \infty$ (Wintgen and Richter 1994, Tanner *et al.* 1996). They are universal also for contributions to the multiply infinite sums. The two coefficients and the remainder terms R are, however, non–universal and contain information about the chaotic near–core dynamics. The coefficients are obtained by fitting (15) to the periodic orbit data (the remainder was approximated by a polynomial in 1/n, i.e. $R_c = \sum_{i=1}^{4} c_i n^{-i}$). The coefficients for the leading periodic orbit family are listed in table 1.

Table 1. The coefficients in Eqs. (15) for the leading periodic orbit family.

	0	2/3	1	2	3	4
s	-0.43709	-0.55993	0.23681	-0.18997	0.07163	-0.08425
σ	0.60982	-1.89786	1.88144	-1.28564	1.40698	-0.83832
l	0.12234	0.25174	-1.3446	3.16088	-3.86478	2.10299

It could be shown (Tanner and Wintgen 1995) that the fundamental term $1 - \sum_{n=1}^{\infty} t_n$ in (14) (after analytic continuation into the negative imaginary z–plane) already builds up the gross structure of the spectrum. Its zeros form a Rydberg series approaching the thresholds $z = N$ or equivalently $E_N = -Z^2/2N^2$. The shadowed terms in (14) are important to describe the overlap of different Rydberg series (Tanner and Wintgen 1995, Tanner *et al.* 1996). We will concentrate now on the fundamental term alone and analyze its structure by approximating the infinite sum itself by stationary phase. This will lead us to an extension of the EBK–formalism for intermittency and provides us with a semiclassical quantum defect theory to describe the spectrum near the various thresholds.

The fundamental term can be transformed into a sum over integrals using Poisson summation. Inserting the identity $\sum_{n=-\infty}^{\infty} \delta(x-n) = \sum_{r=-\infty}^{\infty} \exp(2\pi i r x)$, we can write

$$\sum_{n=1}^{\infty} t_n(z) = \frac{t_1(z)}{2} + \sum_{r=-\infty}^{\infty} \int_1^{\infty} dn\, e^{2\pi i r n}\, t_n(z), \tag{16}$$

where n is now a continuous variable. The stationary phase condition yields

$$\frac{\partial}{\partial n} \left(z\tilde{S}(n) - (r + \frac{1}{2})n - (m + \frac{1}{2})\sigma(n) \right) = 0. \tag{17}$$

The solutions $n_0(z)$ of equation (17) tend to infinity for integer z values, i.e. at the ionization thresholds $E_N = -Z^2/(2N^2)$, $N = r + m + 1$. Approximating the spectral determinant by the fundamental term only leads to the quantization condition $D(E) \approx 1 - \sum_{n=1}^{\infty} t_n \stackrel{!}{=} 0$. The following quantization of the phases alone must then hold, i.e.

$$z\tilde{S}(n_0) - (r + \frac{1}{2})\, n_0 - (m + \frac{1}{2})\, \sigma(n_0) = l + \frac{1}{8} \tag{18}$$

with $l \geq 0$ integer. Equations (17), (18) correspond exactly to an EBK quantization as it has e.g. been carried out for the stable collinear configuration $\theta = 0^o$ (Wintgen and Richter 1994). The quantum numbers r, l are linked to a quantization of the inner and outer electron respectively, m corresponds to a quantization of the θ-degree of freedom. The appearance of intermittency and the non-closed tori structures in phase space leads to the unusual action quantization condition $l + 1/8$ for the motion of the outer electron. The term $1/8$ enters formally through the stationary phase approximation. Combining (17) and (18) leads to a quantization of the "winding number" n, i.e.

$$\begin{aligned} &\left[(r + \frac{1}{2}) + (m + \frac{1}{2})\frac{\partial \sigma}{\partial n}\right] \tilde{S}(n) - \\ &\left[(r + \frac{1}{2})\, n - (m + \frac{1}{2})\, \sigma(n) + (l + \frac{1}{8})\right] \frac{\partial \tilde{S}}{\partial n}(n) = 0. \end{aligned} \tag{19}$$

The quantum energies are now obtained from the Rydberg–like formula

$$E_{r,l,m} = -\frac{2\tilde{S}^2(n_{r,l,m})}{\left[(r + \frac{1}{2})n_{r,l,m} + (l + \frac{1}{8}) + (m + \frac{1}{2})\sigma(n_{r,l,m})\right]^2}. \tag{20}$$

The simple quasi – EBK (QEBK) quantization already reproduces the (real parts) of full 3-dimensional quantum calculations surprisingly well, see table 2. From the stationary phase approximation, we expect (20) to give best results in the large l limit. We find indeed agreement within 2% with respect to the *mean level spacing* near the various thresholds. The error increases for lower l, but remains below 5% even for small l-values, and already gives a good estimate for the full helium ground state (N,l)=(1,0). Note, that there is no free parameter, the coefficients in (15) are determined by the periodic orbit data alone. The deviations of the QEBK–results form the full quantum calculation increases considerably around the first perturber (N, l) =(5,0). The QEBK – approach can not resolve the interaction between different Rydberg series, and higher terms in the expansion (14) have to be taken into account to describe the influence of perturber states (Tanner and Wintgen 1995).

A sensitive test of the QEBK – formula is performed by comparing the quantum defects at the threshold energies obtained from Eqn. (20) with MQDT – fits (Friedrich 1991) estimated from the quantum data available.

Table 2. Quantum energies of the bound Rydberg states $N = 1$ and the resonance series $N = 2$, $N = 4$ are compared with semiclassical quasi EBK results in ${}^1S^e_{m=0}$ – helium.

l	N=1		N=2		N=4	
	QM	QEBK	QM	QEBK	QM	QEBK
0	2.903721	2.784751	0.777868	0.758714	0.200990	0.198877
1	2.145974	2.149065	0.589895	0.592375	0.165734	0.166685
2	2.061272	2.061619	0.544882	0.545636	0.150824	0.151101
3	2.033587	2.033629	0.526687	0.526975	0.142602	0.142656
4	2.021177	2.021172	0.517641	0.517774	0.137685	0.137657
5	2.014563	2.014552	0.512513	0.512584	0.134551	0.134485
6	2.010626	2.010615	0.509332	0.509374	0.132452	0.132358
7	2.008094	2.008085	0.507225	0.507252	0.130999	0.130868
8	2.006370	2.006363	0.505759	0.505777	0.129993	0.129786
(N=5,l=0) →					0.129323	
9	2.005143	2.005138		0.504710	0.128777	0.128976
10	2.004239	2.004235		0.503913	0.128262	0.128354
11	2.003555	2.003551		0.503303	0.127816	0.127868
12	2.003023	2.003021		0.502825	0.127446	0.127479
∞	2.0	2.0	0.5	0.5	0.125	0.125

The quantum defect $\nu_{N,m}(E)$ is defined as the smooth function through the points (Friedrich 1991)

$$E_{N,m,n} = I_N - \frac{1}{2(n - \nu(E_{N,m,n}))^2}, \tag{21}$$

where $n \geq N$ counts the states successively up to the threshold energy $I_N = -Z^2/(2N^2)$. The QEBK threshold values $\nu_{N,m}(I_N)$ are given by the simple formula,

$$\nu_{N,m}(I_N) = (1 + s_0)\, N - (m + \frac{1}{2})\,\sigma_0 - \frac{1}{8} + n_{\text{per}}, \tag{22}$$

where n_{per} counts the number of perturber states in the N series. Information about e^- – e^- interaction in the near core region enters the quantum defect through leading coefficients s_0, σ_0, (see table 1). In table 3, we compare Eqn. (22) with MQDT – data and find good agreement up to $N = 7$. (A determination of quantum defects for $N \geq 8$ is impossible from the quantum data available due to the large number of perturber states.) We are thus able to predict quantum defects from purely classical input together with unusual phase 1/8, which arise from the non–closure of the tori in phase space.

The present quantization of intermittency regroups the periodic orbit sum in a way which accounts for the quasi–regular motion in the separable limit

Table 3. Quantum defects at the various threshold energies $I_N = -2/N^2$.

N	1	2	3	4	5	6	7
MQDT	0.140	0.685	1.272	2.785	3.410	4.953	6.391
QEBK	0.133	0.696	1.259	2.822	3.385	4.948	6.510

of the Hamiltonian. The Rydberg structure is correctly described (including interference) and the spectrum follows an EBK–quantization scheme near the thresholds. Electron – electron interaction is semiclassically included through the action function and even the groundstates in each series, $n \approx N$ or $l = 0$, are obtained reasonably well in a quantization of the asymptotic tori.

In the following sections we will approach the problem from a complementary point of view which supplements the periodic orbit results. Starting from a semiclassical construction of integrated cross sections for two-electron escape at positive energies in the next section, we will come back to to $E < 0$ in section 4 and formulate a cross section in the case of extremely overlapping resonances, i.e. for $n \approx N$ and $N \to \infty$.

3 Complete Fragmentation from Scattering Orbits

In this section we consider positive total energies $E > 0$ where *no* periodic orbits exist. This can be rigorously shown by proving that the hyperradius $\mathcal{R} = \sqrt{r_1^2 + r_2^2}$ of any two–electron (actually many electron) trajectory has at most a single minimum as a function of time for $E > 0$ (Rost 1995). For this reason, no resonances exist. Instead, a new phenomenon occurs, complete fragmentation into the nucleus and two electrons, observable in double-photoionization experiments (Dörner *et al.*, 1996)) or through electron impact ionization which we will discuss here. Both processes can be described semiclassically using an S-matrix formulation (Miller 1975, Rost 1994, 1995).

3.1 The Semiclassical S-Matrix in Reduced Phase Space

For integrated cross sections, for instance the fragmentation cross section, no angular information is needed explicitly. Hence, we use the same restricted collinear phase space as described in the previous section. For moderate energies $E \geq 0$ it turns out that the collinear configuration with the electrons on opposite sides of the nucleus (lower interelectronic repulsion) is most important. Furthermore, for qualitative considerations, we can restrict the discussion to the $L = 0$ case. This yields even quantitative results for $E \to 0$ since all partial cross sections σ_L do not depend on L as $E \to 0$ which follows again from the scaling properties of a Coulomb system (Rost 1995). We are thus left with the collinear Hamiltonian (7) now for $E \geq 0$ in a two dimensional configurations space (r_1, r_2). Trajectories of the three characteristic scattering processes are shown in figure 4. The active target electron is bound with

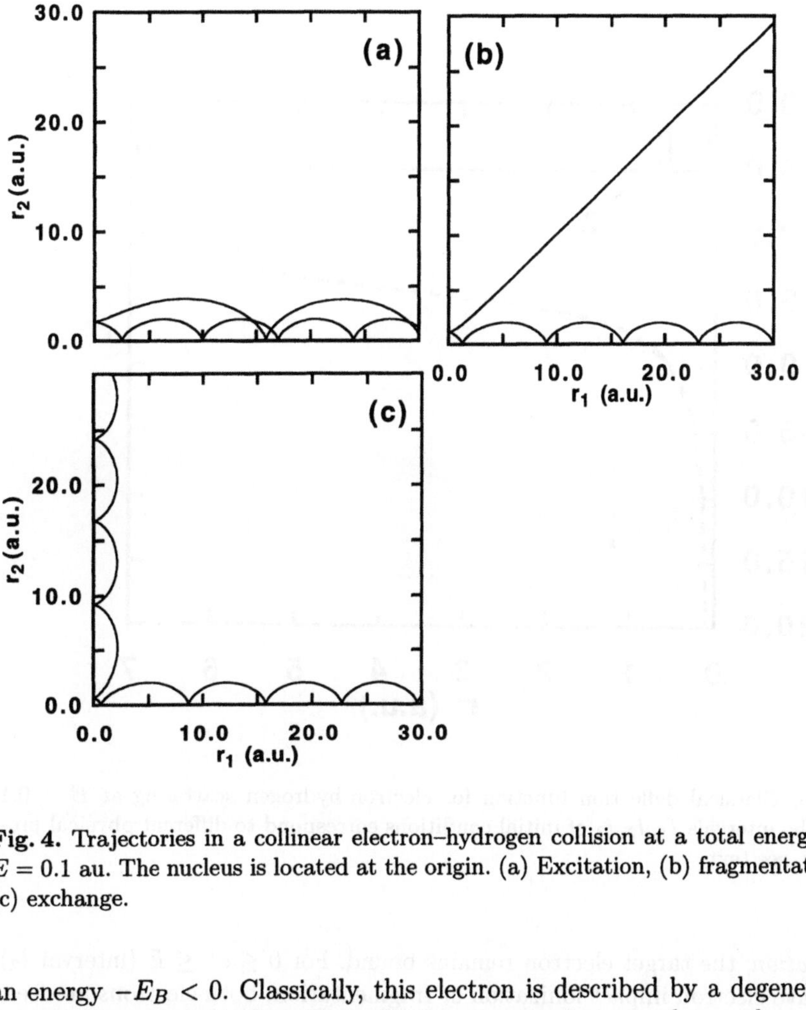

Fig. 4. Trajectories in a collinear electron–hydrogen collision at a total energy of $E = 0.1$ au. The nucleus is located at the origin. (a) Excitation, (b) fragmentation, (c) exchange.

an energy $-E_B < 0$. Classically, this electron is described by a degenerate (eccentricity zero) Kepler ellipse. Without restriction, we choose the outer turning point of the bound electron as a starting point for the coordinate r_2. The only initial phase space variable which is a priori not fixed is the distance of the free projectile electron r_1, which we write as $r_1 = r^- + r_0$. Here, r_0 is a fixed distance large enough to put the system into the asymptotic region before scattering (we use $r_0 = 1000$ au). The momentum p_1 is in the asymptotic regime given by $p_1^- = \sqrt{2E + 2E_B}$. With $\pm$ we indicate in what follows quantities in the limit $t \to \pm\infty$. The observable after the collision is the energy ϵ^+ of one of the electrons. This situation parallels the well known case of elastic scattering where the initial variable is the impact parameter and the final observable the scattering angle. There and here, the relevant quantity is the *deflection function* $\theta^+(b^-)$ or in our case $\epsilon^+(r^-)$. It is shown in figure 5. For $\epsilon^+ < 0$ (initial conditions r^- from interval I_1) we have

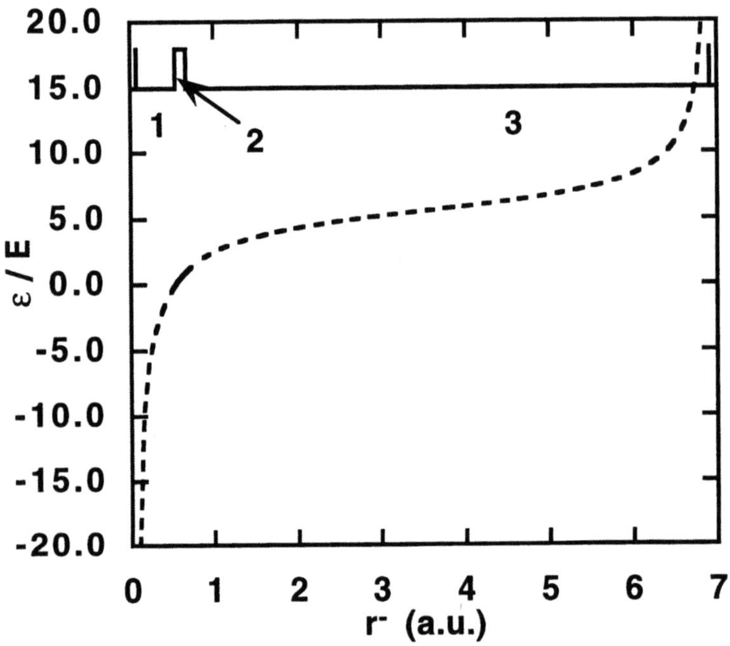

Fig. 5. Classical deflection function for electron-hydrogen scattering at $E = 0.1$ au. The intervals I_1, I_2, I_3 of initial conditions correspond to different physical processes, see text.

excitation, the target electron remains bound. For $0 \leq \epsilon^+ \leq E$ (interval I_2) we have electron impact ionization or fragmentation, both electrons are free. For $\epsilon^+ > E$ (interval I_3) the projectile electron must have negative energy to conserve total energy and it has become bound in *exchange* for the target electron. The classical probability to find an electron with energy ϵ^+ after the collision is given by the integral over the initial phase space (Rost 1996)

$$P(\epsilon^+, E) = \frac{1}{R} \int dr^- \delta(\epsilon^+ - \epsilon(r^-)) = \frac{1}{R} \sum_j \left| \frac{dr_j^-}{d\epsilon} \right|_{\epsilon=\epsilon^+} \equiv \sum_j \mathcal{P}_j(\epsilon^+, E), \quad (23)$$

where R is a normalization of the probability corresponding to the distance the projectile travels during one complete revolution of the target electron. In the following we will suppress the " + " sign in ϵ^+. Since the deflection function is monotonic (Fig. 5) — which is related to the single minimum in the hyperradius of each trajectory — there is only a single trajectory j for each ϵ emerging from a unique initial value of r_j^- and the sum in (23) collapses to one term. This peculiar phenomenon is a property of the Coulomb potential (the same arguments about a single trajectory apply to Rutherford scattering — in this case the hyperradius $\mathcal{R}$ is the ordinary radius r). It becomes very important in the semiclassical context, since the semiclassical scattering matrix is defined by

$$\mathcal{S}(\epsilon, E) = \sum_j \sqrt{\mathcal{P}_j(\epsilon, E)} e^{iS_j - i\nu_j\pi/2} , \tag{24}$$

where S_j is the classical action of trajectory j and ν_j is the Maslov index. Equation (24) is obtained by interpreting the quantum mechanical S-matrix $\mathcal{S}_{fi} = \lim_{|t|\to\infty}\langle f|\exp(iHt/\hbar)|i\rangle$ as a path integral and by evaluating it in stationary phase approximation for small $\hbar$ (Gutzwiller 1990). Essentially, the semiclassical Green's function of (2) is the Fourier transform of the semiclassical propagator in time as used in this section.

3.2 Pauli Principle and Fragmentation Cross Section

Coming back to our problem of electron impact ionization we realize that the semiclassical expression (24) will yield the same probabilities as the classical expression (23) if only a *single* trajectory contributes. However, there is one additional twist in our problem: the electrons are indistinguishable so that the scattering amplitude of (24) must be symmetrized with respect to the exchange of electrons. This introduces a second trajectory and the possibility of interferences (in close analogy to the transition from Rutherford scattering to Mott-scattering, for instance for $\alpha - \alpha$ scattering (Rost and Heller 1994)). Surprisingly, the action $S(\epsilon, E)$ of the direct and the exchanged path, $S(E - \epsilon, E)$ are equal, so that no interference occurs, and the final semiclassical reaction probability may be written as

$$P^{\pm}(\epsilon, E) = \tfrac{1}{2}\left(\sqrt{\mathcal{P}(\epsilon, E)} \pm \sqrt{\mathcal{P}(E-\epsilon, E)}\right)^2 , \tag{25}$$

where " + " refers to ^1S and " − " to ^3S symmetry. The ionization cross section is now easily obtained by integrating over the energy of the electron,

$$P^{\pm}_{ion}(E) = \int_0^{E/2} P^{\pm}(\epsilon, E) d\epsilon. \tag{26}$$

In Fig. 6 we show the integrated ionization cross section for electron-hydrogen scattering together with data from a pioneering experiment by McGowan and

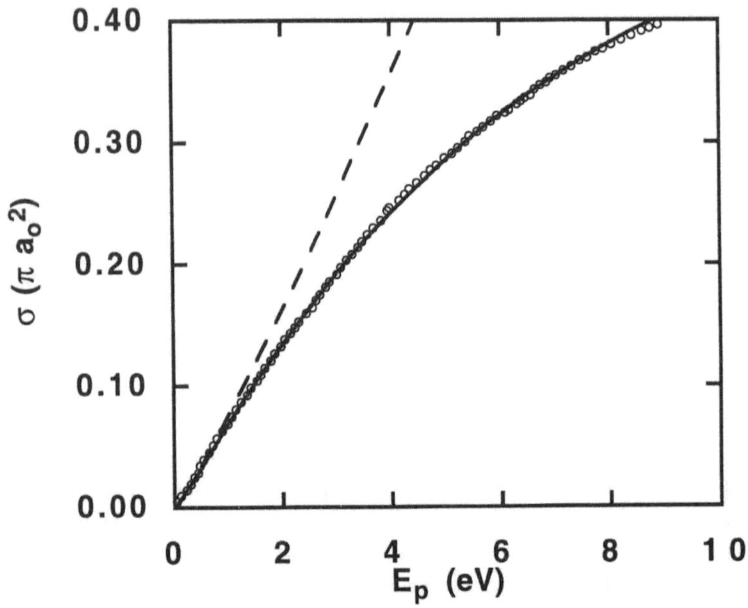

Fig. 6. Electron impact ionization of hydrogen; open circles: from the experiment by McGowan and Clarke (1968); solid line: present theory, normalized to the experimental cross section at 5.82eV; dashed line: Wannier cross section $\sigma \propto E^{1.127}$.

Clarke (1968). The semiclassical S-matrix calculation provided the first theoretical curve describing the entire range of the experiment from threshold to about 10 eV excess energy (Rost 1994). The dashed line indicates Wannier's result $\sigma(E) \propto E^{\alpha}$ with $\alpha = 1.127$ derived classically as a perturbation about $E = 0$ (Wannier 1953). More recently, two other calculations have been published which confirm the earlier experimental and theoretical results from the semiclassical S-matrix approach (Macek and Ovchinnikov 1995, Kato and Watanabe 1995).

In another pioneering experiment Cevjanović and Read (1974) measured very slow ionized electrons following electron impact on helium at energies close to the ionization threshold of 24.6 eV, or for excess energies $E \approx 0$. The technique used is nowadays known as ZEKE spectroscopy (Zero Electron Kinetic Energy). Interestingly, this experiment scanned an energy range above and *below* the fragmentation threshold (Fig. 7). The energy sharing $P(\epsilon, E)$ of the continuum electrons after ionization is constant as a function of ϵ within 5% for $0 \leq E \leq 2$ eV excess energy (Read 1984, Rost 1994). Hence, the measurement of the electrons with $\epsilon \approx 0$ can be related to the total ionization cross section through $P(\epsilon \approx 0, E) \propto \sigma(E)/E$. Consequently,

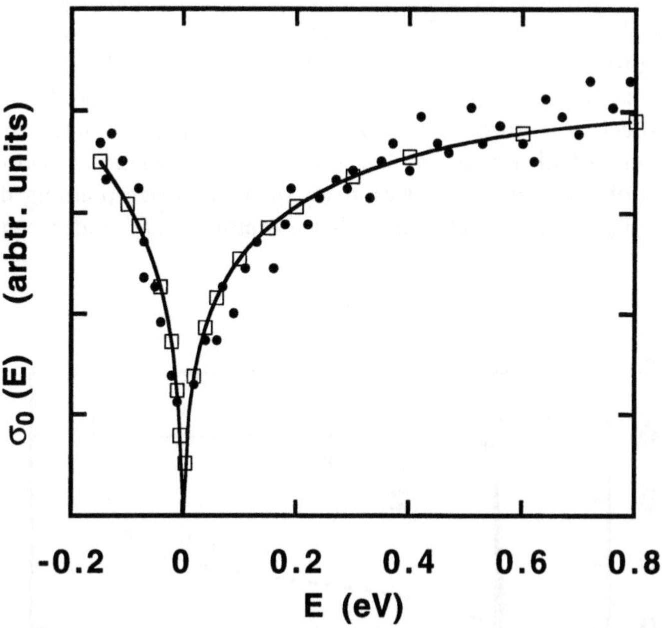

Fig. 7. Cross section for the production of slow electrons following electron impact on atoms close to the fragmentation threshold. The black points are experimental data for a He(1s^2) target from Cvejanović and Read (1974). The squares are the theory for a hydrogen target H(1s) from Rost (1994). The solid line is a fit of the theory according to (28).

one sees in Fig. 7 a dip with $P(\epsilon \approx 0, E) \propto E^{0.127}$ and the ionization cross section behaves as expected for $E > 0$. However, it is not clear off hand how to interpret the measured signal for negative energies where He^- doubly-excited states will show up as resonances, i.e. where we reach energetically a regime which has been discussed in Sect. 2. The highly doubly excited states of He^- behave very much like those of H^- where He^+ serves as a passive core (just like the proton) for the two excited electrons (Rost and Briggs 1988). Tuning the projectile energy down so that the total energy E remains below threshold we are able to simulate this experiment in our collinear phase space by semiclassical scattering theory.

4 Collective Excitation Below the Fragmentation Threshold

For negative energies the classical dynamics of the three–body Coulomb system is quite different and much more complicated compared to $E > 0$ as we have seen in section 2. Multiple Rydberg series exist and begin to perturb each other if the threshold $E = 0$ is approached from below. This energy regime is governed by resonances and the backbone of their semiclassical description is the invariant set of periodic orbits which has been used to calculate spectral features in section 2. Now we ask, how does this region below threshold look like in the framework of a semiclassical *scattering* approach? To obtain a first orientation we show the corresponding deflection function in Fig. 8. As expected we still see *excitation* (I_3) and *exchange* (I_1).

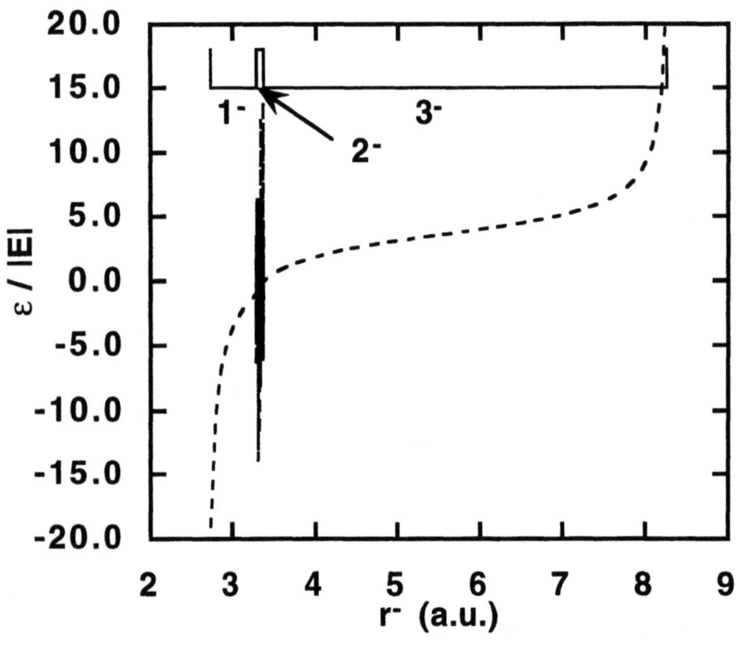

Fig. 8. Same as Fig. 5 but for $E = -0.1$au.

However, instead of ionization which is energetically not possible we see a structure (I_2) in Fig. 8 which turns out to be a fractal (see enlargements in Figs. 9a and b). The attractor of this chaotic deflection function is the set of periodic orbits discussed in 2. The selfsimilar structure has the same origin as the existence of a binary symbolic dynamics, see section 2.3 and can be understood looking at characteristic trajectories. As shown in Fig. 10, the two electrons try to escape as they would do for an initial condition r^- and $E > 0$ leading to fragmentation. However, there is now a potential boundary reflecting the trajectory and two electron escape can not occur. Eventually

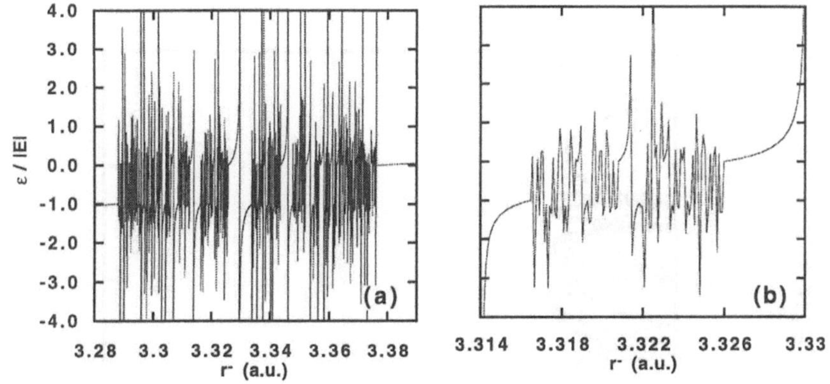

Fig. 9. Stepwise magnification of Fig. 8 demonstrating the selfsimilar, fractal structure of the deflection function.

many re-scattering events happen before finally one of the electrons escapes. The same behaviour can be studied in the Poincaré surface of section, Fig. 3 (here for helium). An electron coming from infinity enters the chaotic sea near the nucleus. One of the electrons leaves the system again after a time period depending sensitively on the initial conditions when entering the near-core dynamics. Compared to a direct scattering trajectory the multiple scattering leads to a time-delay which is the classical signature of a resonance.

4.1 Quantitative Scattering Information from the Fractal

Coming back to the experiment shown in Fig. 7 we have to calculate the ZEKE cross section below threshold. This means summing over all trajectories in (23) which lead to a final energy $\epsilon = 0$. Clearly, this is not possible by brute force. Rather, we will proceed in a similar way as we did for $E > 0$. We approximate the ZEKE cross section by $P_{res}(E)/E$ in analogy to $P_{frag}(E)/E$ above threshold. Furthermore, we circumvent the problem of the explicit summation of all the trajectories by taking the classical limit of Eq. 23. For $E \to 0$ this is justified once again by the scaling $S(E) = \tilde{S}/\sqrt{|E|}$ of the classical action, see section 2.3 Eqn. (10). The action enters the semiclassical scattering amplitude (24) such that the limit $\hbar \to 0$ corresponds to $|E| \to 0$. In this limit the phase in (24) oscillates infinitely rapid averaging out all oscillations in an observable.

For $P_{res}(E)$ we then get

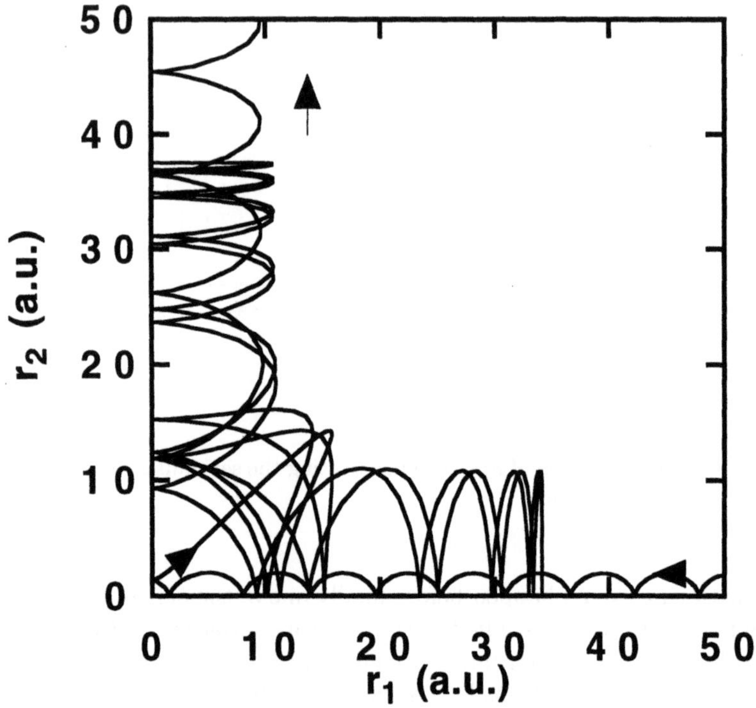

Fig. 10. Chaotic trajectory with an initial condition r^- from interval I_2.

$$
\begin{aligned}
P_{res}(E) &= \int P(\epsilon, E)d\epsilon = \frac{1}{R}\sum_j \int \left|\frac{dr^-}{d\epsilon}\right| d\epsilon \\
&= \frac{1}{R}\sum_j \int_{\{r^-:r^-\in I_2)\}} dr^- = \frac{1}{R}\sum_i \delta_i r^- = \frac{\Delta r^-(I_2)}{R}, \qquad (27)
\end{aligned}
$$

where the δ_i are the intervals of r^- over one branch in Fig. 9. $P_{res}(E)$ can interpreted as a the probability for *collective* resonance excitation of a doubly excited complex during the collision. Individual resonances can not been resolved by averaging over all phase oscillations in calculating $P_{res}(E)$. This has been done in section 2 by explicitly summing over families of periodic orbits and their associated phases. However, $P_{res}(E)$ is a useful quantity to interpret spectra of low resolution or in a region of strongly overlapping resonances. Indeed, the black points in Fig. 7 are calculated according to (27) below threshold and with the corresponding classical expression $P^{cl}_{frag}(E) = \Delta r^-(I_2)/R$ above threshold where I_2 refers to the interval of ionizing initial conditions in Fig. 5. Due to the close analogy of P_{res} and P_{frag} we expect both probabilities to contain the algebraic behavior $|E|^\alpha$. This is indeed the case, the solid line in Fig. 7 is a fit with the function

$$\sigma_\pm(E, \epsilon \approx 0) = s_\pm |E|^{\alpha-1}(1 + a|E|^{\frac{1}{2}} + bE), \tag{28}$$

where $\alpha = 1.127$ and $\sigma_+ \propto P_{frag}/E$ while $\sigma_- \propto P_{res}/E$ (Rost and Wintgen 1996). Due to experimental circumstances (gas pressure etc.) the constants $s_\pm$ have to be determined separately for $E < 0$ and $E > 0$ and the ratio s_+/s_- does not match the theoretical one (Cevjanović *et al.* 1995). This issue needs further clarification in the future.

4.2 Overlapping Resonances Towards Threshold?

A question, which we have not addressed yet, concerns the smoothness of the cross section in Fig. 7 for $E \to 0$ from below which might be caused by lack of resolution or by overlapping resonances. In calculating P_{res}, we have explicitly excluded the description of individual resonances. Also, the periodic orbit description of section 2 does not yet reach into this regime of resonances for $N \approx n \gg 1$. Hence, we have calculated a series of suitable resonances ($N = n$) quantum mechanically and we will try to extrapolate the energy dependence of the widths $\Gamma_N(E)$ for $E \to 0$. Doubly excited states of He^- in the limit of $E \to 0$ behave like those of H^- (Rost and Briggs 1988). Furthermore, the class of resonances which come closest to a collinear configuration are the lowest states in each manifold N with $n = N$ as it has become clear from the discussion in section 2.2.

We find from our full quantum calculation using the method of complex rotation that the $N = n$ resonances in H^- are isolated enough to be determined with sufficient accuracy up to $N = 8$. When fitting their widths with (28), replacing $\sigma_\pm$ by $\Gamma_N(E)/E_N$, we find an exponent of $\alpha = 1.127$. To check the sensitivity of the exponent we performed the fit again with a fixed linear exponent $\alpha = 1$. The relative errors of the widths for both fits shown in Fig. 11 clearly demonstrate that $\alpha = 1.127$ reproduces $\Gamma_N(E)$ significantly better than $\alpha = 1$. The energy dependence of the widths with the Wannier exponent $\Gamma_N(E) \propto |E|^{1.127}$ reveals once more the dominance of the classical mechanics for phenomena close to the fragmentation threshold. It also shows the intimate connection between fragmentation and resonance formation as already pointed out in section 2.2. More importantly however, any dependence $\Gamma_N(E) \propto |E|^\alpha$ with $\alpha < 1.5$ leads to overlapping resonances since the spacing of the $N = n$ resonances is Coulombic with $E_N - E_{N-1} \propto 1/N^3 \propto |E|^{1.5}$.

Our result points to a regime of extremely overlapping resonances for $E \to 0^-$. This is consistent with a recently predicted approximately linear behavior in the widths of helium (Burgdörfer *et al.* 1995, 1997), where $\alpha = 1.056$ is

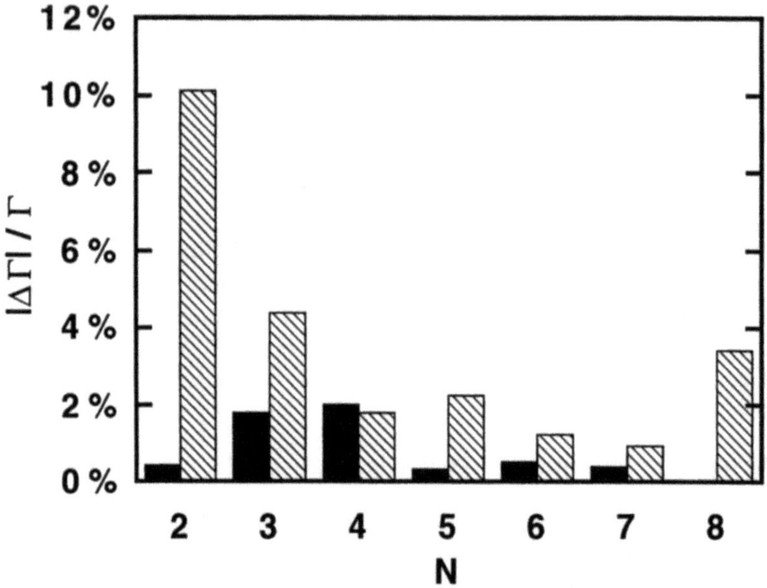

Fig. 11. Relative error of the fit (28) for ${}^1S^e$ saddle resonances in H^- shown with dark columns For comparison the error for a fit with a linear exponent $\alpha = 1$ is shown with light columns.

expected. However, there is a discrepancy to all earlier results which predict a variety of dependencies $\Gamma_N(E) \propto |E|^\alpha$, but all of them with $\alpha > 1.5$, i.e. *non* overlapping resonances (see e.g. Feagin and Macek, 1984; Heim and Rau, 1995).

5 Conclusion

We can summarize our present understanding of the dynamics of two electrons in the limit of high correlation close to the fragmentation threshold as follows: For each fixed ionization threshold $E_N = -Z^2/(2N^2)$, we have three groups of Rydberg series: (a) approximately collinear configurations with the electrons on opposite sides of the nucleus. These resonances emerge from classically chaotic dynamics. (b) approximately collinear configurations with the electrons on the same side of the nucleus. These resonances emerge from classically integrable (regular) dynamics. (c) Energetically in between, there are Rydberg series which strongly perturb each other and whose underlying classical dynamics is not restricted to a lower dimensional (collinear) subspace. Our understanding of these resonances is still rather limited, however, many indications point to a chaotic behavior of this class of resonances as well.

Collective excitations near the fragmentation threshold $|E| \to 0$ can be described in the framework of semiclassical S-matrix theory as presented here. We have seen that resonances with $N \approx n$ will strongly overlap for energy approaching $E = 0$ from below. We expect thus an essentially smooth and structureless cross section from the experiment even for E smaller but close to 0. Only with extremely good resolution it might become possible in the future to resolve the remaining tails of the Rydberg series which are characterized by $n \gg N \gg 1$ and whose members have very small widths. These tails have been the starting point for the periodic orbit description of intermittency as presented here.

In the future we will have to bridge the two gaps which are apparent in the present situation: we need a good semiclassical description for excitation modes which are at the boundary of regular Rydberg behavior and collective excitation. This concerns already the dynamics which can be restricted to the collinear configuration space. In addition we need to understand the irregular behavior of resonances formed in the full configuration space.

Experimentally (e.g. in photo absorption experiments with synchrotron radiation), it will be most interesting in the future to see if it is possible to observe the transition to collective excitation and to see if Ericson fluctuations can be observed as they are well known from other areas of physics in similar situations (Ericson 1960).

Acknowledgment. We are grateful for financial support by the Deutsche Forschungsgemeinschaft, in Freiburg within the SFB 276 and the Gerhard-Hess-Programm.

References

Artuso R., Aurell E. and Cvitanović P., Nonlinearity **3**, 325 and 361 (1990).
Balian R. and Bloch C. , Ann. Phys. (N.Y.) **69**, 75 (1972); **85**, 514 (1974).
Berry M. V. and Tabor M. , Pro. Roy. Soc. London A **349**, 101 (1976); J. Phys A **10**, 371 (1977).
Berry M. V. and Keating J. P., J. Phys. A **23**, 4839 (1990); Berry M. V. and Keating J. P., Proc. R. Soc. A **437**, 151 (1992).
Blümel R. and Reinhardt W. P., in *Quantum Chaos* ed. G. Casati and B. V. Chirikov (Cambridge Univ. Press), p 301 (1995).
Bohigas O., Tomsovic S. and Ullmo D., Phys. Rep. **223**, 44 (1993).
Born M, *Vorlesung über Atommechanik* (Springer, Berlin, 1925).
Bürgers A., *Doppelt angeregte S–Zustände in Helium*, Phd – thesis, Freiburg (1995).
Bürgers A., Wintgen D. and Rost J. M., J. Phys. B **28**, 3163 (1995).
Burgdörfer J., Yang X. , and Müller J., Chaos, Sol. & Fractals **5**, 1235 (1995).
Burgdörfer J., Qiu Y. , and Müller J. (1997), p 304 of this volume.
Cvejanović S. and Read F. H., J. Phys. B **7**, 1841 (1974).

Cvejanović S., Shiell R. C., and Reddish T. J., J. Phys. B **28**, L707 (1995).

Cvitanović P., Phys. Rev. Lett. **61**, 2729 (1988); Cvitanović P. and Eckhardt B. Phys. Rev. Lett. **63**, 823 (1989).

Dörner R. *et al.*, Phys. Rev. Lett., **76**, 2654 (1996).

Domke M., Schulz K., Remmer G., Gutierrez A., Kaindl G., and Wintgen D., Phys. Rev. A **51**, R4309 (1995).

Draeger M., Handke G., Ihra W. and Friedrich H., Phys. Rev. A **50**, 3793 (1994).

Ericson T., Phys. Rev. Lett. **5**, 430 (1960).

Ezra G. S., Richter K., Tanner G., and Wintgen D. J. Phys. B **24**, L413 (1991).

Feagin J. M. and Macek J., J. Phys. B **17**, L245 (1984).

Friedrich H. and Wintgen D., *Phys. Rep.* **183**, 37 (1989).

Friedrich H., *Theoretical Atomic Physics* (Springer, Berlin, 1991).

Grujić P. V. and Simonović N. S., J. Phys. B **28**, 1159 (1995).

Gutzwiller M. C., *Chaos in Classical and Quantum Mechanics* (Springer, New York, 1990).

Handke G., Draeger M. and Friedrich H., Physica A **197**, 113 (1993).

Handke G., Draeger M., Ihra W. and Friedrich H., Phys. Rev. A **48**, 3699 (1993).

Heim, T. A. and Rau A. R. P., J. Phys. B **28**, 5309 (1995).

Herrick D. E., Adv. Chem. Phys. **52**, 1 (1983).

Ho Y. K., Phys. Rev. A **34**, 4402 (1986).

Ihra W., Draeger M., Handke G. and Friedrich H., Phys. Rev. A **52**, 3752 (1995).

Kato D., and Watanabe S., Phys. Rev. Lett. **74**, 2443 (1995).

McGowan J. W. and Clarke E. M., Phys. Rev. **167**, 43 (1968).

Miller W. H., Adv. Chem. Phys. **30**, 77 (1975).

Müller J. and Burgdörfer J., Phys. Rev. Lett. **70**, 2375 (1993).

Ozorio de Almeida A. M. and Hanney J. H., J. Phys. A **20**, 5873 (1987).

Macek J. H. and Ovchinnikov S. Yu., Phys. Rev. Lett. **74**, 4631 (1995).

Read F. H., J. Phys. B **17**, 3965 (1984).

Richter K., G. Tanner G. and Wintgen D., Phys. Rev. A **48**, 4182 (1993).

Rost J. M., Greenland P. T., and Briggs J. S., J. Phys. B **B22**, L353 (1989).

Rost J. M. and Briggs J. S., J. Phys. B **24**, 4293 (1991).

Rost J. M., Phys. Rev. Lett. **72**, 1998 (1994).

Rost J. M., J. Phys. B **27**, 5923 (1994).

Rost J. M., Heller E J, J. Phys. B **27**, 1287 (1994).

Rost J. M., J. Phys. B **28**, 3003 (1995).

Rost J. M., Phys. Rev. A **53**, R640 (1996).

Rost J. M., and Wintgen D., Europhys. Lett. **35**, 19 (1996).

Schuster H. G., *Deterministic Chaos* (VCH, Weinheim, 1989).

Sieber M. and Steiner F., Phys. Rev. Lett. **67**, 1941 (1991).

Tanner G., Scherer P., Bogomolny E., Eckhardt B. and Wintgen D., Phys. Rev. Lett. **67**, 2410 (1991).

Tanner G. and Wintgen D., Phys. Rev. Lett. **75**, 2928 (1995).

Tanner G., Hansen K. T. and Main J., *The semiclassical resonance spectrum of hydrogen in a constant magnetic field.* Nonlinearity **3**, 1641 (1996).

Tanner G., *How chaotic is the Stadium Billiard? – A semiclassical Analysis*, J. Phys. A (in press).

Voros A., J. Phys. A **21**, 685 (1988).

Wannier G. H., Phys. Rev. **90**, 817 (1953).

Wintgen D., K. Richter K., and Tanner G., CHAOS **2**, (1992).
Wintgen D., Bürgers A., Richter K. and Tanner G., Prog. Theo. Phys. **116**, 121 (1994).
Wintgen D. and Richter K., Comm. At. Mol. Phys. **29**, 261 (1994).

Periodic Orbit Spectra of Simple Atoms

Joachim Burgdörfer[1,2], Yanghui Qiu[1], and Jörg Müller[1]

[1] Department of Physics, University of Tennessee, Knoxville, TN 37996-1200
[2] Oak Ridge National Laboratory, Oak Ridge, TN 37831-6377

Abstract. We discuss the correspondence and interplay between classical and quantum dynamics in simple atoms within the framework of the periodic-orbit (PO) expansion of quantum spectra. The semiclassical expression for the spectral density (Gutzwiller's trace formula) can be viewed as a Fourier expansion in terms of classical (quasi) periodic orbits. Inverse Fourier transform of the full quantum spectral density therefore not only gives the PO spectrum in the semiclassical limit, but may also provide a tool to characterize the quantum spectrum. We study the scaled energy Fourier transform spectrum of the hydrogen atom as the prototype example of a regular, highly degenerate system and doubly excited resonances of helium exemplifying simple atomic systems that display soft chaos. Fundamental classical periodic orbits can be identified which form the basic building blocks of the quantum spectrum. This provides a tool for analyzing and interpreting electron correlation effects in atoms. We analyze the mechanism for the suppression of chaos due to the presence of Rydberg series in systems with long-range Coulomb interaction and discuss signatures of Ericson fluctuations in highly excited atoms close to the double ionization threshold.

1 Introduction

The most celebrated of all dynamical problems, the "Problem of Three Bodies" (Whittaker, 1944), has been studied for more than two centuries because of its importance in Celestial Mechanics, one-electron molecules and, most recently, in the investigation of doubly excited states in atoms. In spite of the simplicity of the Hamiltonian function, it has been shown by Bruns (1887) and Poincare (1890) that the "classical integrals," namely the integrals of motion of the center of gravity, the integrals of angular momentum, and the integral of energy are the only independent algebraic integrals of the problem of three bodies. Thus the classical phase space displays soft chaos, i.e. intermingled regions of regular and chaotic motion in phase space, and constants of motion can only be found locally.

The structure of the classical phase space, in particular the properties of periodic orbits and their semiclassical (i.e., Bohr-Sommerfeld) quantization, played an important role in the "old quantum theory" of the H_2^+ molecule (Pauli, 1922; see also Müller et al., 1995) and of helium (van Vleck, 1922). Various periodic orbits of the three-body Coulomb problem were employed in the Bohr-Sommerfield quantitization of the action along periodic orbits, $\oint p(r) \cdot dr = 2\pi\hbar n$. The failure to yield a stable H_2^+ molecule and to predict the ground state energy of He led to the development of modern quantum mechanics and the demise of the old quantum theory. With the benefit of the knowledge of recent developments of semiclassical theory, one may state in retrospect that the pioneers of quantum mechanics "did the right thing for the wrong reason." The failure of the old quantum theory was not due to the semiclassical approximation underlying Bohr-Sommerfeld-type quantization itself but due to its incomplete implementation. Specifically, essential ingredients of modern semiclassical theory such as the Maslov indices for conjugate points of classical trajectories and the proper relationship between periodic orbits and individual quantum states were not known. Clearly, the role of the latter could be appreciated only after semiclassical mechanics was understood as the short-wave limit of wave mechanics.

In the last twenty years, considerable progress has been made in the semiclassical theory of atomic spectra. Semiclassical theory has provided the link between observable quantum spectra and the classical dynamics of non-separable and chaotic systems. Atomic systems have proven to be the ideal experimental and theoretical testing ground for the notion of "quantum chaos," that is, how and to what extent properties of quantum systems reflect the chaotic phase space of its classical counterpart. In the following article we will elucidate the correspondence between classical periodic orbits and the quantum spectrum by investigation of the periodic orbit (PO) spectrum.

Semiclassical quantization starts with classical dynamics by identifying families of classical periodic orbits which are then used in the Gutzwiller trace formula (Gutzwiller, 1990) for the spectral density (or variants thereof for the spectral determinant) to calculate an approximate quantum spectrum. This approach is plagued by difficulties stemming from the exponential proliferation of contributing orbits and from poor convergence behavior of the trace formula. Nevertheless, considerable progress along those lines has been made (see e.g., Tanner and Wintgen, 1995) and is reviewed in the contribution by Rost and Tanner (1997) in this volume.

Our point of departure is, to some extent, complementary. We employ the trace formula "in reverse". Starting from the quantum spectrum, we determine the dominant classical periodic orbit in the semiclassical expression by inverse scaled Fourier transform. The resulting periodic orbit spectrum serves to identify its classical or semiclassical "content," i.e., the periodic orbits of the underlying classical dynamics that support the quantum spectrum. This method has been implemented in the past for many systems that display soft chaos with Dieter Wintgen being one of the prime movers in this field (see, e.g., Friedrich and Wintgen (1989); Holle et al. (1986); Pullen and Edmonds (1981), Kim and Ezra (1991), Eichmann et al. (1988), Du and Delos (1987)). The most notable example is the photoabsorption spectrum of atoms in a strong magnetic field. The quasi-Landau "modulation" in the photoabsorption spectrum of barium in a strong magnetic field observed in the early 1970s (Garton and Tomkins (1969); Edmonds (1970)) appears to be one of the first cases where this approach has been utilized. They were initially misidentified as quasi-Landau "resonances" (Fano and Rau, 1986). Only later it was realized that these modulations were the fundamental Fourier component of an exceedingly complicated quantum spectrum near the ionization threshold, associated with the unstable periodic cyclotron (Landau) orbit in the simultaneous presence of the Coulomb and the magnetic field. For atoms in external fields, the periodic orbit spectrum could be studied at fixed scaled energy by simultaneously varying both the energy and the external field. In the following we study atoms at zero field. Here the (scaled) energy is the only available control parameter.

Our analysis is primarily motivated by recent experimental advances in the study of doubly excited resonances near the double-ionization threshold (Seng *et al.* 1995; Seng and Sandner, 1996). With multi-step laser excitation and up to six lasers it is now possible to produce high-lying resonances in quasi-two-electron atoms (e.g., strontium) very close to the double-ionization threshold. In this region, many Rydberg series strongly overlap and the width of resonances becomes comparable to the spacing between the levels. The identification of individual resonances is, for principal reasons, no longer possible irrespective of the attainable experimental resolution. All well-established classification schemes that have proven successful for low to intermediate excited states are expected to fail since the non-separability manifests itself in the strong mixing of spectral series rendering approximate quantum numbers meaningless. The Fourier transform of the measured spectrum and the identification of Fourier components in terms of classical periodic orbits appears to be one (if not the only one) meaningful way to interpret and classify the spectrum. Our analysis focuses in the following on the periodic orbit spectrum rather than the recurrence spectrum. Quantum mechanically, this corresponds to the study of the density of states rather than the local density of states near the nucleus. The reason is that the pathway to doubly excited states via sequential excitation deposits the energy into the atomic system in a more delocalized manner compared to one-photon-excitation from the ground state. In the latter case, orbits contribute that emanate from and return to the region near the nucleus.

Many properties of the periodic orbit spectrum are not yet well understood. Most notably, the behavior of the spectrum for a mixed phase space is an open question since the trace formula addresses "pure" cases when the phase space is either regular or chaotic. Furthermore, the influence of the finite width of the resonances of the PO spectrum is of interest since the direct semiclassical calculation of the resonance width, unlike energy position, has proven difficult. The latter is of particular significance in the regime of overlapping resonances, where the quantum spectrum is characterized by Ericson fluctuations.

In Sect. 2, we briefly review the semiclassical expression for the spectral density. The application to the hydrogen atom (Sect. 3) not only serves as an instructive example for periodic-orbit analysis but reveals important features ubiquitously present in all atomic systems governed by long-range Coulomb forces. The periodic-orbit analysis of doubly excited helium is presented in Sect. 4. The suppression of signatures of chaos in atomic systems and the role of Ericson fluctuations near the multiple-particle break-up threshold are discussed in Sect. 5.

2 Periodic Orbits Expansion of the Spectral Density

The density of states of a quantum system

$$d(E) = \sum_n \delta(E - E_n) = \lim_{\eta\to 0} \frac{1}{\pi} \sum_n \frac{\eta}{(E - E_n)^2 + \eta^2} \tag{1}$$

can be written in terms of the trace of the Green's function $G(E)$,

$$d(E) = \frac{-1}{\pi} \operatorname{Im} Tr(G(E)) , \tag{2}$$

where the spectral representation of the causal Green's function is given by

$$G(E) = \sum_n \frac{|n\rangle\langle n|}{E - E_n + i\eta} . \tag{3}$$

Eqs. (1-3) remain valid for resonances embedded in the continuum upon replacing the infinitesimal imaginary part η by a finite half-width, $\eta \to \Gamma/2$, of the resonance.

The trace can now be written in terms of the coordinate space representation of G as

$$Tr(G(E)) = \int d^{\mathrm{N}}q\langle q|G(E)|q\rangle \tag{4}$$

with N the number of degrees of freedom. The matrix element $\langle q'|G(E)|q\rangle$ can be viewed as the propagator from the initial coordinate q to q' on the energy shell E. In the trace (Eq.(4)), i.e. $q' \to q$, only closed orbits contribute.

In the primitive semiclassical (PSC) approximation to Eq.(2), which assumes the validity of the stationary phase approximation in performing the integral over all paths as well as the trace, the spectral density is given by the Gutzwiller trace formula (Gutzwiller, 1990),

$$d(E) \to d^{PSC}(E) = d_{av}(E) + \frac{1}{2\pi} \sum_{j\in PPO's} T_j \sum_{k=-\infty}^{\infty} \frac{1}{\sqrt{|det(M_j^k - I)|}} e^{i(kS_j - \sigma_{jk}\pi/2)} . \tag{5}$$

The index j runs over all classical primitive periodic orbits (PPO) of the system, k counts the number of their repetitions. S_j is the action integral of the jth primitive orbit, $T_j = dS_j/dE$ its period, M_j^k the monodromy matrix, and σ_{jk} the Maslov index of its kth repetition. While the action for the kth repetition is a multiple of that of periodic orbit, the Maslov index may depend in a non-trivial manner on k (see e.g., Brack and Jain, 1995; Mao *et al.* 1992). $d_{av}(E)$ is the smoothed state density, given for bound states by the Thomas-Fermi formula,

$$d_{av}(E) = \frac{1}{(2\pi\hbar)^{\mathrm{N}}} \int d^{\mathrm{N}}q d^{\mathrm{N}}p \quad \delta(E - H(p,q)) \ , \tag{6}$$

which gives the change of volume of classical phase space per unit energy in units of the volume of the Planck cell $(2\pi\hbar)^{\mathrm{N}}$. The Gutzwiller trace formula is, in general, valid for manifolds with isolated periodic orbits and is not convergent for real E. Therefore, variants employing resummation and renormalization schemes have been proposed (see, e.g., Artuso *et al.*, 1990; Voros, 1988; Eckhardt and Aurell, 1989). Since we will not attempt to directly calculate the spectral density from Eq. (5), we will not dwell on this aspect. Our point of departure is the interpretation of Eq. (5) as a Fourier expansion of the full quantum level density (Eq.(2)) in terms of classical periodic orbits of the system with each periodic orbit (the primitive orbit as well as its repetitions) corresponding to one Fourier component. No a priori assumption of the validity of the PSC approximation is required. Instead, Eq.(5) provides a means to uncover the "classical content" underlying the quantum spectrum.

Within the PSC, the oscillating part of the spectral density, $\Delta d(E) = d(E) - d_{av}(E)$ can be written for systems with two degrees of freedom and *isolated* stable orbits as (Gutzwiller, 1990)

$$\Delta d(E) = \sum_{j\in\{PPO's\}} \frac{T_j}{2\pi i} \sum_{\substack{k=-\infty \\ k\neq 0}}^{\infty} \frac{e^{i(S_j-\mu_j\pi/2)k}}{2\sin(k\pi\nu_j)} \ , \tag{7}$$

where ν_j is the winding number (frequency ratio) and μ_j is the number of conjugate points of the jth primitive orbit. For unstable periodic orbits Eq. (5) becomes

$$\Delta d(E) = \sum_{j\in\{PPO's\}} \frac{T_j}{2\pi i} \sum_{\substack{k=-\infty \\ k\neq 0}}^{\infty} \frac{e^{i(S_j-\mu_j\pi/2)k}}{2\sinh(k\lambda_j/2)} \tag{8}$$

for hyperbolic orbits and

$$\Delta d(E) = \sum_{j\in\{PPO's\}} \frac{T_j}{2\pi} \sum_{\substack{k=1-\infty \\ k\neq 0}}^{\infty} \frac{e^{i(S_j-\mu j\pi/2)k}}{2\cosh(k\lambda_j/2)} \tag{9}$$

for inverse hyperbolic orbits. For systems with hard chaos, periodic orbits are isolated. For mixed systems, regular stable orbits are generally not isolated but may form continuous families. Degeneracies within islands lead to rational winding numbers and thereby to singularities in Eq.(7). Nevertheless, the trace formula still may yield the correct density of states if a regularization process is employed, as will be shown below for hydrogen.

The calculation of the oscillatory component $\Delta d(E)$ from the quantum spectrum requires the accurate determination of the averaged spectral density d_{av}. One technique to obtain d_{av} uses the homogeneity of the potential. For a system with homogenous potentials V of degree ζ,

$$V(aq) = a^{\zeta} V(q) \ , \tag{10}$$

where a is a scale factor and $\zeta = -1$ for Coulomb potentials, Eq. (6) is reduced to

$$d_{av}(E) \ = \ CE^{(1/2+1/\zeta)\mathrm{N}-1} \tag{11}$$

with C as an energy-independent constant. It turns out, however, that in the case of a Coulomb interacting system with a divergent density of states near threshold, the direct evaluation of

$$d_{av} \ = \ \frac{d\langle n(E)\rangle}{dE} \tag{12}$$

from the spectrum, in general, is more accurate (Qiu *et al.*, 1996).

The key ingredient for scaled periodic orbit spectroscopy is the scaling property of the Hamiltonian function for a homogeneous potential. For an atom with only Coulomb interactions present, the equations of motion are invariant under the following non-canonical scaling transformation of the canonical variables (q, p) and the time

$$Q = |E/E_0|^{-1} q \tag{13}$$

$$P = |E/E_0|^{1/2} p \tag{14}$$

$$T = |E/E_0|^{-3/2} t \ , \tag{15}$$

where the lower case variables denote the values at the reference energy E_0 taken to be in the following as $E_0 = -1$. Accordingly, the classical action scales as

$$S_j(E) = s_j |E|^{-1/2} \equiv s_j z \ , \tag{16}$$

where $s_j = S_j(E_0 = -1)$ is the scaled action at the reference energy E_0. $z = |E|^{-1/2}$ is the transformed energy variable with respect to which the inverse Fourier transform of the quantum spectrum is evaluated. The action s_j is the corresponding conjugate variable of the transform.

Since the bound state spectrum as well as the spectrum of resonances extends only over a finite energy window, effects due to the integration over a finite domain of z must be taken into account,

$$f(s) = \int_{z_0-\Delta/2}^{z_0+\Delta/2} dz\, e^{-isz}\, \Delta d(z) \tag{17}$$

$$= \sum_{j,k} a_{jk}\, e^{i(s_{jk}-s)z_0}\, g_\Delta(s_{jk}-s), \tag{18}$$

where the peak shape function

$$g_\Delta(x) = \frac{\sin(x\Delta/2)}{x} \underset{\Delta\to\infty}{\longrightarrow} \pi\delta(x) \tag{19}$$

reduces to a δ function in the limit of an infinite window. The spectral density as a function of the variable z is given by

$$\Delta d(z) = \Delta d(E(z)) \frac{d|E|}{dz} \quad . \tag{20}$$

If the semiclassical limit holds, $f(s)$ represents the periodic orbit spectrum with amplitudes a_{jk} given by

$$a_{jk} = \begin{cases} \frac{t_j e^{-i\mu_j \pi k/2}}{2\pi\alpha i \sin(k\pi\nu_j)} & \text{(elliptic)} \\ \frac{t_j e^{-i\mu_j \pi k/2}}{2\pi\alpha i \sinh(k\lambda_j/2)} & \text{(hyperbolic)} \\ \frac{t_j e^{-i\mu_j \pi k/2}}{2\pi\alpha \cosh(k\lambda_j/2)} & \text{(inverse hyperbolic),} \end{cases} \tag{21}$$

where all classical orbital parameters (for example, t_j,) refer to the reference energy E_0. The PO spectrum yields a (infinite) set of peak positions and complex amplitudes $\{s_{jk}, a_{jk}\}$ which permit a semiclassical interpretation of a segment of the quantum spectrum. However, its significance extends well beyond the validity of the trace formula itself. The PO spectrum also allows a meaningful characterization of the full quantum spectrum in cases where corrections to the semiclassical approximation are expected to be important. For example, contributions from "ghost orbits" (see e.g., Main *et al.*, 1994; Main, 1997), which get lost in the stationary phase approximation of the non-uniform ("primitive") semiclassical quantization, manifest themselves as peaks in the periodic orbit spectrum. Likewise, the absence of pronounced peaks and their replacement by the broad "noisy" PO spectrum signals the presence of transient resonances whose lifetimes are shorter than the classical orbital period.

It is convenient to introduce the PO power spectrum defined as

$$F(s) = |f(s)|^2$$

$$= \sum_{j,k}\sum_{j'k'} a_{jk}\, a^*_{j'k'}\, e^{iz_0(s_{jk}-s_{j'k'})} g_\Delta(s_{jk}-s) \times g_\Delta(s_{j'k'}-s) \quad . \tag{22}$$

The cross terms in Eq. (22) can be interpreted as due to interference between different periodic orbits. The variation of the size of spectral window, Δ, allows one, at least in principle, to control the influence of interference on the power spectrum. In the case that the spectral window is sufficiently large as to resolve all individual periodic orbits, $|\Delta(s_{jk}-s_{j'k'})| \gg 1$, the power spectrum is simply given by an incoherent sum over periodic orbit contributions

$$F(s) \simeq \sum_{jk} |a_{jk}|^2 \left(g_\Delta(s_{jk}-s)\right)^2 \ . \tag{23}$$

In this limit, the power spectrum depends only on the set of parameters $\{|a_{jk}|^2, s_{jk}\}$. The PO power spectrum is therefore independent of phases of the amplitude, i.e. Maslov indices do not contribute. The significance of this property for Rydberg series will be discussed below.

Because of the exponential proliferation of orbits in chaotic systems, the condition $|s_{jk}-s_{j'k'}|\Delta \gg 1$ will be eventually violated for any finite Δ. However, the decomposition of $F(s)$ into an incoherent sum over individual orbits (Eq. (23)) will still be valid for short orbits since the exponential increase occurs for long orbits with $s_{jk} \to \infty$.

3 Periodic Orbit Spectrum of Hydrogen

The PO analysis of hydrogen can serves as a pedagogical example for a regular, highly degenerate system. For this exactly solvable textbook example the systematic study of properties of the PO power spectrum, in particular of spectral window effects as well of effects of level broadening due to finite lifetimes, is possible (Qiu *et al.*, 1996). More importantly, quasi-hydrogenic Rydberg series are ubiquitously present in (almost) all atomic spectra. An understanding of their influence is a prerequisite for an analysis of the power spectrum in more complicated cases.

The density of states for a given angular momentum eigenstate of hydrogen $\{|nlm>, \quad n=l+1,\ldots\}$ is given by

$$d(E) = \sum_{n=l+1}^{\infty} \delta(E+\frac{1}{2n^2}) \ . \tag{24}$$

The Fourier integral Eq. (17) can be written as

$$\begin{aligned} \int dz e^{isz}\, d(z) &= \int d\left(\frac{1}{\sqrt{|E|}}\right) e^{-is/\sqrt{|E|}}\, d(E)\, 2|E|^{3/2} \\ &= \int dz\, e^{-isz} \sum_n \delta(z-\sqrt{2}n). \end{aligned} \tag{25}$$

Expressed in terms of the energy E, the Fourier integral (Eq. (25)) is similar to, but not identical to, the previously proposed generalized Fourier transform for Coulomb systems (Blümel and Reinhardt, 1991) where E rather that $\frac{1}{\sqrt{|E|}}$ was used as the variable of integration. While the peak position remains unaffected by this transformation, the amplitude of the Fourier components can no longer be identified with Fourier components since s is no longer the corresponding conjugate variable.

Evaluation of Eq. (25) yields

$$f(s) = \sum_{n_1}^{n_2} e^{-is\sqrt{2}n} \tag{26}$$

where n_1 is the smallest integer $\geq (z_0 - \Delta/2)/\sqrt{2}$ and n_2 is the largest integer $\leq (z_0 + \Delta/2)/\sqrt{2}$. If we neglect window effects, Eq. (25) can be conveniently evaluated by use of the Poisson sum formula

$$\begin{aligned} f(s) &= \frac{1}{\sqrt{2}} \int dz\, e^{-isz} \sum_k e^{2\pi i k z/\sqrt{2}} \\ &= 2\pi \sum_k \frac{1}{\sqrt{2}} \delta(s - \sqrt{2}\pi k) \ . \end{aligned} \tag{27}$$

The Fourier transform consists of equidistantly spaced δ peaks ("picket fence" spectrum) at the actions

$$s_k = \sqrt{2}\pi k \tag{28}$$

or, equivalently, at scaled periods

$$t_k = \frac{\pi}{\sqrt{2}} k, \qquad (k = 1, 2, \ldots) \tag{29}$$

where k labels the repetitions of the primitive orbit (Fig. 1). The Fourier amplitude becomes

$$a_k = \sqrt{2} = \frac{2t}{\pi} \ , \tag{30}$$

where $t = t_1$ is the scaled primitive period, i.e. the period of the Kepler orbit. This expression can be interpreted as the effective one-dimensional (1-D) limit of the amplitude in the trace formula (Eq. (21)). Comparing Eq. (30) with Eq. (21) shows that in the 1-D case the factor in the denominator of Eq.(21) should be removed, i.e. $2i\sin(\nu\pi) \to 1$. We arrive at the remarkable conclusion that, apart from window effects, the exact hydrogenic quantum spectrum (not only in the limit of large n) is completely and uniquely determined by just one classical parameter, the period of the Kepler orbit.

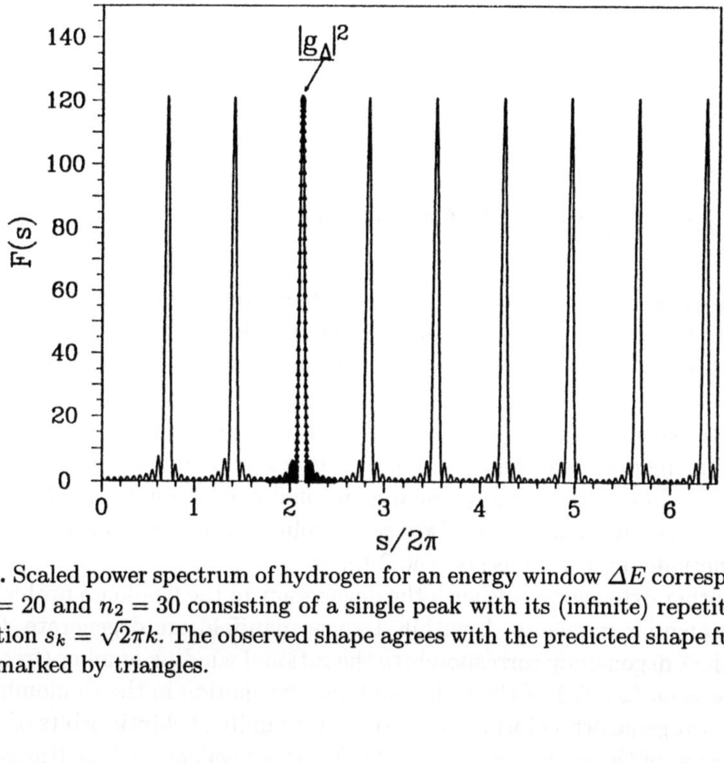

Fig. 1. Scaled power spectrum of hydrogen for an energy window ΔE corresponding to $n_1 = 20$ and $n_2 = 30$ consisting of a single peak with its (infinite) repetitions at the action $s_k = \sqrt{2}\pi k$. The observed shape agrees with the predicted shape function $|g_\Delta|^2$ marked by triangles.

The influence of the "window size", i.e. the finite length of the spectrum entering the Fourier transform, is illustrated in Fig. 1. The scaled power spectrum $F(s) = |f(s)|^2$ is shown for an energy window ΔE corresponding to $n_1 = 20$ and $n_2 = 30$. As expected we find a single peak with its (infinite) repetitions at the action $s_k = \sqrt{2}\pi k$, all with the same peak height. While the peak position is independent of the window, the peak shape (height and width) are strongly dependent on the spectral window. The observed shape agrees with the predicted shape function $|g_\Delta|^2$ (Eq. (23)). Evidently, the scaled power spectrum consists of an incoherent superposition of individual orbits. Accordingly, the peak height is given by $\Delta^2/2 = (n_2 - n_1 + 1)^2$.

Two extensions of this result are of interest: One refers to quasi-hydrogenic Rydberg series in atoms other than hydrogen. In the presence of a non-hydrogenic core potential the homogeneity relation of the Coulomb potential, and therefore the classical scaling invariance, is broken. Nonetheless, the scaled Fourier transform is still applicable as an approximation when we exploit the fact that the electron spends most of its time in the Coulomb field outside the core and encounters with the core occur on a time scale short compared to the orbital period. This is precisely the reasoning underlying quantum defect theory (Friedrich, 1991). The spectral density is given in terms of the quantum defect μ by

$$d(E) = \sum_n \delta(E + \frac{1}{2(n-\mu)^2}) \ . \tag{31}$$

In Eq.(31) μ is assumed to be independent of n. The Fourier transformation of the above expression, neglecting window effects, leads to

$$f(s) = 2\pi \sum_k \frac{1}{\sqrt{2}} \ e^{2\pi i \mu k} \ \delta(s - \sqrt{2}\pi k) \ . \tag{32}$$

The existence of a non-zero quantum defect only manifests itself as a phase shift of the amplitude. The physical interpretation of Eq. (32) is that each encounter with the core corresponds to an elastic scattering event resulting in a phase shift by $2\pi\mu$ without, however, changing the modulus. Consequently, the power spectrum (Eq. (23)) is unaffected by a quantum defect analogously to the independence of the Maslov index. It is remarkable that the PO power spectrum "looks" completely one-dimensional even though the electron orbits, prescessing at a rate of $2\pi\mu$ per revolution, are no longer closed but quasi-periodic and lie dense on the 2-torus.

Another extension deals with the degeneracy of the Coulomb problem. All orbital angular momenta, l, within a given manifold are degenerate. Classically, the ℓ degeneracy corresponds to the rational winding number (frequency ratio) $\nu = \omega_r/\omega_\theta$=1:1 of the radial and angular motion in the Coulomb problem. Hydrogenic orbits form a one-parameter family of elliptic orbits of a fixed orientation of the major axis $a = 1/(2|E|)$, or equivalently of the Runge-Lenz vector (Englefield, 1972), in space with l as the parameter. The conservation of the Runge-Lenz vector is associated with the dynamical $O(4)$ symmetry of the Coulomb problem. As a continuous family, periodic orbits are not isolated as has been assumed in the derivation of the trace formula for stable PO (Eq.(7)). Furthermore, Eq.(7) becomes singular for rational winding numbers. Nevertheless, the PO spectrum can be calculated for the degenerate case using a regularization scheme in which we take the rational winding number as the limit of an irrational number

$$\nu(\delta) = \frac{1+\delta}{1-\delta} \qquad \delta \to 0^+ \tag{33}$$

to relate the Fourier transform to amplitudes of the Gutzwiller trace formula. In Eq. (33) ν is assumed to be a diophantine irrational number (Brack and Jain, 1995). Consider all $m_\ell = 0$ states with the n-fold l degeneracy ($\ell = 0, \ldots n-1$).

The quantum density of states is

$$d(E) = \sum_n n \quad \delta\left(E + \frac{1}{2n^2}\right) \ . \tag{34}$$

Scaling transformation of Eq. (34) yields

$$d(z) = \sum_n \frac{z}{\sqrt{2}} \delta(z - \sqrt{2}n) \ . \tag{35}$$

Obviously, the Fourier transform of Eq. (35) is singular, being related to the singularity of $\{\sin[\pi\nu(\delta)k]\}^{-1}$ as $\delta \to 0^+$ in Eq. (7).

With help of the Poisson sum formula, the Fourier transform can be written as

$$f(s) = \frac{1}{2} \sum_k \int dz z \, e^{i(\sqrt{2}\pi k - s)z} \ . \tag{36}$$

The integrand of the divergent Fourier integral can now be written, with the help of Eq. (33), as the $\delta \to 0$ limit of the sum of two terms:

$$\frac{1}{2} z \, exp[i(\sqrt{2}\pi k - s)z] = \lim_{\delta \to 0} \frac{1}{\sqrt{2}} \ [\frac{(-1)^k exp[i(\sqrt{2}\pi k(1+\delta) - s)z]}{2i \sin[k\pi\nu(\delta)]} + \frac{(-1)^k exp[i(\sqrt{2}\pi k(1-\delta) - s)z]}{2i \sin[k\pi\nu(\delta)^{-1}]}] \tag{37}$$

Consequently,

$$f(s)|_{\delta \to 0} = \frac{1}{\sqrt{2}} \sum_k \left[\frac{(-1)^k}{2i \sin[k\pi\nu(\delta)]} 2\pi\delta(s - \sqrt{2}\pi k(1+\delta)) + \frac{(-1)^k}{2i \sin[k\pi\nu(\delta)^{-1}]} 2\pi\delta(s - \sqrt{2}k\pi(1-\delta)) \right] . \tag{38}$$

Eq. (38) can be recognized as the Fourier transform of a 2-degree of freedom system with two isolated periodic orbits (and its repetitions) with frequencies $\omega_{r,\theta} = 2\sqrt{2}(1 \pm \delta)$ and fundamental periods $t_{r,\theta} = \frac{\pi}{\sqrt{2}}(1 \mp \delta)$. The periodic orbits correspond to the purely radial motion $(0 \le r \le r_{max})$ with zero angular momentum $\ell = 0$ and to the circular orbit $\ell = \ell_{max}$ with $\dot{r} = 0$. The continuous family of all closed orbits on the torus are then recovered in the limit $\delta \to 0$. The fundamental period agrees with that of the 1-D - case (see Eq. (29)). The further extension of this method to the full shell quantum degeneracy $\sim n^2$ is straightforward.

Excited atomic states form resonances in the continuum. The spectral density is therefore not a sequence of δ functions but a sequence of Lorentzian peaks, (Eq. (1) with $\eta \to \Gamma_n/2$). The influence of the finite width on the power spectrum depends on the ratio of the width of the resonance to the level spacing, $\beta = \Gamma_n/|E_n - E_{n+1}|$. Fig. 2 shows the hydrogenic spectrum with a finite decay width given by $\Gamma_n = Cn^{-3}$. This n dependence applies to both the radiative width as well as non-radiative width due to autoionization along an unperturbed Rydberg series in multi-electron systems. As long as $\beta = C << 1$, the scaled Fourier spectrum remains essentially unchanged. For larger β's, peaks for longer orbits are exponentially damped. Note that

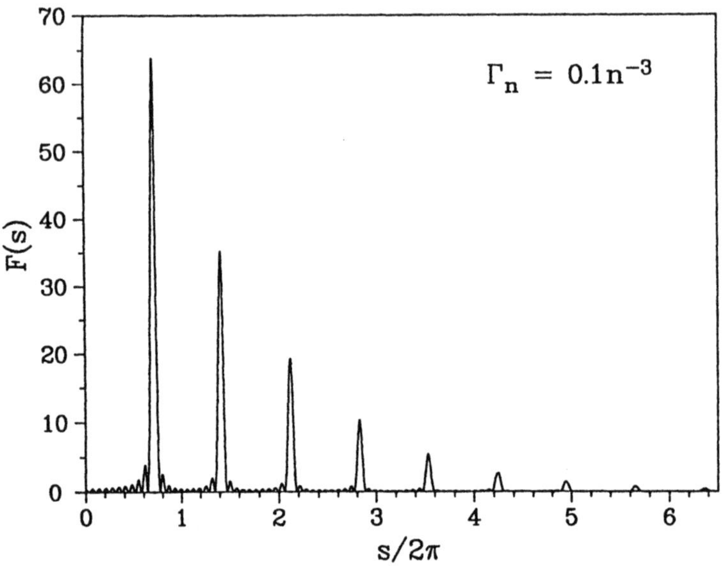

Fig. 2. Hydrogenic spectrum with a finite decay width given by $\Gamma_n = Cn^{-3}$. The PO peaks for longer orbits are exponentially damped.

the scaled action axis can be interpreted as a scaled time axis. Fig. 2 can therefore be visualized as a damped oscillator in the time domain.

For doubly excited resonances of helium the n^{-3} dependence of Γ breaks down when a large number of Rydberg series overlap and, instead, an approximate scaling for the (average) width of the resonances $\Gamma_n = Cn^{-2}$ holds (Burgdörfer *et al.*, 1995, see below). For n^{-2} scaling and for an arbitrary value of C the region of overlapping resonances $\beta \approx 1$ will eventually be reached for sufficiently large n. Fig. 3 shows the effect of overlapping resonances on the PO spectrum for a hydrogenic model case where only one periodic orbit and its repetitions are present.

A peak at the action $s_{k=1}$ of the primitive orbit is still observable but repetitions are damped out. With increasing length of the orbit, the width of the levels prevents the resolution of repetitions. The immediate consequence is that only short orbits, if any, remain observable in the regime of overlapping resonances, which is generic for non-separable systems exhibiting Ericson fluctuations. Therefore, characterization of the spectrum in terms of a few short orbits is likely the most appropriate approach in this spectral region. Moreover, the PO spectrum now possesses intensity below the action of the fundamental periodic orbit, indicating the presence of low-frequency "noise" predicted for Ericson fluctuations.

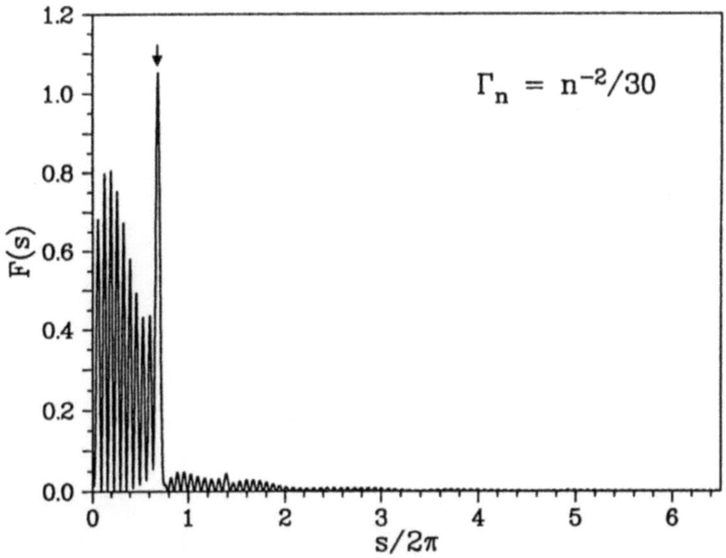

Fig. 3. Hydrogenic spectrum for an energy window ΔE corresponding to $n_1 = 20$ and $n_2 = 30$. A finite decay width is given by $\Gamma_n = Cn^{-2}$ with $C = \frac{1}{30}$, chosen such that $\Gamma_n \sim |E_{29} - E_{30}|$. A peak at the action $s_{k=1}$ of the primitive orbit is still observable, but the repetitions are damped out.

4 Periodic Orbit Spectrum of Doubly Excited Resonances of Helium

We now apply the PO analysis to ^1S doubly excited resonances in helium. This system represents a non-separable three degree of freedom problem with mixed phase space. It is of considerable interest from both an experimental and a theoretical point of view. Experiments using synchrotron radiation (Domke *et al.*, 1995) and sequential excitation by laser photons (Seng *et al.*, 1995) permit the investigations of the spectrum in the immediate vicinity of the three-body break-up threshold which is a cluster point of Rydberg series.

For low-lying resonances the classification in terms of approximate O(4) group-theoretical "collective" quantum numbers (NnKTL) (Wulfman, 1971; Herrick and Kellman, 1980; Kellman and Herrick, 1980; Lin, 1984) has proven successful. In view of the chaotic classical phase space, the existence of approximate quantum numbers indicating quasi-separability appears surprising. Two circumstances are key to the appearance of approximate quantum numbers for low-lying states. Firstly, different Rydberg series of the outer electron n for different N of the inner electron do not overlap (Fig. 4), thereby rendering the quantum numbers N and n approximately conserved. To zeroth order in the electron-electron interaction the resonance positions along the Rydberg series ($N \ll n$) are given by

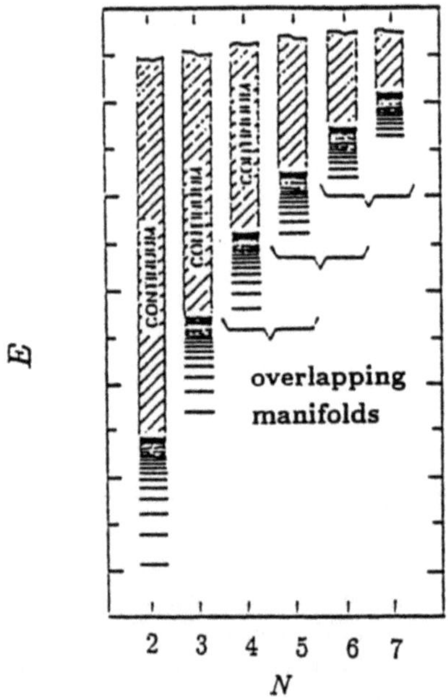

Fig. 4. Spectrum of doubly excited helium (schematically).

$$E_{Nn}^{(0)} = -\frac{Z^2}{2N^2} - \frac{(Z-1)^2}{2n^2} \ . \tag{39}$$

For helium ($Z = 2$), these energies remain, up to $N = 4$, separated from the adjacent intershell $(N + 1)$ manifold

$$E_{N+1}^{(0)} = -\frac{(Z-\sigma)^2}{(N+1)^2} \tag{40}$$

where $\sigma \approx 0.17$ is an effective screening constant. States with $N = n$ are referred to as intrashell resonances while $N < n$ as intershell resonances. As long as the approximate separability holds, intrashell resonances arise from symmetric radial excitations of the two electrons with $\langle r_1 \rangle \simeq \langle r_2 \rangle$ while asymmetric excitations $\langle r_1 \rangle \ll \langle r_2 \rangle$ correspond to intershell resonances. For higher N, series begin to increasingly overlap and for such highly excited resonances principal quantum numbers become meaningless. The second circumstance is related to the high degree of degeneracy within each (N, n) manifold due to the $O(4) \times O(4)$ symmetry of two hydrogenic electrons. The electron-electron interaction can be approximately diagonalized within each subspace by constructing an irreducible representation of the $O(4) \subset O(4) \times O(4)$ subgroup. The basis states of the irreducible representation of the $O(4)$ subgroup are labelled by K ($T = 0$ for $L = 0$),

$$E_{Nn}^{(1)} = E_{Nn}^{(0)} + \langle NnK|\frac{1}{v_{12}}|NnK\rangle \ . \tag{41}$$

It is important to realize that the physical meaning of the O(4) subgroup label K is different for intershell ($N << n$) and intrashell ($N = n$) resonances, even though the same label $K = N-1, N-3, \ldots -(N-1)$ is customarily used (Herrick, 1983). For asymmetric excitations $N << n$, K represents the projection of dipole moment of the inner electron onto the $\mathbf{r}$ vector of the outer electron

$$< \cos\theta_{12} > = -\frac{K_d}{N} \tag{42}$$

and we denote this Stark-like quantum number by K_d. For intrashell resonances the corresponding quantum number is denoted by K_v which is associated with the angular ("bending") vibrational excitation quantum number v of the two electrons in the interelectronic angle θ_{12},

$$N - K_v - 1 = v \ . \tag{43}$$

The vibrational ground state ($K_v = N-1$) corresponds to two electrons localized on opposite sides of the nucleus ($\cos\theta_{12} = -1$) with the angular motion confined to the zero-point fluctuation about this position. Resonances with decreasing K correspond to an increase of excitation of the bending vibration. Near $K = K_{\max}$ classifications in terms of K_v and K_d are equivalent. For small K values, however, they are completely different which has led to some confusion in the literature. This is not surprising since for small K and energy-unfavored configurations, the coupling to the adjacent Rydberg series is large and, therefore, the break-down of approximate quantum numbers is expected to be most pronounced.

PO analysis provides an alternative means of classification of high-lying resonances where simple classification schemes based on quasi-separability break down. The spectrum can be mapped onto a few "fundamental" periodic orbits in the case of the three-body Coulomb problem. The point to keep in mind is that the mapping between orbits and individual resonances is not one-to-one (even though some resonant wavefunction carry pronounced "scars " of individual orbits). Instead a stretch of the spectrum can be associated with families of orbits. Nevertheless, a simple physical picture of the underlying correlated electronic motion i.e. of electron correlation, emerges. A few fundamental orbits that play an important role in the PO analysis are displayed in Fig. 5. Figure 5(a) shows the asymmetric stretch (AS) orbit, sometimes also referred to as hyperspherical orbit (Watanabe, 1987), because the hyperradius $R = \sqrt{r_1^2 + r_2^2}$ is approximately conserved. This orbit corresponds to an out-of-phase purely, radial motion of two electrons on opposite sides of the nucleus ($\cos\theta_{12} = -1$) without bending vibration. Fig. 5(b) shows the Langmuir (L) orbit (Langmuir, 1922; Müller *et al.*, 1992) which corresponds to maximum vibrational bending and only minimal (in-phase)

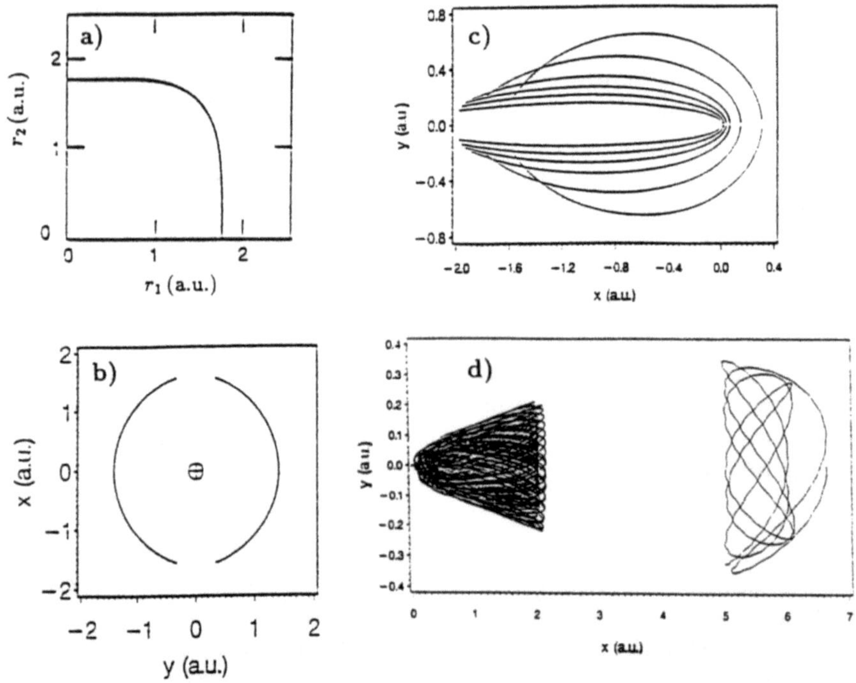

Fig. 5. Periodic orbits for the classical two-electron problem with symmetric energy sharing: (a) The weakly unstable asymmetric stretch (or hyperbolic orbit, Watanabe, 1987)); (b) the stable Langmuir orbit (Langmuir, 1922); (c) the unstable asymmetric Langmuir orbit (Müller and Burgdörfer, 1993); (d) quasi-periodic orbit with asymmetric energy sharing: frozen planetary orbit (Wintgen *et al.*, 1992).

radial motion. Fig. 5(c) shows its asymmetric counterpart, the asymmetric Langmuir (AL) orbit, in which the concentric semicircles degenerate into semi-ellipses (Müller *et al.*, 1993; Müller and Burgdörfer, 1993). Fig. 5(d) shows the "frozen planet" orbit (P) representing the correlated motion of an outer electron in the field of the nucleus and of the inner orbit (Richter and Wintgen, 1991; Wintgen *et al.*, 1992). In this orbit, the two electrons are localized on the same side of the nucleus and the angular motion of the outer electron is confined to small angles θ_{12} around zero ("hindered rotor"). The inner electron executes a Stark-like motion on a precessing and breathing Kepler orbit. Note that the frozen planet orbit corresponds to an asymmetric excitation with $\langle r_1 \rangle = \langle r_2 \rangle/3$, unlike all other orbits listed in Fig.5 which represent symmetric excitations with $\langle r_1 \rangle \simeq \langle r_2 \rangle$.

In order to facilitate the identification of PO's we perform the Fourier transform of sections of the spectrum, using the approximate labels K_d or K_v as guides. We first focus on the so called "Wannier ridge" resonances which correspond to states with $K = K_{\max}$. This term was originally coined from the notion of a quantum state associated with the classical "Wannier orbit" of an in-phase symmetric stretch motion of the two electrons. Recent calculations (Ezra *et al.*, 1991) have indicated, however, that the underlying orbit is the asymmetric (out-of phase) stretch mode. The enhanced density near the Wannier ridge then results from the slowing down of the motion across the ridge near the top of the barrier.

As input for the PO analysis we use positions and widths of resonances from recent accurate quantum calculations (Bürgers, 1995; Bürgers *et al.*, 1995) for resonances up to $N = 10$ and $n = 17$, similar calculations (Müller *et al.*, 1994; Müller, 1993) as well as a simple empirical analytic formula for ^1S resonances positions (Müller, 1993; Qiu *et al.*, 1996),

$$E(N, n, K_v) = \frac{(Z - \frac{\lambda^2-\mu^2}{4\lambda})^2}{2N^2}\left(1 + \frac{N^2}{n^2}\right) \tag{44}$$

with

$$\lambda^2 = \mu^2 + \frac{2(1-\mu^2)}{\omega},$$

$$\mu^2 = (1-4\sigma_{n0})k^2(\frac{N}{n})^3(1-(\frac{N}{n})^2)(1+(\frac{N}{n})^2(1-2\omega(1-\frac{1}{2Z})-\frac{4}{3}(\frac{Z-1}{Z})^2\frac{K}{n})),$$

$$\omega = 1 - \cos(\pi k + \arccos(1 - \frac{1}{8\sigma_{l0}}(1-k)^2)),$$

$$k = \frac{1}{2}(1 + \frac{K}{N}),$$

$$\sigma_{n0} = \frac{1}{2}(1 - \frac{2}{\pi}),$$

$$\sigma_{l0} = (0.18 \ln Z + 0.53)^2,$$

where K is the quantum number of bending vibration and $Z = 2$ is the nuclear charge. This formula fits almost all known resonance positions for $N \leq 10$ (with the exception of the frozen-planet resonances) to within $\simeq 1.5\%$, and converges to the semiclassical limit as $N \to \infty$ for the asymmetric stretch and Langmuir states. The formula permits an extrapolation to very high N as required for a large energy window in the Fourier transform. Its accuracy for large but finite N and overlapping Rydberg series is, however, uncertain.

The present treatment is an extension of earlier work by Kim and Ezra (1991) and by Ezra *et al.* (1991) who applied the scaled-energy Fourier transform to the one-dimensional collinear model of helium, which approximates the $K = K_{max}$ manifold quite well, and also of similar work by Dräger *et al.* (1994) for the s-wave helium model. Fig.6a gives the PO power spectrum resulting from the empirical formula Eq. (44), while Fig. 6b shows the same spectrum resulting from the *ab initio* quantum calculation for the level positions (Bürgers, 1995). Both PO spectra represent the Fourier transform of the spectral density within the same window $-5.469 \leq E \leq -1.150eV$ $(-0.202 \leq E \leq -0.0423a.u.)$. The two spectra agree remarkably well in both the peak positions and overall distribution of intensities. A few deviations in peak shape are, however, noticeable.

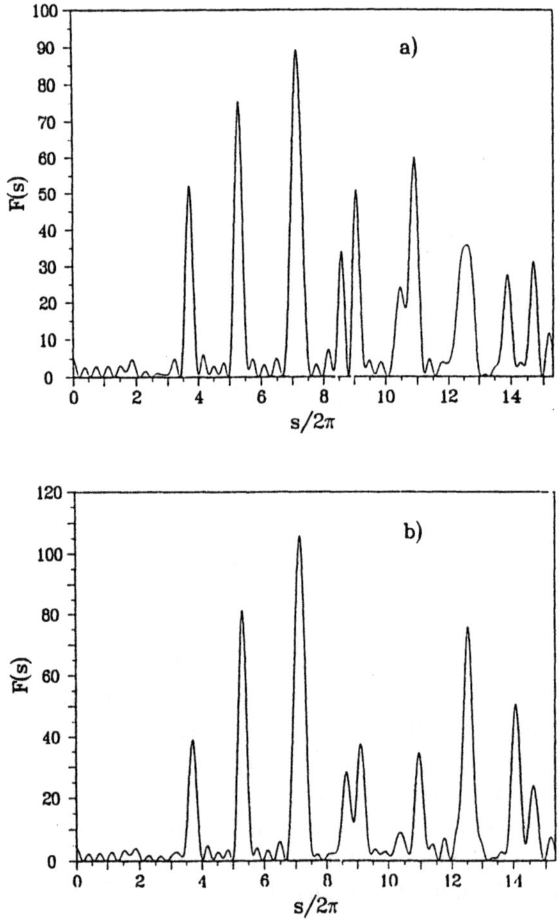

Fig. 6. Fourier transform of $K = K_{max}$ resonances for an energy window -5.469 $\leq E \leq -1.150$ eV. (a) The PO power spectrum resulting from the analytical formula Eq.(44) and (b) the same spectrum resulting from the *ab initio* quantum calculation for the level positions (Bürgers, 1995).

From the analysis of window-size effects (Eq. (22)), it follows that only those orbits whose actions satisfy the condition $|s_{jk} - s_{j'k'}|\Delta >> 1$ can be resolved. For the energy window $\Delta(\frac{1}{\sqrt{|E|}}) = 0.88$, for which the *ab initio* calculation is available, only orbits whose differences in action satisfy $|s_{jk} - s_{j'k'}|/2\pi >> 0.2$ can be resolved. The broad and strong peak near $s/2\pi \simeq 7$ is therefore expected to contain contributions from more than one periodic orbit. This hypothesis can be confirmed by extending the energy window using Eq.(44). Fig. 7 displays the PO spectrum for actions $s/2\pi$ up to 15 calculated for an energy window $-0.276 \leq E \leq -0.042$ eV ($-0.01016 \leq E \leq -0.001560$ a.u.). This interval was chosen in accord with the energy window of recent experimental data (Seng et al., 1995). A comparison between Fig. 6 and 7 illustrates that, with increased resolution, additional periodic orbits can be uncovered. For example, the broad peak mentioned above breaks up into two well-separated peaks with actions $s/2\pi = 6.99$ and $s/2\pi = 7.28$.

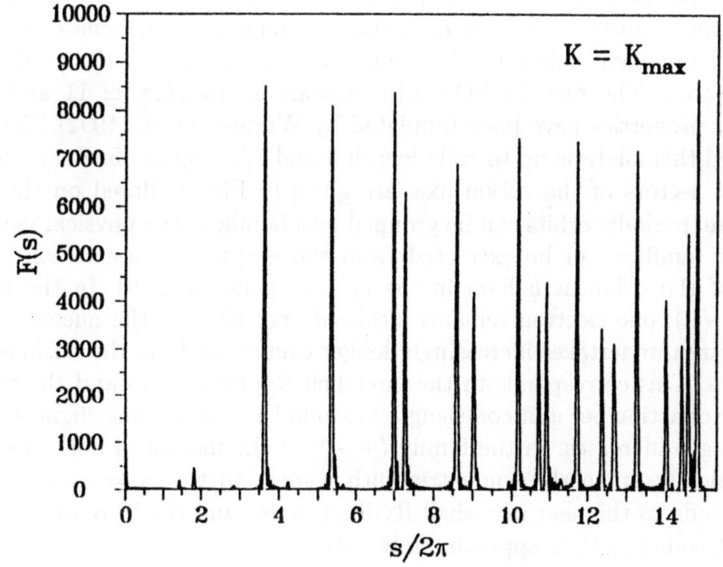

Fig. 7. PO power spectrum of $K = K_{max}$ resonances of helium for an energy window -0.276 $\leq E \leq -0.042$ eV.

Several families of periodic orbits can be identified. Fig. 8a-c give the peak identification for selected regions of the PO power spectrum. In analogy to the Fourier spectrum of hydrogen (see Sect. 3), we find (Fig. 8a) an equidistant sequence of peaks corresponding to the period of the Coulomb orbits labeled by R (for Rydberg). However, the actions s_k of their peak positions are twice as large as for hydrogen, i.e.

$$\frac{s_k}{2\pi} = \frac{Z}{\sqrt{2}}k \qquad (k = 1, 2, \ldots) \tag{45}$$

with $Z = 2$. The important observation is that in two-electron spectra only the periodic orbit of the inner electron is visible. This result, counter-intuitive at first glance, follows from the fact that $z = 1/\sqrt{|E|}$ rather than $|E|$ is the fundamental variable of the Fourier transform. From Eq. (39) we have for the spectral density on the z scale

$$d(z) = \sum_{N,n} \delta\left(z - \frac{N\sqrt{2}}{Z}\left(1 + \left(\frac{Z-1}{Z}\frac{N}{n}\right)^2\right)^{-1/2}\right) . \tag{46}$$

For a Rydberg series with $n >> N$, all peaks are clustered near $N\sqrt{2}/Z$ which leads, upon Fourier transformation, to dominant peaks at s_k given by Eq. (45).

A large number of additional orbits can be identified in terms of periodic orbits of the classical collinear Hamiltonian. In this subspace with two degrees of freedom all unstable periodic orbits can be enumerated in terms of a binary code of subsequent collisions of the nucleus with the same (+) or the other (−) electron. The first 22 PO's with actions up to $s/2\pi \simeq 11$ and their stability properties have been tabulated by Wintgen *et al.* (1992). We have extended this analysis up to code length 8 and $s/2\pi$ up to 15. Examples for different sectors of the action axis are given in Fig. 8. Based on the code structure, periodic orbits can be grouped into families. The physical meaning of these families can be extracted from the graphical presentation of the orbits of the collinear helium in the $r_1 - r_2$ plane (Fig. 9). In the family $(+^{n-2}, --)$, one electron remains localized very close to the nucleus, while the electron undertakes increasingly longer excursions from the nucleus as n increases. This corresponds to the intershell Rydberg series and the typical increase of action per unit code length is around 1.5, decreasing slightly as the code length increases. In the family $(+, -^{n-1})$ the motion of both electrons is completely or nearly symmetric with respect to the $r_1 = r_2$ axis. This corresponds to the near intrashell Rydberg series and the increase of action per unit code length is approximately 1.83.

It is remarkable that all peaks of the quantum PO spectrum for $K = K_{\max}$ can be identified with a PO of the classical collinear helium model. The converse is obviously not true. Instead, there is a strong correlation between the stability index χ and the fraction of observed PO's. The PO's missing from the spectrum have large stability indices, more precisely, large Lyapunov exponents $\lambda = \chi/(s/2\pi) \gtrsim 0.50$. For the same reason, the symmetric stretch (or Wannier) orbit is absent.

The fact that only a subset of unstable periodic orbits is reflected in the quantum spectrum raises the question, to which extent the proliferation of unstable periodic orbits is mirrored in quantum mechanics. Exponential proliferation of unstable periodic orbits is the hallmark of a fully chaotic phase

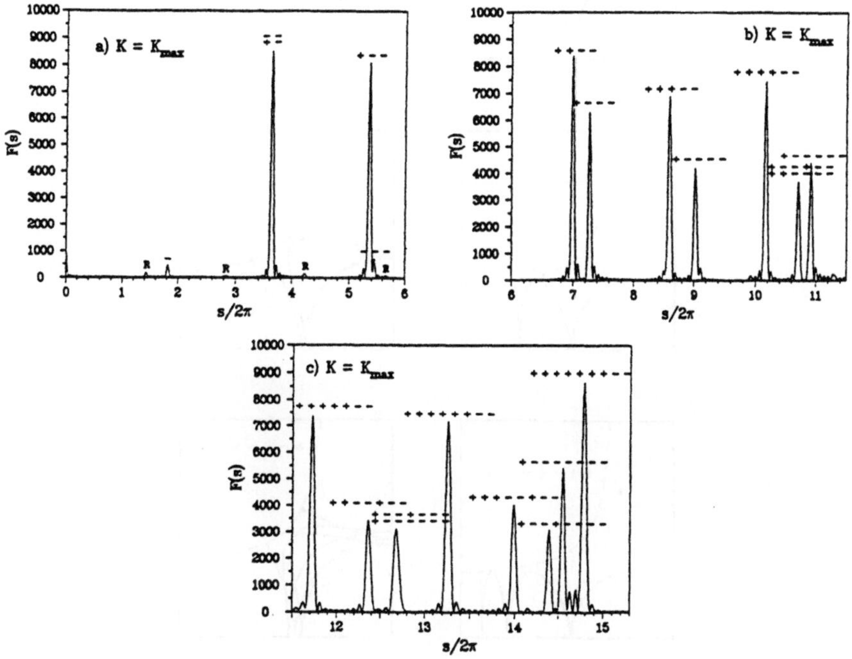

Fig. 8. Identification of several families of periodic orbits for selected regions of the PO power spectrum of Fig. 7.

space. In Fig. 10 we compare the number of PO's of the classical collinear helium, N_{cl}, as a function of the action s with the corresponding number of PO's, N_{qm}, identified in the quantum spectrum of Fig. 8 (Fourier spectrum from the empirical formula (Eq. (44)). We find the number proliferates exponentially as a function of the action which is itself a measure of the length of the orbit and obeys the numerical relation,

$$N_{cl} = N_0 \, e^{0.28(s/2\pi)} \quad . \tag{47}$$

However, this system is not a C-(or Anosov) system, i.e. the Lyapunov exponents are not bounded from below (Lichtenberg and Liebermann, 1982).

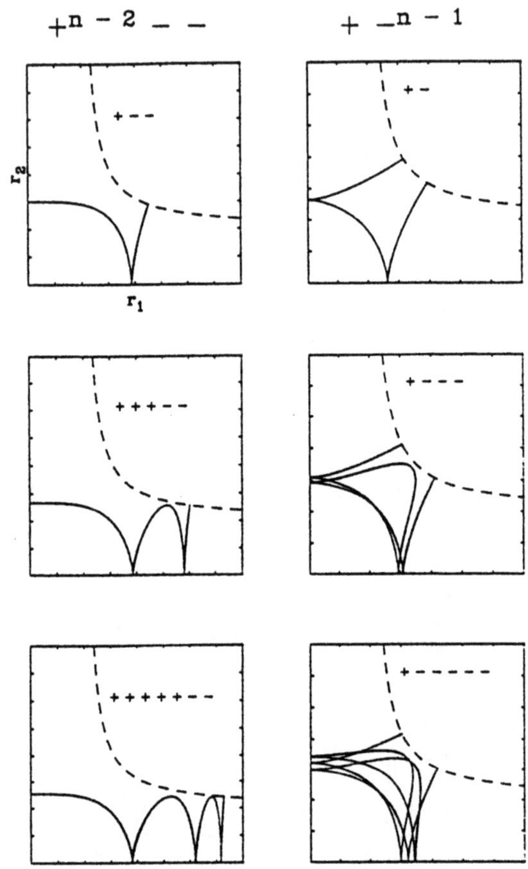

Fig. 9. PO trajectories for families $+^{n-2}--$ (left column) and $+-^{n-1}$ (right column) of the collinear helium. Also shown in each plot is the potential barrier.

Therefore, the exponential proliferation includes the birth of long, marginally unstable orbits with very small Lyapunov exponents. Elements of the family $(+^{n-2},--)$ with large Rydberg-like excursions from the nucleus are a prime example. Consequently, the quantum PO spectrum which is most sensitive to marginally unstable orbits is capable of mirroring the exponential proliferation. Their number N_{qm} grows exponentially, however with a smaller slope than in classical dynamics. It is the presence of Rydberg-like orbits, characteristics of Coulomb-interacting systems, which "controls" the PO spectrum and the extent to which atomic spectra can reflect classical chaos. We will come back to this point in Sect. 5.

Other fundamental orbits can be identified in the sections with smaller K_d (or K_v) $\simeq 0$ and $K \approx K_{\min} = -(N-1)$. Figure 11 shows the PO

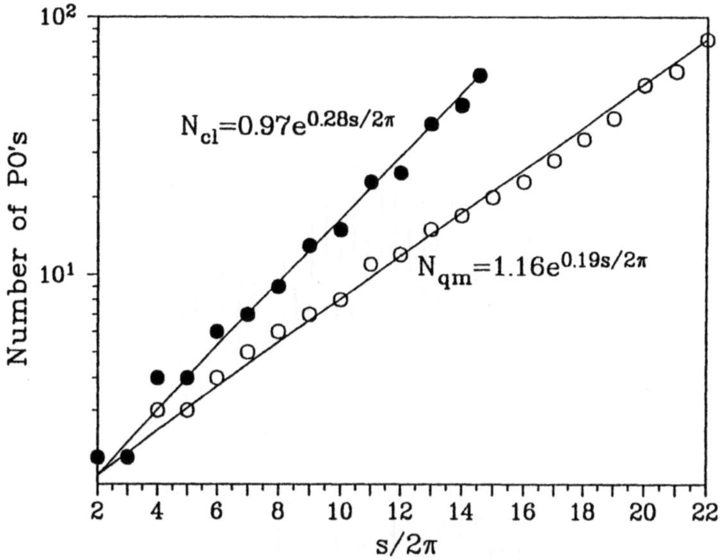

Fig. 10. Exponential proliferation of the number of PO's with the action $s/2\pi$ in collinear subspace ($\theta = \pi$). •, number of PO's, N_{cl}, including the Wannier orbit and the orbit with binary code + for which one electron is removed to infinity where the second electron is moving on a degenerate Kepler ellipse; ∘, number of PO's, N_{qm}, identified in the quantum PO power spectrum.

power spectrum for all $K_d = -(N-1)$ and $-(N-3)$ resonances taken from the calculation of Bürgers (1995) for $4 \leq N \leq 10$. Distinct series of peaks can be observed. The peak position labeled P associated with the $K_d = -(N-1)$ resonances agrees with the action of the planetary orbit ($s_k/2\pi \approx 1.49$) (Fig. 5d). In the $K_d = -(N-3)$ sector two peak series can be identified, one with actions $s_k/2\pi = 1.45k$ ($k = 1, 2 \ldots$) and another one with $s_k/2\pi = 1.43k (k = 1, 2, \ldots)$. Both of them are remarkably close to, but not identical to, the Coulomb orbit of the inner electron (Eq. 45), possibly signifying perturbed Rydberg series.

In the classification scheme with K_v representing bending vibrations, the Langmuir orbit should be associated with the maximum bending vibration excitation $K_v = -(N-1)$. The angle θ_{12} between the two electrons varies from 180° to the outer turning point, 24.05^o, with the average of $\langle\cos\theta_{12}\rangle$=0.059. The hyperradius R varies between 1.99 and 2.26 with an average of $\langle R \rangle$ =2.1. The action of the symmetric Langmuir orbit is $s/2\pi = 1.35$ which lies below the action of the Coulomb (Rydberg) orbit of the inner electron (Eq.45). This indicates an important difference to other orbits discussed above. Contributions from this periodic orbit can only be found in the region of overlapping Rydberg series. This type of symmetric excitation with $N-1$ nodes along θ_{12} appears as a shape resonance above the Rydberg threshold ($n \to \infty$) belonging to the quantum number N rather than a Feshbach resonance be-

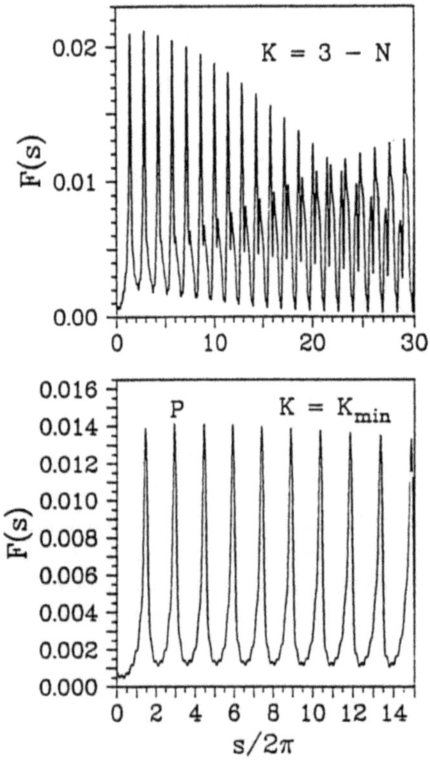

Fig. 11. PO power spectrum for all $K_d = -(N-1)$ and $-(N-3)$ resonances taken from the calculation of Bürgers (1995) for $4 \leq N \leq 10$. P stands for the planetary orbit.

low this threshold. Resonant wavefunctions tracing the Langmuir orbit have, indeed, been found in large scale complex scaling CI calculations (Müller and Burgdörfer, 1993), an example of which, labeled as ($N = 10$, $K_v = -9$) is shown in Fig. 12. The wavefunction is projected onto three planes: the hyperradius $R - \theta_{12}$ plane, the hyperangle $\alpha = \tan^{-1} r_2/r_1 - \theta_{12}$ and the $r_1 - r_2$ plane. The density is strongly enhanced near the classical turning point ($\theta = 24^o$ of the Langmuir orbit). Furthermore the distortion of the semi-circular orbit due to the interelectronic Coulomb repulsion, which leads to a slight increase of the hyperradius R near the turning point, is accurately reproduced by the quantum density distribution (see the $R - \theta_{12}$-plane). Another feature in the $\alpha - \theta_{12}$-plane is worth noting: while the dominant part of the probability density, which is located at $\theta_{12} < 90^o$, is peaked at $\alpha = 45^o$ (i.e. $\langle r_1 \rangle = \langle r_2 \rangle$), it spreads out over $\alpha = 45^o \pm \delta$ for $\theta_{12} > 90^o$

with $\delta \simeq 15^o$. This resonance with 9 nodes along the θ_{12} direction and an energy of $E = -0.0192$ lies embedded in the overlapping $N = 11, 12$, and 13 Rydberg manifolds.

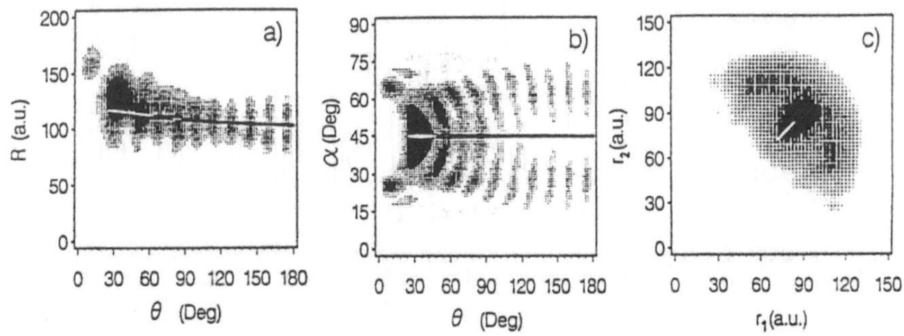

Fig. 12. Resonance tracing the Langmuir orbit ($K_v = -9$) for helium projected onto **a)** $R - \theta$, **b)** $\alpha - \theta$, and **c)** $r_1 - r_2$ planes. The solid line shows the projection of the classical Langmuir orbit.

It is now interesting to inquire into where the Langmuir orbit appears in a classification scheme using K_d instead. Since $\langle \cos\theta_{12} \rangle$ is small and positive we take from the numerically calculated resonance positions (Bürgers, 1995) the subset of "intershell" resonances ($N < n$) with small negative $K_d = 0$ or -1. The resulting PO spectrum (Fig. 13) displays a series of peaks with actions

$$s_k/2\pi = 1.36\,k \quad (k = 1\ldots) \tag{48}$$

in close proximity to that of the Langmuir orbit and its repetitions. In other words, the symmetric Langmuir orbit is in this classification scheme "embedded" in the manifolds of intershell resonances. This example illustrates that scaled energy Fourier transform analysis permits a meaningful characterization of the quantum spectrum in terms of contributing periodic orbits which is independent of the assignment of approximate quantum numbers (or labels). This feature becomes increasingly important when with increasing N as a large number of Rydberg series begin to overlap and simple classification schemes based on quasi-separability fail.

The Langmuir orbit resides in a stable island of classical phase space. Very close to the double ionization threshold the size of this island is sufficient to support individual quantum states. Rigorous semiclassical EBK quantization

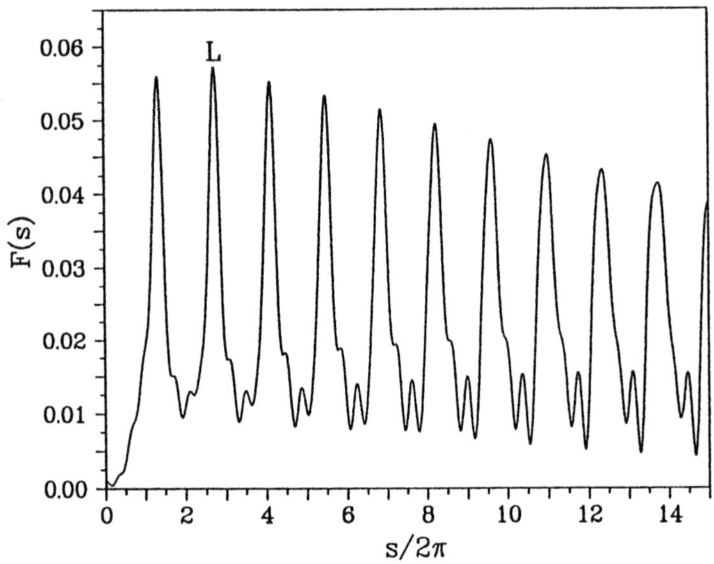

Fig. 13. PO power spectrum of helium resonances from the calculation of Bürgers (1995) with $n \sim [1/\sqrt{2(|E_{NNK_v}| - 2/N^2)}]$, $K_d = 0$, or -1 (see the text). The spectrum shows a peak at action $s/2\pi = 1.36$, in agreement with the Langmuir orbit (L).

(Müller *et al.*, 1992) shows that only for energies corresponding to $N \gtrsim 500$ will at least one quantum state reside on this island. In this case the correspondence between orbit and state is one-to-one.

The asymmetric Langmuir orbit (Fig. 5c) only appears for two electron atoms with larger nuclear charge. The latter is a direct signature of the bifurcation of the classical phase space structure (Fig. 14). At $Z = 5.60$ the Langmuir orbit bifurcates into the asymmetric Langmuir orbit where the electrons follow to semi-elliptic trajectories. In the limit $Z \rightarrow \infty$, this asymmetric Langmuir orbit degenerates to a "symmetric stretch" mode, however with both electrons on the same side of the nucleus. In this limit, degenerate perturbation theory for intrashell mixing of the underlying the group-theoretical O(4) basis states becomes increasingly accurate. Fig. 15 displays the $\theta - \alpha$ projection of the probability density of all states $K = N-1, N-3\ldots -(N-1)$ in the $N = 10$ manifold. With decreasing $K = K_v$ an increasing number of nodes along the θ direction, signifying the excitation of bending vibrations, becomes visible. However, at the top of the manifold $K \simeq -(N-1)$, the vibronic pattern all but disappears and is replaced by strong localization at $\alpha = 45^o$ or $\theta = 0^o$, in complete accord with predictions of classical dynamics for the $Z \rightarrow \infty$ limit of the asymmetric Langmuir orbit. This motion could also properly be described as "riding on the Wannier ridge", while the motion

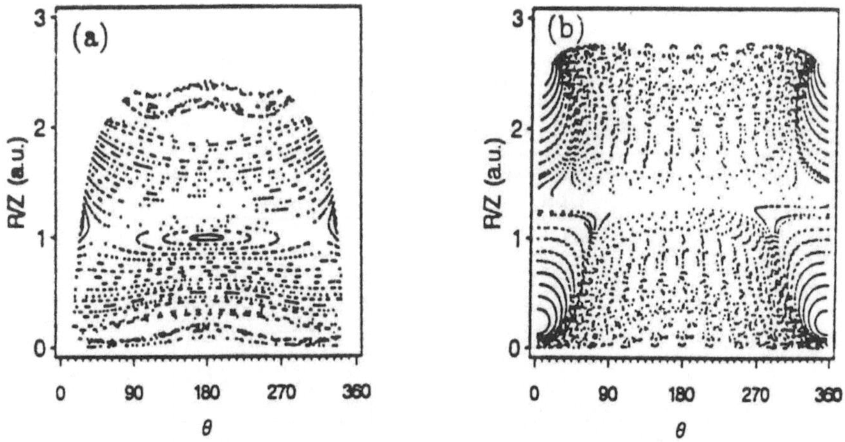

Fig. 14. Bifurcation of Langmuir orbit in Poincare surface of section ($\theta_{12} - R$ plane) as a function of Z. (a) Z=2; (b) Z=20. At $Z = 5.6$ the stable island breaks up into two islands.

perpendicular to the ridge (along the α- direction) is reduced to zero-point fluctuations.

5 Suppression of Chaos and Ericson Fluctuations

The PO analysis discussed in the previous sections provides clear evidence for the influence of (mostly) unstable, periodic orbits on the quantum spectrum. Nevertheless, the energy positions of resonances for $N \lesssim 10$ display a high degree of order and regularity defying any notion of disorder or chaos. The approximate validity of the simple analytical formula (Eq. (44)), as well as the fact that values of approximate quantum labels K_v or K_d could be associated with the prevalence of certain classical periodic orbits, illustrates this fact. It has become clear that the orderly picture is closely tied to the ubiquitous presence of (possibly locally perturbed) Rydberg series. As was recently shown (Burgdörfer *et al.*, 1995), this apparent suppression of classical chaos is a specific feature of Coulomb interacting systems and is therefore expected to be operative in all atomic systems.

Because of the long-range behavior of the Coulomb field, an electron with an energy close to a Rydberg threshold spends most of its orbital period, T_{Coulomb}, on a Kepler orbit in the Coulomb-dominated separable region while visiting intermittently the 'core' region of the "inner" electron(s) for short periods, T_{core}. Only this segment of the orbit accounts for the non-vanishing

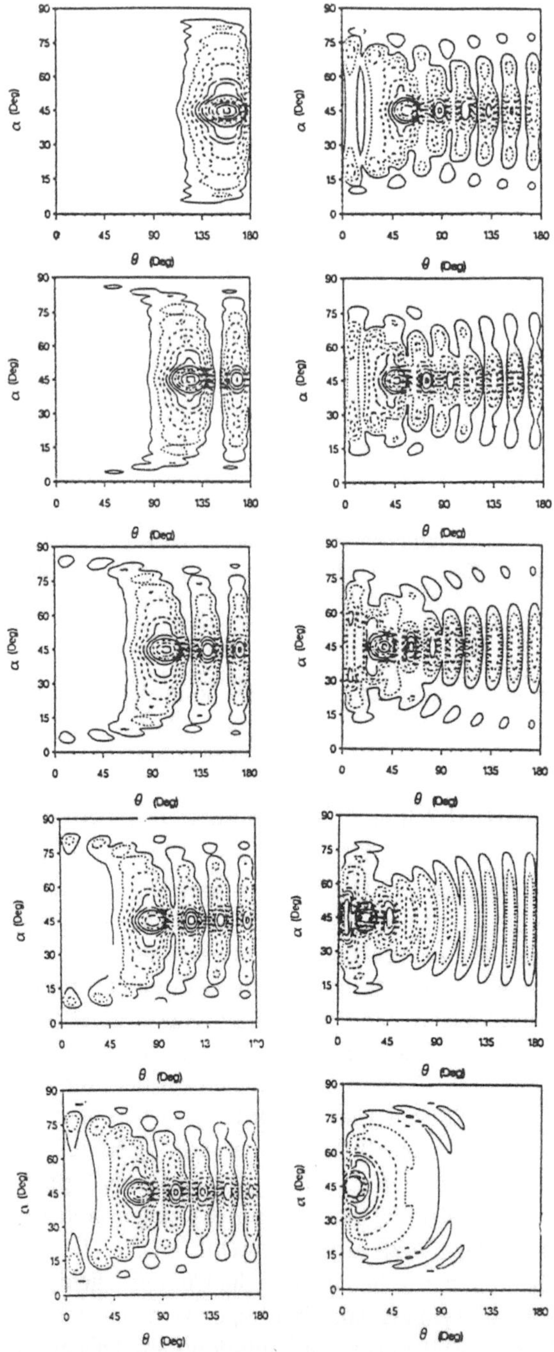

Fig. 15. Intrashell resonances in the limit $Z \to \infty$, projected onto the $\alpha - \theta_{12}$ plane, K= 9, 7, …, -9 left column (top to bottom): K= 9,7,…, 1; right column: K = -1,-3, … -9).

Lyapunov exponent, and hence for classical chaotic scattering (see Fig. 9). The disparity in time scales $T_{\text{Coulomb}} >> T_{\text{core}}$ can be expressed in terms of the density of states. Referring to the separable limit of the Hamiltonian (Eqs. (39,40)), the orbital period of the outer electron can be related to the density of states as

$$d_{Ryd}(n) = (n-\mu)^3 \simeq T_{\text{Coulomb}}/2\pi \ . \tag{49}$$

On the other hand, the interaction time with the core can be related to the orbital period, and hence, the density of states of the inner electron

$$d_{\text{core}}(N) = N^3 \simeq T_{\text{core}}/2\pi \ . \tag{50}$$

Consequently, sufficiently close to a Rydberg threshold there always exists a region with

$$d_{\text{Ryd}}(n) \gg d_{\text{core}}(N) \ . \tag{51}$$

In the language of multi-channel quantum defect theory (Friedrich, 1991), we can find spectral regions where the density of perturber states will be negligibly small compared to the density of the main Rydberg series, thereby preserving a quasi-regular Rydberg progression. This amounts to a quantum suppression of classical chaos specific to Coulomb systems. While the interaction in the nonseparable region is sufficient to render the classical orbit unstable, the amount of energy transfer between channels in a classical core collision is small compared to the quantized energy difference to the nearest perturber series, thereby suppressing strong interchannel coupling. The mechanism is also operative in other systems, for example, in hydrogen in a strong magnetic field, where regular Rydberg progressions have been observed well above the first (Landau) ionization threshold (Iu *et al.*, 1989).

As suggested by Blümel and Reinhardt (1992), the width rather than the position of resonances might be a better place to search for signatures of irregular scattering, i.e. transient chaos. Decay of resonant states by autoionization requires the two electrons to exchange energy with each other in a close collision, i.e. in the 'core' region where the classical instability occurs. The width is therefore very likely to probe selectively transient chaos in the core region while being insensitive to the excursion into the outer region of the Kepler orbit.

This picture is confirmed by an analysis of the width along a series where energy positions still closely follow a Rydberg progression. Figure 16, based on data from calculations (Bürgers, 1995; Müller, 1993) for the $K_v = K_d = N-1$ resonances, illustrates the distortion of an orderly sequence of widths as a function of the number of overlapping manifolds. In the absence of a perturber state embedded in a Rydberg manifold, the width Γ follows an approximate n^{-3} scaling, which is a measure of the fraction of the orbital period spent in the 'core' region (Fig. 16(a)). As soon as a perturber level is embedded in a Rydberg series, Γ displays large-scale fluctuations (Fig. 16(b)). With an increasing number of perturbers (Fig. 16(c)) the distribution of Γ's begins to randomize. It is therefore plausible to assume that Γ approaches a statistical distribution (an approximate Porter-Thomas distribution (Brody *et al.*, 1981)) when the density of perturbers becomes comparable to the level density of the manifold within a stretch of the spectrum.

The analytic expression (Eq.(44)) permits an estimate of the N quantum number where this region is reached. The number of perturbers from other Rydberg series ($N' > N$) embedded in the Rydberg series N for fixed $K = N - 1$ scales as

$$N_{\text{pert}} \simeq 0.143N^3 - 0.43N \tag{52}$$

for large N. Therefore, the width distribution will approach a locally random distribution when $N_{\text{pert}} \simeq N$ corresponding to $N \simeq 10$, in good agreement with the results of Fig. 16. A high density of perturber states can, however, only be found within a limited stretch of the spectrum.

Randomization of the width can be considered to be a precursor to the regime of overlapping resonances. The latter requires mean spacings between resonances $\langle \Delta E \rangle = \langle E_{k+1} - E_k \rangle$ to be smaller than the width $\langle \Gamma \rangle$, i.e.

$$\langle \Delta E \rangle / \langle \Gamma \rangle < 1 \ . \tag{53}$$

In this regime, reaction cross sections (such as photoabsorption cross sections) are expected to display Ericson's fluctuations (Ericson and Mayer-Kuckuk, 1966) due to a large number of overlapping resonances. We estimate $\langle \Delta E \rangle \simeq \langle E(N, N+1, K) - E(N, N, K) \rangle_K$ and $\langle \Gamma \rangle \simeq \langle \Gamma(N, N, K) \rangle_K$, where we average in each case over all K quantum numbers of the intrashell manifold, and extrapolate the data for Γ to larger N (Burgdörfer et al., 1995),

$$\langle \Gamma \rangle \simeq \frac{0.03(N-2) + 0.01}{N^3} \ , \tag{54}$$

the criterion for Ericson fluctuations is satisfied for $N = 16$. The onset is considerably lower than for the one-dimensional helium ($N \simeq 26$, Blümel and Reinhardt, 1991). The reason for the difference is that resonances with intermediate values of K, ($K \simeq 0$), possess the largest widths, and such resonances are not present in the one-dimensional helium atom. Width distributions and branching ratios (partial widths) therefore appear as prime candidates for investigations of irregular scattering in two-electron systems in the region which is experimentally accessible in multi-photon and synchrotron photoabsorption studies and theoretically accessible in large-scale solutions of the eigenvalue problem.

The extrapolation of Eq.(54) towards the threshold ionization has interesting implications for the behavior of reaction cross sections just below the threshold. Expressed in terms of the energy E relative to the three-body break-up threshold, Eq.(54) yields

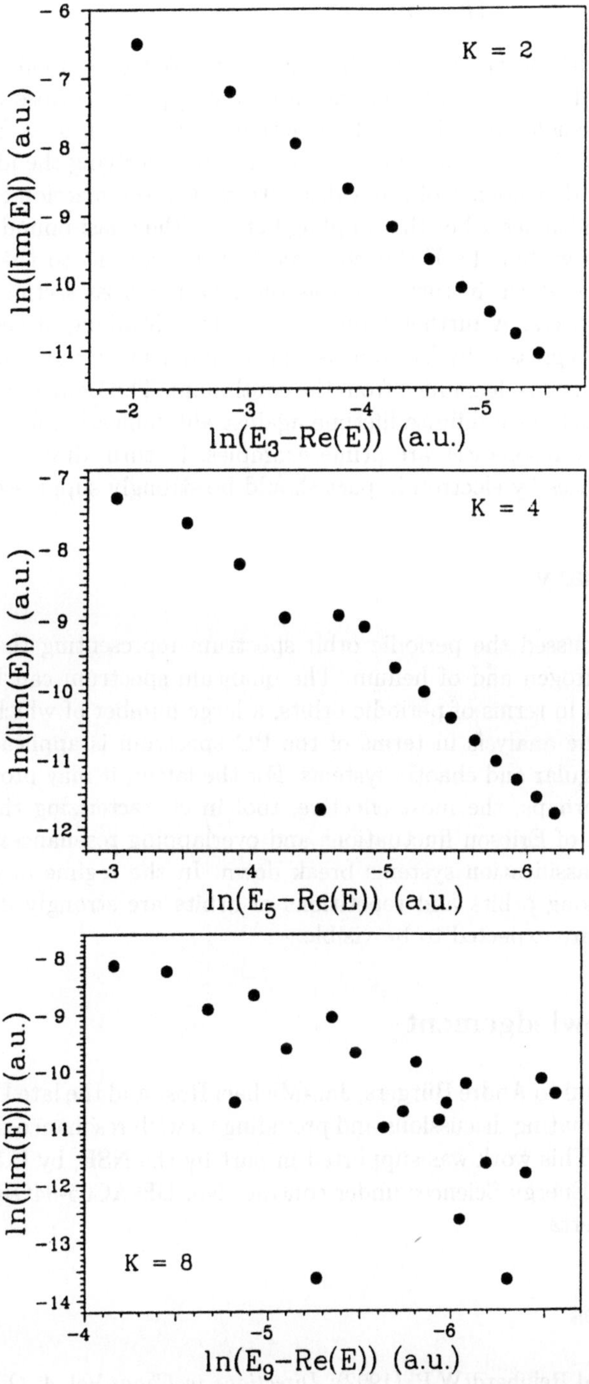

Fig. 16. Width of the Rydberg series of $K = N - 1$ resonances in helium including perturbers (data from Bürgers, 1995; Müller, 1993). (a) $N = 3$; (b) $N = 5$; and (c) $N = 9$.

$$\langle \Gamma \rangle \propto |E|^{1.0} \qquad (E < 0) \tag{55}$$

In parallel with observation for H^- (Rost and Wintgen, 1996), the energy dependence of $\langle \Gamma \rangle$ for He below the threshold appears to closely mimic the Wannier threshold behavior of the reaction cross section $\sigma \sim E^{1.056}$ above the threshold ($E > 0$). The implicit assumption underlying the identification of the energy dependence of the width with that of the reaction cross section is that the width describes the coupling between the quasi-bound state space (the Q space within the Feshbach projection formalism) to the continuum (the P space) which, in turn, controls the reaction cross section in a (half) scattering process. A further implication of the shrinking of the accessible phase space expressed by Eq. (55) is that a significant fraction of the phase space near $E \simeq 0$ is decoupled from the continuum. The Langmuir resonances (Sect. 4), which have infinite lifetime against autoionization for $N \gtrsim 500$, as well planetary resonances are prime examples. In turn, direct excitation of these resonances by electron impact should be strongly suppressed.

6 Summary

We have discussed the periodic orbit spectrum representing the density of states of hydrogen and of helium. The quantum spectrum can be uniquely characterized in terms of periodic orbits, a large number of which have been identified. The analysis in terms of the PO spectrum is applicable to both classically regular and chaotic systems. For the latter, it may provide a powerful, and perhaps, the most effective, tool in characterizing the spectrum in the region of Ericson fluctuations and overlapping resonances, where approximate classification systems break down. In the regime of overlapping resonances, long orbits and repetitions of orbits are strongly damped and short orbits are expected to be visible.

7 Acknowledgement

We are grateful to Andre Bürgers, Jan-Michael Rost and the late Dieter Wintgen for illuminating discussions and providing us with resonance data prior to publication. This work was supported in part by the NSF; by U.S. DOE Office of Basic Energy Sciences under contract No. DE-AC05-84OR21400 with Martin Marietta.

References

Blümel R. and ReinhardtẆ.P. (1992): *Directions in Chaos* Vol. 4, *Quantum Non-Integrability*, ed. D.H. Feng and J.-M. Yuan (World Scientific, Singapore) 245.

Brack M. and Jain S., (1995): Phys. Rev. A **51** 3462.

Bohigas O., (1991): *Chaos and Quantum Physics*, (North-Holland, Amsterdam), 87.
Brody T., Flores J., French J., Mello P., Pandey A., and Wong S., (1981): Rev. Mod. Phys. **53**, 385.
Bruns H., (1887): Berichte der Kg. Sächs. Ges. der Wiss., 1.
Burgdörfer J., Yang X., and Müller J., (1995): Chaos, Solitons, Fractals **5**, 1235.
Bürgers A., (1995); Ph.D. Thesis, U. Freiburg (unpublished).
Bürgers A., Wintgen D., and Rost J. (1995); J. Phys. B **28**, 3163.
Domke M. and Schulz K., Remmers G., Guiterrez A., Kaindl, G., and Wintgen D., (1995): Phys. Rev. A **51**, R4309.
Dräger M., Handke G., Ihra W., and Friedrich H. (1994): Phys. Rev. A **50**, 3793.
Du L., and Delos J. (1987): Phys. Rev. Lett. **58**, 1731.
Eckhardt B. and Aurell E. (1989): Europhys. Lett. **9**, 509.
Edmonds A., (1970): J. Phys. (Paris) Colloq. **31**, C4-71.
Eichmann U., Richter K., Wintgen D., and Sandner W., (1988): Phys. Rev. Lett. **61**, 2438.
Englefield M., (1972): *Group Theory and the Coulomb Problem* (Wiley, New York).
Ericson T. and Mayer-Kuckuk T., (1966): Ann. Rev. Nucl. Sci. **16**, 183.
Ezra G., Richter K., Tanner G. and Wintgen D., (1991): J. Phys. B. **24**, L413.
Fano U., and Rau A., (1987) *Atomic Collisions and Spectra* (Academic Press, NY).
Friedrich H. (1991): *Theoretical Atomic Physics*, (Springer-Verlag, New York).
Friedrich H. and Wintgen W. (1989): Phys. Rep. **183**, 37.
Garton W., and Tomkins F., (1969): Astrophys. J., **158**, 839.
Gutzwiller M. (1990): *Chaos in Classical and Quantum Mechanics.* Springer.
Herrick D.R., and Kellman M., (1980): Phys. Rev. A **21**, 418.
Herrick D.R., (1980): Phys. Rev. A **22**, 1346.
Herrick D.R. , and Kellman M., and Poliak D., (1980): Phys. Rev. A **22**, 1517.
Holle A., Wiebusch G., Main J., Hager B., Rottke H., and Welge K. (1986): Phys. Rev. Lett. **56**, 2594.
Herrick D.R., (1983): Adv. Chem. Phys. **52**, 1.
Iu C., Welch G., Kash M., Hsu K., and Kleppner D., (1989): Phys. Rev. Lett. **63**, 1133.
Kellman M., and Herrick D., (1980): Phys. Rev, A **22**, 1536.
Kim J. and Ezra G. (1991): *Proceedings of the Adriatico Research Conference on Quantum Chaos, Trieste* (World Scientific, Singapore) p. 436.
Langmuir I. (1921): Phys. Rev. **17**, 339.
Lichtenberg A. and Lieberman M. (1982); *Regular and Stochastic Motion* (Springer-Verlag, NY)
Lin C.D., (1984): Phys. Rev. A **29**, 1019.
Main J., Wiebusch G., Welge K., Shaw J., and Delos J. (1994): Phys. Rev. A **49**, 847.
Main J. (1997): contribution to this volume.
Mao J.-M., Shaw J., and Delos J.B., (1992): J. Stat. Phys. **68**, 51.
Müller J., (1993): Ph.D. thesis, Universität of Frankfurt am Main, Reihe Physik (Verlag Deutsch, Frankfurt, 1993), **19**, 1-218.
Müller J., Burgdörfer J., and Noid D., (1992): Phys. Rev. A **45**, 1471.
Müller J., Yang X., and Burgdörfer J., (1994): Phys. Rev. A **49**, 2470.
Müller J., and Burgdörfer J., (1993): Phys. Rev. Lett. **70**, 2375.

Müller J., Burgdörfer J., and Noid D., (1995); J. Chem. Phys. **103**, 12.
Pauli W. (1922): Ann. Phys. **11**, 177.
Poincaré H. (1890): Acta Math., **XIII**, 259.
Pullen R. and Edmonds A. (1981): J. Phys. A **14**, L477.
Qiu Y., Müller J., and Burgdörfer J., (1996): Phys. Rev. A **54**, 1922.
Richter K., and Wintgen D. (1991): J. Phys. B **24**, 565.
Rost J., and Tanner G., (1997): this volume.
Rost J., and Wintgen D. (1996): Europhys. Lett. **35**, 19.
Seng M., Halka M., Heber K.-D., and Sandner W., (1995): Phys. Rev. Lett. **74**, 3344.
Seng M., and Sandner W., (1996): private communication.
Tanner G., and Wintgen D., (1995): Phys. Rev. Lett. **75**, 2928.
Van Vleck J. (1922): Phil. Mag., **44** 842.
Watanabe S. (1987): Phys. Rev. A **36**, 1566.
Whittaker E., (1944): *A Treatise of the Analytical Dynamics of Particles and Rigid Bodies.* Cambridge University Press.
Wintgen D., Richter K., and Tanner G., (1992): Chaos **2**, 19.
Wulfman C., (1971): *Group Theory and Its Application*, (Academic, New York).

Subject Index

Lecture Notes in Physics

**For information about Vols. 1–455
please contact your bookseller or Springer-Verlag**

Vol. 456: H. B. Geyer (Ed.), Field Theory, Topology and Condensed Matter Physics. Proceedings, 1994. XII, 206 pages. 1995.

Vol. 457: P. Garbaczewski, M. Wolf, A. Weron (Eds.), Chaos – The Interplay Between Stochastic and Deterministic Behaviour. Proceedings, 1995. XII, 573 pages. 1995.

Vol. 458: I. W. Roxburgh, J.-L. Masnou (Eds.), Physical Processes in Astrophysics. Proceedings, 1993. XII, 249 pages. 1995.

Vol. 459: G. Winnewisser, G. C. Pelz (Eds.), The Physics and Chemistry of Interstellar Molecular Clouds. Proceedings, 1993. XV, 393 pages. 1995.

Vol. 460: S. Cotsakis, G. W. Gibbons (Eds.), Global Structure and Evolution in General Relativity. Proceedings, 1994. IX, 173 pages. 1996.

Vol. 461: R. López-Pen˜a, R. Capovilla, R. Garcı´a-Pelayo, H. Waelbroeck, F. Zertuche (Eds.), Complex Systems and Binary Networks. Lectures, México 1995. X, 223 pages. 1995.

Vol. 462: M. Meneguzzi, A. Pouquet, P.-L. Sulem (Eds.), Small-Scale Structures in Three-Dimensional Hydrodynamic and Magnetohydrodynamic Turbulence. Proceedings, 1995. IX, 421 pages. 1995.

Vol. 463: H. Hippelein, K. Meisenheimer, H.-J. Röser (Eds.), Galaxies in the Young Universe. Proceedings, 1994. XV, 314 pages. 1995.

Vol. 464: L. Ratke, H. U. Walter, B. Feuerbach (Eds.), Materials and Fluids Under Low Gravity. Proceedings, 1994. XVIII, 424 pages, 1996.

Vol. 465: S. Beckwith, J. Staude, A. Quetz, A. Natta (Eds.), Disks and Outflows Around Young Stars. Proceedings, 1994. XII, 361 pages, 1996.

Vol. 466: H. Ebert, G. Schütz (Eds.), Spin – Orbit-Influenced Spectroscopies of Magnetic Solids. Proceedings, 1995. VII, 287 pages, 1996.

Vol. 467: A. Steinchen (Ed.), Dynamics of Multiphase Flows Across Interfaces. 1994/1995. XII, 267 pages. 1996.

Vol. 468: C. Chiuderi, G. Einaudi (Eds.), Plasma Astrophysics. 1994. VII, 326 pages. 1996.

Vol. 469: H. Grosse, L. Pittner (Eds.), Low-Dimensional Models in Statistical Physics and Quantum Field Theory. Proceedings, 1995. XVII, 339 pages. 1996.

Vol. 470: E. Martı´nez-González, J. L. Sanz (Eds.), The Universe at High-z, Large-Scale Structure and the Cosmic Microwave Background. Proceedings, 1995. VIII, 254 pages. 1996.

Vol. 471: W. Kundt (Ed.), Jets from Stars and Galactic Nuclei. Proceedings, 1995. X, 290 pages. 1996.

Vol. 472: J. Greiner (Ed.), Supersoft X-Ray Sources. Proceedings, 1996. XIII, 350 pages. 1996.

Vol. 473: P. Weingartner, G. Schurz (Eds.), Law and Prediction in the Light of Chaos Research. X, 291 pages. 1996.

Vol. 474: Aa. Sandqvist, P. O. Lindblad (Eds.), Barred Galaxies and Circumnuclear Activity. Proceedings of the Nobel Symposium 98, 1995. XI, 306 pages. 1996.

Vol. 475: J. Klamut, B. W. Veal, B. M. Dabrowski, P. W. Klamut, M. Kazimierski (Eds.), Recent Developments in High Temperature Superconductivity. Proceedings, 1995. XIII, 362 pages. 1996.

Vol. 476: J. Parisi, S. C. Müller, W. Zimmermann (Eds.), Nonlinear Physics of Complex Systems. Current Status and Future Trends. XIII, 388 pages. 1996.

Vol. 477: Z. Petru, J. Przystawa, K. Rapcewicz (Eds.), From Quantum Mechanics to Technology. Proceedings, 1996. IX, 379 pages. 1996.

Vol. 479: H. Latal, W. Schweiger (Eds.), Perturbative and Nonperturbative Aspects of Quantum Field Theory. Proceedings, 1996. X, 430 pages. 1997.

Vol. 480: H. Flyvbjerg, J. Hertz, M. H. Jensen, O. G. Mouritsen, K. Sneppen (Eds.), Physics of Biological Systems. From Molecules to Species. X, 364 pages. 1997.

Vol. 481: F. Lenz, H. Grießhammer, D. Stoll (Eds.), Lectures on QCD. VII, 276 pages. 1997.

Vol. 482: X.-W. Pan, D. H. Feng, M. Vallières (Eds.), Contemporary Nuclear Shell Models. Proceedings, 1996. XII, 309 pages. 1997.

Vol. 483: G. Trottet (Ed.), Coronal Physics from Radio and Space Observations. XVII, 226 pages. 1997.

Vol. 484: L. Schimansky-Geier, T. Pöschel (Eds.), Stochastic Dynamics. XVIII, 386 pages. 1997.

Vol. 485: H. Friedrich, B. Eckhardt (Eds.), Classical, Semiclassical and Quantum Dynamics in Atoms. VIII, 341 pages. 1997.

Vol. 486: G. Chavent, P. C. Sabatier (Eds.), Inverse Problems of Wave Propagation and Diffraction. XV, 379 pages. 1997.

Vol. 488: B. Apagyi, G. Endrédi, P. Lévay (Eds.), Inverse and Algebraic Quantum Scattering Theory. XV, 385 pages. 1997.

Vol. 489: G. M. Simnett, C. E. Alissandrakis, L. Vlahos (Eds.), Solar and Heliospheric Plasma Physics. VIII, 278 pages, 1997.

Vol. 491: O. Boratav, A. Eden, A. Erzan (Eds.), Turbulence Modeling and Vortex Dynamics. XII, 245 pages, 1997.

New Series m: Monographs

Vol. m 1: H. Hora, Plasmas at High Temperature and Density. VIII, 442 pages. 1991.

Vol. m 2: P. Busch, P. J. Lahti, P. Mittelstaedt, The Quantum Theory of Measurement. XIII, 165 pages. 1991. Second Revised Edition: XIII, 181 pages. 1996.

Vol. m 3: A. Heck, J. M. Perdang (Eds.), Applying Fractals in Astronomy. IX, 210 pages. 1991.

Vol. m 4: R. K. Zeytounian, Mécanique des fluides fondamentale. XV, 615 pages, 1991.

Vol. m 5: R. K. Zeytounian, Meteorological Fluid Dynamics. XI, 346 pages. 1991.

Vol. m 6: N. M. J. Woodhouse, Special Relativity. VIII, 86 pages. 1992.

Vol. m 7: G. Morandi, The Role of Topology in Classical and Quantum Physics. XIII, 239 pages. 1992.

Vol. m 8: D. Funaro, Polynomial Approximation of Differential Equations. X, 305 pages. 1992.

Vol. m 9: M. Namiki, Stochastic Quantization. X, 217 pages. 1992.

Vol. m 10: J. Hoppe, Lectures on Integrable Systems. VII, 111 pages. 1992.

Vol. m 11: A. D. Yaghjian, Relativistic Dynamics of a Charged Sphere. XII, 115 pages. 1992.

Vol. m 12: G. Esposito, Quantum Gravity, Quantum Cosmology and Lorentzian Geometries. Second Corrected and Enlarged Edition. XVIII, 349 pages. 1994.

Vol. m 13: M. Klein, A. Knauf, Classical Planar Scattering by Coulombic Potentials. V, 142 pages. 1992.

Vol. m 14: A. Lerda, Anyons. XI, 138 pages. 1992.

Vol. m 15: N. Peters, B. Rogg (Eds.), Reduced Kinetic Mechanisms for Applications in Combustion Systems. X, 360 pages. 1993.

Vol. m 16: P. Christe, M. Henkel, Introduction to Conformal Invariance and Its Applications to Critical Phenomena. XV, 260 pages. 1993.

Vol. m 17: M. Schoen, Computer Simulation of Condensed Phases in Complex Geometries. X, 136 pages. 1993.

Vol. m 18: H. Carmichael, An Open Systems Approach to Quantum Optics. X, 179 pages. 1993.

Vol. m 19: S. D. Bogan, M. K. Hinders, Interface Effects in Elastic Wave Scattering. XII, 182 pages. 1994.

Vol. m 20: E. Abdalla, M. C. B. Abdalla, D. Dalmazi, A. Zadra, 2D-Gravity in Non-Critical Strings. IX, 319 pages. 1994.

Vol. m 21: G. P. Berman, E. N. Bulgakov, D. D. Holm, Crossover-Time in Quantum Boson and Spin Systems. XI, 268 pages. 1994.

Vol. m 22: M.-O. Hongler, Chaotic and Stochastic Behaviour in Automatic Production Lines. V, 85 pages. 1994.

Vol. m 23: V. S. Viswanath, G. Müller, The Recursion Method. X, 259 pages. 1994.

Vol. m 24: A. Ern, V. Giovangigli, Multicomponent Transport Algorithms. XIV, 427 pages. 1994.

Vol. m 25: A. V. Bogdanov, G. V. Dubrovskiy, M. P. Krutikov, D. V. Kulginov, V. M. Strelchenya, Interaction of Gases with Surfaces. XIV, 132 pages. 1995.

Vol. m 26: M. Dineykhan, G. V. Efimov, G. Ganbold, S. N. Nedelko, Oscillator Representation in Quantum Physics. IX, 279 pages. 1995.

Vol. m 27: J. T. Ottesen, Infinite Dimensional Groups and Algebras in Quantum Physics. IX, 218 pages. 1995.

Vol. m 28: O. Piguet, S. P. Sorella, Algebraic Renormalization. IX, 134 pages. 1995.

Vol. m 29: C. Bendjaballah, Introduction to Photon Communication. VII, 193 pages. 1995.

Vol. m 30: A. J. Greer, W. J. Kossler, Low Magnetic Fields in Anisotropic Superconductors. VII, 161 pages. 1995.

Vol. m 31: P. Busch, M. Grabowski, P. J. Lahti, Operational Quantum Physics. XI, 230 pages. 1995.

Vol. m 32: L. de Broglie, Diverses questions de mécanique et de thermodynamique classiques et relativistes. XII, 198 pages. 1995.

Vol. m 33: R. Alkofer, H. Reinhardt, Chiral Quark Dynamics. VIII, 115 pages. 1995.

Vol. m 34: R. Jost, Das Märchen vom Elfenbeinernen Turm. VIII, 286 pages. 1995.

Vol. m 35: E. Elizalde, Ten Physical Applications of Spectral Zeta Functions. XIV, 228 pages. 1995.

Vol. m 36: G. Dunne, Self-Dual Chern-Simons Theories. X, 217 pages. 1995.

Vol. m 37: S. Childress, A.D. Gilbert, Stretch, Twist, Fold: The Fast Dynamo. XI, 410 pages. 1995.

Vol. m 38: J. González, M. A. Martín-Delgado, G. Sierra, A. H. Vozmediano, Quantum Electron Liquids and High-T_c Superconductivity. X, 299 pages. 1995.

Vol. m 39: L. Pittner, Algebraic Foundations of Non-Commutative Differential Geometry and Quantum Groups. XII, 469 pages. 1996.

Vol. m 40: H.-J. Borchers, Translation Group and Particle Representations in Quantum Field Theory. VII, 131 pages. 1996.

Vol. m 41: B. K. Chakrabarti, A. Dutta, P. Sen, Quantum Ising Phases and Transitions in Transverse Ising Models. X, 204 pages. 1996.

Vol. m 42: P. Bouwknegt, J. McCarthy, K. Pilch, The W_3 Algebra. Modules, Semi-infinite Cohomology and BV Algebras. XI, 204 pages. 1996.

Vol. m 43: M. Schottenloher, A Mathematical Introduction to Conformal Field Theory. VIII, 142 pages. 1997.

Vol. m 44: A. Bach, Indistinguishable Classical Particles. VIII, 157 pages. 1997.

Vol. m 45: M. Ferrari, V. T. Granik, A. Imam, J. C. Nadeau (Eds.), Advances in Doublet Mechanics. XVI, 214 pages. 1997.

Vol. m 46: M. Camenzind, Les noyaux actifs de galaxies. XVIII, 218 pages. 1997.

Vol. m 48: P. Kopietz, Bosonization of Interacting Fermions in Arbitrary Dimensions. XII, 259 pages. 1997.

Vol. m 49: M. Zak, J. B. Zbilut, R. E. Meyers, From Instability to Intelligence. Complexity and Predictability in Nonlinear Dynamics. XIV, 552 pages. 1997.

MIX
Papier aus verantwortungsvollen Quellen
Paper from responsible sources
FSC® C105338

If you have any concerns about our products,
you can contact us on
ProductSafety@springernature.com

In case Publisher is established outside the EU,
the EU authorized representative is:
Springer Nature Customer Service Center GmbH
Europaplatz 3, 69115 Heidelberg, Germany

Printed by Libri Plureos GmbH
in Hamburg, Germany